ADVANCED POLYMERIC MATERIALS FOR SUSTAINABILITY AND INNOVATIONS

ADVANCED POLYMERIC MATERIALS FOR SUSTAINABILITY AND INNOVATIONS

Edited by

Sajith Thottathil
Sabu Thomas
Nandakumar Kalarikkal
Didier Rouxel

AAP | APPLE ACADEMIC PRESS

Apple Academic Press Inc. | Apple Academic Press Inc.
3333 Mistwell Crescent | 9 Spinnaker Way
Oakville, ON L6L 0A2 | Waretown, NJ 08758
Canada | USA

© 2019 by Apple Academic Press, Inc.

First issued in paperback 2021

Exclusive worldwide distribution by CRC Press, a member of Taylor & Francis Group
No claim to original U.S. Government works

ISBN 13: 978-1-77-463074-7 (pbk)
ISBN 13: 978-1-77-188633-8 (hbk)

Library and Archives Canada Cataloguing in Publication

Advanced polymeric materials for sustainability and innovations / edited by Sajith Thottathil, Sabu Thomas, Nandakumar Kalarikkal, Didier Rouxel.

Includes bibliographical references and index.
Issued in print and electronic formats.
ISBN 978-1-77188-633-8 (hardcover).--ISBN 978-1-315-10243-6 (PDF)

1. Polymers. I. Thomas, Sabu, editor II. Kalarikkal, Nandakumar, editor III. Rouxel, Didier, editor IV. Thottathil, Sajith, editor

TA455.P58A42 2018   620.1'92   C2018-904504-3   C2018-904505-1

CIP data on file with US Library of Congress

Apple Academic Press also publishes its books in a variety of electronic formats. Some content that appears in print may not be available in electronic format. For information about Apple Academic Press products, visit our website at **www.appleacademicpress.com** and the CRC Press website at **www.crcpress.com**

CONTENTS

ABOUT THE EDITORS

Sajith Thottathil, MTech, is an assistant professor at the International and Interuniversity Center for Nanoscience& Nanotechnology, Mahatma Gandhi University, Kottayam, Kerala, India. He was previously a research scholar at the same university. He completed his BTechdegreein mechanical engineering from Kerala University, Kerala, India, and went on to earn his MTech innanotechnology from the National Institute of Technology, Calicut, Kerala, India. His MTech project was "Effect of Graphene Oxides on Mechanical and Tribological properties of Epoxy nanocomposites." His research areas include nanotechnology, mechanical engineering, and polymer nanocomposites. He won an award for his poster presentation at the International Conference on Polymer Processing and Characterization, 2016, ICPPC 16.

E-mail: sajithrahman87@gmail.com

Sabu Thomas, PhD, is the Pro-Vice Chancellor of Mahatma Gandhi University and Founding Director of the International and Inter University Center for Nanoscience and Nanotechnology, Mahatma Gandhi University, Kottayam, Kerala, India. He is also a full professor of polymer science and engineering at the School of Chemical Sciences of the same university. He is a fellow of many professional bodies. Professor Thomas has (co-) authored many papers in international peer-reviewed journals in the area of polymer science and nanotechnology. He has organized several international conferences. Professor Thomas's research group has specialized in many areas of polymers, which includes polymer blends, fiber-filled polymer composites,

particulate-filled polymer composites and their morphological characterization, ageing and degradation, pervaporation phenomena, sorption and
diffusion, interpenetrating polymer systems, recyclability and reuse of
waste plastics and rubbers, elastomeric crosslinking, dual porous nanocomposite scaffolds for tissue engineering, etc. Professor Thomas's
research group has extensive exchange programs with different industries
and research and academic institutions all over the world and is performing
world-class collaborative research in various fields. Professor Thomas's
Center is equipped with various sophisticated instruments and has established state-of-the-art experimental facilities, which cater to the needs of
researchers within the country and abroad.

Professor Thomas has published over 750 peer-reviewed research papers,
reviews, and book chapters and has a citation count of 31,574. The H index
of Prof. Thomas is 81, and he has six patents to his credit. He has delivered
over 300 plenary, inaugural, and invited lectures at national/international
meetings over 30 countries. He is a reviewer for many international journals. He has received MRSI, CRSI, nanotech medals for his outstanding
work in nanotechnology. Recently Prof. Thomas has been conferred an
Honoris Causa (DSc) by the University of South Brittany, France, and
University Lorraine, Nancy, France.
E-mail: sabuchathukulam@yahoo.co.uk, sabupolymer@yahoo.com

Dr. Nandakumar Kalarikkal is the Joint Director
at the International and Inter University Centre
for Nanoscience and Nanotechnology and Asso-
ciate Professor at the School of Pure and Applied
Physics, Mahatma Gandhi University, Kottayam,
Kerala, India. His research activities involve
applications of nanostructured materials, laser
plasma, phase transitions, etc. He is the recipient
of research fellowships and associateships from
prestigious government organizations such as the
Department of Science and Technology and the Council of Scientific and
Industrial Research of the Government of India. He has active collabora-
tion with national and international scientific institutions in India, South
Africa, Slovenia, Canada, France, Germany, Malaysia, Australia, and US.

He has more than 130 publications in peer-reviewed journals. He has also co-edited nine books of scientific interest and co-authored many book chapters.He has supervised 8 PhD theses, and currently 12 PhD students are working under him.
E-mail: nkkalarikkal@mgu.ac.in, drkalarikkal@gmail.com

Didier Rouxel was born in Lyon, France, on March 5, 1965. He received an Engineer degree (5-year degree) in materials in 1989 from the Ecole Supérieure des Sciences et Techniques de l'Ingénieur de Nancy, Vandoeuvre, France. He received his Ph.D. in material sciences and engineering from the University of Nancy I, France in 1993. He is now fulltime professor in the Université de Lorraine and he leads the team "Micro and Nano Systems" in Institut Jean Lamour in Nancy, France. His research has been mainly focused on physical chemistry of surfaces and thin films, piezoelectric materials, elastic properties of materials studied by Brillouin spectroscopy, and development of micro-devices based on polymer–nanoparticles nanocomposite materials, in particular for biomedical applications. He was expert for the French agency ANSES for the topic "Nanomaterials and Health" and Member of the Year of the French Society of Nanomedicine in 2014.

LIST OF CONTRIBUTORS

Hussein Ali
Department of Polymer Engineering, Maharashtra Institute of Technology, Kothrud,
Pune 411038, India
E-mail: hussien_a2004@yahoo.com

K. Aruna
Assistant professor, Department of Mechanical Engineering, Sri Venkateswara University,
Tirupati 517502, India
E-mail: karuna@svuniversity.ac.in

Salifu T. Azeko
Nigerian Turkish Nile University, Abuja, Nigeria

Vishalakshi B.
Department of Chemistry, Mangalore University, Mangalagangotri, India

B. Haritha Bai
Department of Mechanical Engineering, Sri Venkateswara University, Tirupati 517502, India

I. Barandiaran
Materials and Technologies Group, Universidad del País Vasco/Euskal Herriko Unibertsitatea
UPV/EHU, Plaza Europa 1, 20018 Donostia, Spain

Anjali Bishnoi
Department of Polymer and Rubber Technology, Shroff S. R. Rotary Institute of Chemical
Technology, Block no: 402, At: Vataria, Ankleshwar-Valia road, Ta: Valia, Dist: Bharuch,
Gujarat 393003, India; Department of Chemistry, Indian Institute of Technology Hauz Khas,
New Delhi 110016, India
E-mail: anjali. bishnoi@srict.in, anjali_bishnoi@yahoo.com

Vibha C.
Biomedical Technology Wing, Sree Chitra Tirunal Institute for Medical Sciences and Technology,
Poojappura, Thiruvananthapuram 695012, India

Harichandra U. Chandekar
Department of Mechanical Engineering, Goa College of Engineering, Farmagudi, Ponda 403401,
Goa, India
E-mail: harichandekar@gmail.com

Vikas V. Chaudhari
Department of Mechanical Engineering, BITS Pilani, K. K. Birla Goa Campus, Zuarinagar 403726,
Goa, India
E-mail: vikas@goa.bits-pilani.ac.in

J. Dhanalakshmi
Department of Chemistry, Kamaraj College of Engineering and Technology, K. Vellakulam 625701,
India

J. M. Gadgil
Polymer and Rubber Consultants, Pune, India

Narendra N. Ghosh
Department of Chemistry, BITS Pilani, K. K. Birla Goa Campus, Zuarinagar 403726, Goa, India
E-mail: nghosh@goa.bits-pilani.ac.in

Vikas V. Gite
Department of Polymer Chemistry, School of Chemical Sciences, North Maharashtra University,
Jalgaon 425001, Maharashtra, India
Phone No.: +91 257 2257432, Fax No.: +91 257 2257406, E-mail: vikasgite123@gmail.com

M. K. Gupta
Department of Mechanical Engineering, Motilal Nehru National Institute of Technology Allahabad
211004, India

Sandeep Gupta
Indian Institute of Technology, Jodhpur 342011, Rajasthan, India

D. Jeyabala
Department of Polymer Technology, Kamaraj College of Engineering and Technology,
K. Vellakulam 625701, India

H. V. Joshi
M. E. Polymer Engineering Department, MIT, Pune, India

P. S. Joshi
Polymer and Rubber Consultants, Pune, India

Sunil Kumar
Department of Polymer and Rubber Technology, Shroff S. R. Rotary Institute of Chemical
Technology, Block no: 402, At: Vataria, Ankleshwar-Valia road, Ta: Valia, Dist: Bharuch 393003,
Gujarat, India; Centre for Polymer Science and Engineering, Indian Institute of Technology
Hauz Khas, New Delhi 110016, India

Amrita Kaurwar
Indian Institute of Technology, Jodhpur 342011, Rajasthan, India

G. Kortaberria
Materials and Technologies Group, Universidad del País Vasco/Euskal Herriko Unibertsitatea
UPV/EHU, Plaza Europa 1, 20018 Donostia, Spain
E-mail: galder.cortaberria@ehu.eus

M. B. Kulkarni
Department of Polymer Engineering, Maharashtra Institute of Technology, Kothrud, Pune 411038,
India
E-mail: malhari.kulkarni@mitpune.edu.in

P. P. Lizymol
Biomedical Technology Wing, Sree Chitra Tirunal Institute for Medical Sciences and Technology,
Poojappura, Thiruvananthapuram 695012, India
Tel: 0471-2520329, +91-9447216561, E-mail: lizymol@rediffmail.com
E-mail: mkgupta@mnnit.ac.in, mnnit.manoj@gmail.com

Ankit Mishra
Department of Polymer and Rubber Technology, Shroff S. R. Rotary Institute of Chemical Technology, Block no: 402, At: Vataria, Ankleshwar-Valia road, Dist: Bharuch 393003, Gujarat, India

Raghvendra Kumar Mishra
International and Inter University Centre for Nanoscience and Nanotechnology (IIUCNN), Mahatma Gandhi University, Kottayam 686560, Kerala, India
E-mail: raghvendramishra4489@gmail.com

B. R. P. Nayak
Faculty, Department of Mechanical Engineering, SIETEK, Puttur, A. P., India

Mahendra R. Nevare
Department of Polymer Chemistry, School of Chemical Sciences, North Maharashtra University, Jalgaon 425001, Maharashtra, India

B. C. Pai
CSIR-National Institute for Interdisciplinary Science and Technology, Trivandrum 695019, Kerala, India

Jerin K. Pancrecious
CSIR-National Institute for Interdisciplinary Science and Technology, Trivandrum 695019, Kerala, India

Fathima Parveen
Department of Engineering, Amity University, Dubai

P. V. Patil
Polymer and Rubber Consultants, Pune, India

Anand Plappally
Indian Institute of Technology, Jodhpur 342011, Rajasthan, India

Nandini R.
Department of Chemistry, MITE, Moodabidri 574226 (DK), Karnataka, India

S. Radhakrishnan
Department of Polymer Engineering, Maharashtra Institute of Technology, Kothrud, Pune 411038, India

Sandeep Rai
Department of Polymer and Rubber Technology, Shroff S. R. Rotary Institute of Chemical Technology, Block no: 402, At: Vataria, Ankleshwar-Valia road, Ta: Valia, Dist: Bharuch 393003, Gujarat, India; GRP Limited, GIDC Estate, Panoli, Bharuch, Gujarat, India

Ramya Rajan
CSIR-National Institute for Interdisciplinary Science and Technology, Trivandrum 695019, Kerala, India

T. P. D. Rajan
CSIR-National Institute for Interdisciplinary Science and Technology, Trivandrum 695019, Kerala, India E-mail: tpdrajan@rediffmail.com, tpdrajan@niist.res.in

B. Sudheer Reddy
Faculty, Department of Mechanical Engineering, SIETEK, Puttur, A. P., India
E-mail: sudheerb121@gmail.com

Raj Kumar Satankar
Indian Institute of Technology, Jodhpur 342011, Rajasthan, India

Winston O. Soboyejo
Princeton University, Princeton, NJ 08544, USA

Alfred B. O. Soboyejo
The Ohio State University, Columbus, OH 43210, USA

R. K. Srivastava
Department of Mechanical Engineering, Motilal Nehru National Institute of Technology
Allahabad 211004, India

C. S. Sunandana
School of Physics, University of Hyderabad, Hyderabad 500046, Telangana, India
E-mail: sunandana@gmail.com

Sarah B. Ulaeto
CSIR-National Institute for Interdisciplinary Science and Technology, Trivandrum 695019,
Kerala, India

K. Usha
ACESSD, M G University, Kottayam 686560, Kerala, India

C. T. Vijayakumar
Department of Polymer Technology, Kamaraj College of Engineering and Technology,
K. Vellakulam 625 701, India, E-mail: ctvijay22@yahoo.com

Sachin D. Waigaonkar
Department of Mechanical Engineering, BITS Pilani, K.K.Birla Goa Campus,
Zuarinagar 403726, Goa, India
E-mail: sdw@goa.bits-pilani.ac.in

S. A. Yawalkar
M. E. Polymer Engineering Department, MIT, Pune, India
E-mail: sharayu1@gmail.com

PREFACE

This book, *Advanced Polymer Material for Sustainability and Innovations,* discusses the recent advancements in the research and developments in synthesis, characterization, processing, morphology, structure, and properties of advanced polymeric materials. This book has a special focus on eco-friendly polymers, polymer composites, nanocomposites, blends, and materials for traditional and renewable energy.

In this book the relationship between processing, morphology, property, and applications of the polymeric materials is well established. Recent advances in the synthesis of new functional monomers have shown strong potential in generating better property polymers from renewable resources. Fundamental advances in the field of nanocomposites, blends, and nanostructured polymeric materials in automotive, civil, biomedical, and packaging/coating applications are the highlights of this book.

Chapter 1 is a review that discusses recent advances in the synthesis and characterization of nanocomposites based on block copolymers and magnetic nanoparticles. It describes the use of different routes to prepare nanocomposites with magnetic properties. The functionalization of nanoparticles can improve the dispersion of the nanoparticles and thereby final properties.

Development of environment-friendly green and low-cost ionic liquid (IL) for rubber compounding has been described in Chapter 2. IL was synthesized and used as an accelerator in medium acylonitrile cold NBR compounding. Due to nonvolatile nature, IL usage in an industry will lead to almost no emission of volatile organic compounds.

Chapter 3 is an empirical study of green bio-composite beams containing dispersed equine ordure in naturally occurring clayey soil matrix. These materials containing 50% by volume of the clayey matrix have sustained importance in rural households in western Rajasthan, India, for manufacture of cantilever slabs. The implication of this work is toward a new set of bioactive materials that can be locally manufactured to substitute cement-based structures.

The effect of blending polyvinyl chloride (PVC) plasticized with epoxidized soy oil with polybenzoxazine has been reported in Chapter 4.

These studies indicate the compositional dependence of performance of unfilled polybenzoxazine–PVC blends and also the effect of addition of fumed silica on the properties of the material. Thermal properties revealed that the presence of more thermally stable PBZ and FS in the composites enhances the overall thermal stability of the composites.

Chapter 5 discusses on the dry bonding of rubber, which could be achieved by resorcinol formaldehyde resin (R) being used as a methylene acceptor and hexamethoxymethylmelamine (HMMM) as a methylene donor. From the differential scanning calorimetry (DSC) curves, the interaction existing between R and HMMM is explicit. The TG studies of the cured blend B (60% R + 40% HMMM) showed that PB is thermally more stable than the resin R and the rate of degradation of the cured material is comparatively slower than the blend B.

Chapter 6 deals with the newly developed composites from hybrid resins containing strontium, which exhibited good mechanical properties with low polymerization shrinkage. The synergetic effect of the inorganic content within the resin of this particular composition supplements better properties for the composites. So the application of these hybrid resins can be expanded to orthopedic, dental, and bioactive coating applications as it contains polymerizable methacrylate groups which can endure in-situ polymerization to obtain bioactive polymers.

Chapter 7 talks about hybrid jute/sisal fiber-reinforced epoxy composites which are prepared using hand lay-up technique with varying wt.% of fibers. The positive effect of hybridization is observed as increase in dynamic mechanical properties. Glass transition, crystallization, and decomposition temperatures are found to increase due to hybridization as compared to pure jute and sisal composites. The potential applications of jute-based hybrid composites in packaging, automobiles, and constructions are going to increase in the near future.

Chapter 8 illustrates unidirectional jute-fiber-reinforced epoxy composites which are prepared using hand lay-up technique. The composites were subjected to mechanical tests (tensile, flexural, and impact) and thermal tests using DMA and DSC. The mechanical properties of present jute composite were found to be good. No significant change is observed in values of crystallization and glass transition temperatures of jute composites when compared to epoxy resin.

Chapter 9 describes the sodium alginate and sodium carboxymethyl cellulose polymers. Based on the results, it is concluded that the acridine orange is more effective in inducing metachromasy in sodium alginate than in sodium heparinate. Cooperativity in binding is observed to occur due to neighboring interactions among the bound dye molecules at lower P/D ratios leading to stacking. The stacking tendency is enhanced by the easy availability and close proximity of the charged sites.

In Chapter 10 the reclaimed rubber is produced by breaking down the vulcanized structure of rubber using heat, chemicals, and mechanical techniques. Devulcanization and reclaiming of rubber waste is carried out by physical and chemical processes. Physical process involves application of mechanical, thermo-mechanical, microwave, or ultrasound energy to partially devulcanized rubber while chemical reclaiming processes use various chemicals and devulcanizing agents. The chemical system under study was based on natural rubber (NR) waste. Diphenyl disulfide (DPDS) and garlic oil were used as devulcanizing agents. Promising results are obtained for combination of thermal and mechano-chemical modes of devulcanization process.

Chapter 11 illustrates the production of new PMCs with the reinforcement materials of fly ash and SiC by injection molding process. Tests are conducted to find out their impact strength and hardness. Hardness of the PMC is increases with the increase of percentage of reinforcement. It is also established that the new materials produced have the suitable mechanical property with strengths for practical applications.

Chapter 12 gives a brief review on the ion-conducting polymers, which provides a perspective on the problem of ion conduction in polymer electrolytes. It mainly focuses on the properties of polymer that favor ion transport. Contemporary issues in polymer electrolytes are discussed next. This is followed by a discussion of two fundamental experimental probes: electrical conductivity and positron annihilation.

Chapter 13 describes the synergistic effect of LDH and calcium carbonate. The problem regarding weak compatibility between calcium carbonate and PP has been dealt with by modifying calcium carbonate by sodium stearate as organically modified filler that can reduce the aggregation tendency. Moreover, polar and nonpolar functional groups present in sodium stearate can help to improve the interfacial adhesion of hydrophilic filler particles with hydrophobic polymer molecules.

Chapter 14 explains and compares different aerogel fabrication techniques such as supercritical drying, ambient pressure drying, and freeze drying along with their mechanisms, advantages, and disadvantages. The applications of aerogels in various fields are also described in brief.

Chapter 15 describes the incorporation of nanodispersoids in polymeric coatings for improving barrier performance as it influences other properties as well as imparts new properties to the coating system. Dispersing nanodispersoids, instead of larger particles has allowed increase in the "interfacial material" content significantly. The investigated novel coating formulations and their tailored responses have been designed to provide new insight toward understanding the mechanism of corrosion protection of different materials such as aluminum and its alloys, brass, steel and so forth.

Chapter 16 illustrates the study of the processing parameters for jute–PP composites. The thermal analysis has been carried out using TGA and DSC. The TGA study reveals that the thermal stability of the jute–PP composite is better than neat PP. While the DMA investigations show that at high temperatures up to 120°C the reinforcing effect of jute fibers is more pronounced. The tan δ curve indicates lower energy dissipation in the jute–PP composite which suggest a better interfacial bonding.

CHAPTER 1

STILL LOOKING FOR THE MAGIC SPOT: DISPERSING MODIFIED MAGNETIC NANOPARTICLES INTO NANOSTRUCTURED BLOCK COPOLYMERS

I. BARANDIARAN and G. KORTABERRIA[*]

Materials and Technologies Group, Universidad del País Vasco/Euskal Herriko Unibertsitatea UPV/EHU, Plaza Europa 1, 20018 Donostia, Spain

[*]Corresponding author. E-mail: galder.cortaberria@ehu.eus

CONTENTS

ABSTRACT

In this chapter it has been carried out an extended revision of magnetic nanocomposites based on block copolymers and magnetic nanoparticles, analyzing different techniques used by different authors to disperse and selectively place magnetic nanoparticles in block copolymer matrix. These techniques varies from the use of low molar mass molecules, to the use of polymeric brushes to improve the compatibility between nanoparticles and the matrix. In addition, different methods to modify nanoparticle surface with polymer brushes have been studied (*grafting to, grafting through* and *grafting from*), explaining the differences between them and also the advantages and disadvantages of each method.

1.1   INTRODUCTION

Nanocomposites are multiphase solid materials in which one of the phases has one, two, or three dimensions of less than 100 nm, or structures having nanoscale repeating distances between the different phases that make up the material. Different types of nanocomposites can be found. Depending on the presence of polymers, such as matrices, they can be classified in two main groups: polymer-based nanocomposites and nonpolymer-based nanocomposites.

Among polymer-based nanocomposites, organic–inorganic ones can be distinguished, hybrid materials based on polymeric matrices (homopolymer or copolymer) that constitute the organic part, with nanoparticles (NPs) or other nanofillers dispersed on them. The addition of inorganic functional NPs, endows the nanocomposites with specific advantageous optical, conductive, electric, or magnetic properties[1-3] for various fields of application, such as photonic bandgap materials, solar cells, sensors, and high-density magnetic storage devices.[4-7] Among them, hybrid nanocomposites based on block copolymers and magnetic NPs can be mentioned. Magnetic NPs have attracted great interest for researchers from a wide range of disciplines, including magnetic fluids,[8] catalysis,[9] biotechnology,[10] magnetic resonance imaging,[11] data storage,[12] and environmental remediation.[13]

Block copolymers have the ability to self-assemble into continuous nanostructures. This makes them excellent candidates for the self-assembly of inorganic NPs within nanodomains, in order to design periodic arrangements for materials with enhanced properties at nanometer scale.[14] Block copolymers consist of more than one sequence of monomer units with different chemical composition that are covalently bonded. Owing to the thermodynamic incompatibility between blocks, they can self-assemble into a wide variety of nanostructures, covalent linkage among blocks preventing the phase separation at macroscopic scale.[15–17] Morphologies of self-assembled block copolymers depend on several parameters such as copolymer composition (f), Flory–Huggins interaction parameter among blocks (χ), and polymerization degree of the copolymer (N). Most common morphologies obtained for diblock copolymers are either in bulk or in thin films such as spheres, cylinders, gyroids, and lamellas.[18–21]

Preparation of nanocomposites based on block copolymers and magnetic NPs have attracted attention of many researches. In order to successfully prepare them, one of the obstacles that should be overtaken is the good dispersion and selective placement of NPs into the block copolymer, as it will determine the final properties of nanocomposites. For dispersing magnetic NPs, different techniques have been used by different authors, such as the use of small molecules as surfactants or the modification of NP surface by polymer brushes. In the following pages, different works related to the use of different routes to prepare nanocomposites with magnetic properties will be presented.

1.2   LOW MOLAR MASS MOLECULES FOR DISPERSING MAGNETIC NANOPARTICLES IN BLOCK COPOLYMERS

In order to disperse and selectively place inorganic NPs in self-assembled block copolymers, the use of electrostatically stabilized NPs[22,23] or NPs covered with low molar mass ligands[24,25] have been some of the methods; those methods provide the possibility to selectively incorporate them into the desired domain of the block copolymer. In such systems, the enthalpic interactions which will drive the NP incorporation into a block copolymer are hydrogen bonding, ionic, or dipole–dipole interactions acting between the NP surface and polymer chains, providing good selectivity of NPs.

Regarding magnetic NPs, the use of low molar mass ligands has been probed as a successful technique, as it can be found in different works.

For example, Park et al.[26] prepared oleic acid-modified maghemite (γ-Fe_2O_3) NPs and dispersed them into a self-assembled poly (styrene-*b*-isoprene) (PS-*b*-PI) block copolymer. They analyzed the effect of neutral or selective solvents to polyisoprene (PI) block, in the phase behavior of a cylinder-forming PS-*b*-PI diblock copolymer with magnetic γ-Fe_2O_3 NPs. By small-angle X-ray scattering (SAXS) and transmission electron microscopy (TEM) analysis, they observed that when toluene was used as casting solvent (neutral solvent), the hexagonal cylinder morphology of neat block copolymer first transforms into suppressed ordered morphology up to 5 wt.% and well-ordered by-centered cubic phase at 15 wt.%. NP aggregates started at 3 wt.% of NPs, considered provoking morphological transformation. While using hexane as casting solvent, selective to PI block, they observed higher ordered hexagonal morphology, maintained at lower NP loading, transforming into a disordered morphology with extensive aggregation of NPs for nanocomposites with 10 wt.% of NPs. From this work, they concluded that NP dispersion could be improved by using a selective solvent when NPs are marginally selective to one of the blocks, as the use of a selective solvent will improve the selective placement of NPs and prevent aggregates formation until higher concentration.

In the same direction, Char et al.[27] also observed that selective location of magnetic NPs and the size of aggregates can be clearly altered by changing casting solvents. They studied two block copolymers with different solvents, poly (styrene-*b*-4-vinyl pyridine) (PS-*b*-P4VP) and PS-*b*-PI diblock copolymers, and maghemite NPs. In the case of PS-*b*-P4VP-based nanocomposites, at 5 wt.% of NP concentration they analyzed the effect of three solvents: $CHCl_3$ (neutral and fast evaporation), Tetrahydrofuran (THF) (selective to PS and fast evaporation), and toluene (selective to PS and slow evaporation), as it can be seen in Figure 1.1. With the first solvent, they observed that NPs tended to form aggregates, according to their previous work[26] neutral solvents favored aggregates formation. With THF, NPs were well dispersed and selectively placed at PS domains, although they tend to the interfaces between PS and poly (4-vinyl pyridine) (P4VP) domains. With toluene, selective for PS, NP aggregates were formed, as low evaporation rates favored aggregates formation.[26] These differences due to the use of surfactants can be seen in Figure 1.1. While with $CHCl_3$ or toluene NPs tended to aggregate, with THF they were well dispersed and selectively placed. At higher NP concentration using THF as solvent, NP placement at the interfaces was more prominent, significantly disturbing block copolymer morphology.

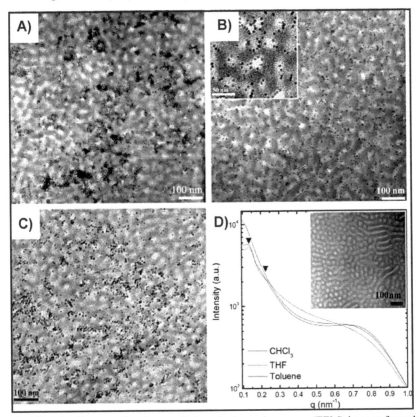

FIGURE 1.1 Cross-sectional transmission electron microscopy (TEM) images for poly (styrene-*b*-4-vinyl pyridine) (PS-*b*-P4VP) (20/40)/nanoparticle (NP) mixtures (5 wt.%) cast with: (A) $CHCl_3$, (B) THF, and (C) toluene. P4VP domains are shown in black by the iodine staining and even darker spots in each TEM micrograph represent dispersed NPs of 6 nm in diameter. Inset of (B) shows the NPs localized at the interfaces between PS and P4VP blocks. The small-angle X-ray scattering (SAXS) profiles of samples cast with three different solvents are also shown in (D). Different NP correlations at around $q = 0.75$ nm^{-1} indicate significantly suppressed NP aggregation for the sample cast with THF compared with toluene and $CHCl_3$. The inverted filled triangles represent the peaks with a ratio of $1:\sqrt{3}$ corresponding to the hex phase and the TEM image of a neat PS-*b*-P4VP (20/40) is also shown in the inset of (D).

Source: Adapted with permission from ref 27. Copyright (2009) Elsevier.

Wu et al.[28] also used oleic acid as dispersing agent for magnetic NPs, but in this case in poly (styrene-*b*-butadiene-*b*-styrene) (SBS) block copolymer. They analyzed the effect of NP size in their dispersion and selective placement into SBS block copolymer. As the octadecene segments on the oleic acid molecules have chemical affinity with polybutadiene (PB) block, NPs tended to locate in these domains. While observing the effect of NP size, they concluded that for NP diameter smaller than the thickness of PB lamella ($d/l \sim 0.5$) with the addition of 9.1 wt.% NPs they were selectively confined into PB domains, but tended to join for higher concentration (25 wt.%). For diameters in the same order of PB lamella thickness ($d/l \sim 1$), NPs were not uniformly dispersed in the SBS matrix, but rather aggregate to form clusters of 100–300 nm in size at all loadings. However, it was observed that, although NP aggregates were formed, those aggregates were still selectively confined into PB domains and, thus, form oriented strings along the PB phase. If d was larger than l, as NPs were larger than the lamellar thickness of the PB domains, they strongly aggregate, forming clusters of hundreds of nanometers to several micrometers. Selective dispersion was not observed due to the fact that PB domains could not accommodate NPs larger than their lamellar thickness.

Horechyy et al.[29] also worked with nanocomposites based on block copolymers and magnetic NPs. In this case, magnetite NPs were synthesized in the presence of a mixture of tri-*n*-octylphosphine oxide (TOPO) and oleylamine surfactants. They analyzed their dispersion into PS-*b*-P4VP and poly (*n*-pentyl methacrylate-*b*-methyl methacrylate) (PPMA-*b*-PMMA) block copolymers. They observed that for systems based on PS-*b*-P4VP, NPs interacted preferentially with P4VP block appearing selectively localized at P4VP cylindrical domains. However, for systems based on PPMA -*b*-PMMA, NPs had no significant preference for the any of the blocks and were segregated at the polymer/substrate interface, provoking evident domain orientation change.

Lastly, Lo et al.[30] used pyridine-tethered nanorods to endow block copolymer with magnetic properties. In this case, they first synthesized Fe_2P nanorods and then modified their surface with pyridine through a ligand exchange process according to Gupta et al.,[31] finally preparing nanocomposites with poly (styrene-*b*-2-vinyl pyridine) (PS-*b*-P2VP) block copolymer. They prepared nanorods with different lengths of 20.9, 40.8, and 100.9 nm. Regarding the dispersion in poly (2-vinyl pyridine)

(P2VP) homopolymer, they found that nanorods with length of 20.9 and 40.8 nm were well dispersed, but the longest nanorods tended to aggregate. On the other hand, in the case of block copolymers, the shortest nanorods were well dispersed and aligned parallel to the interface between PS and P2VP at low particle loading (2%), but at higher nanorod concentration they failed to disperse individually in polymer phase and small aggregates were formed, as it can be seen in Figure 1.2. Despite those aggregates, PS-*b*-P2VP still exhibited lamellar structure. In addition, by SAXS measurements, they found that P2VP lamellas became thicker; this increase in nanodomain size with an increase in rod concentration was associated with the confinement into the referred domains of block copolymer. With 40.8 nm length nanorods, incorporation of longer nanorods in PS-*b*-P2VP caused the dislocation and declination of composite microstructure and those nanorods preferred to locate at the dislocation and declination. The longer nanorods also reduced the selection of nanorods in the P2VP domains by forming aggregates. With longest ones, nanorods formed extensive aggregates. This work demonstrated that modifying nanorods with pyridine groups is an adequate technique to disperse them in P2VP-based block copolymers.

From the analysis of works in which low molar mass ligands were used to disperse magnetic nanofillers, it can be said that these can constitute an adequate option to prepare nanocomposites with magnetic properties. Oleic acid is a commonly used surfactant to disperse magnetic NPs in different block copolymers, improving compatibility with one of the domains of the block copolymer, affinity that can be changed with block copolymer composition. When oleic acid was used to disperse NPs in PS-*b*-PI block copolymer, they were selectively placed at PI domains,[26,27] they tended to PS domains when dispersing in PS-*b*-P4VP,[6] and to PB domains in the case of SBS.[28]

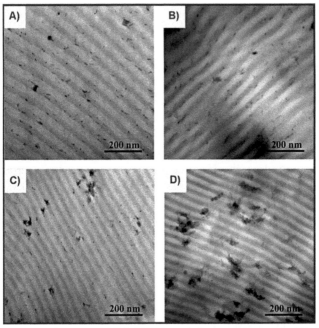

FIGURE 1.2   TEM images of PS-*b*-P2VP/nanorod composites with: (A) 0.5; (B) 1; (C) 2%; and (D) 4 wt.%. The rod length is 20.9 nm.

Source: Adapted with permission from ref 30. Copyright (2013) American Chemical Society.

1.3   POLYMER BRUSHES FOR DISPERSING MAGNETIC NANOPARTICLES IN BLOCK COPOLYMERS

As it was mentioned before, another method to disperse NPs in block copolymers is the modification of their surface with polymer brushes corresponding to one of the blocks, in order to improve compatibility between NP and the corresponding domains of the self-assembled block copolymer.

Modifying NP surface with polymer brushes changes the interactions between NPs and polymeric matrix. With the right polymer brushes, dispersion of NPs in polymeric matrix will be thermodynamically favored. The solubility of polymer brushes in polymeric matrix is determined by the Flory–Huggins interaction parameter.[32] Usually, the Flory–Huggins interaction parameter between polymers has a positive value if polymers are chemically different, in those cases polymeric matrix and polymer brushes are insoluble; in rare cases where χ is negative, polymeric matrix and polymer brushes will be soluble; when $\chi = 0$ both polymeric matrix and polymer brushes are chemically identical polymers, so they will be soluble.[32] Lots of works have been published about solubility of NPs with polymer brushes in polymeric matrices; works have been mainly directed to the study of different parameters that affect solubility when polymer brushes and matrix are chemically identical.[33–36] The parameters affecting the solubility of polymer brushes in a polymeric matrix are the molecular weight of both matrix and polymer brushes, and the grafting density of the brushes at NP surface. Depending on those parameters different states have been identified.[33–36]

As NP dispersion is improved by modifying its surface with polymer brushes chemically identical to the polymeric matrix, the same behavior could be expected when dispersing NPs in block copolymers. Modifying NP surface with polymer brushes chemically identical to one of the blocks will improve the affinity between NP and one of the blocks, and NPs could be dispersed and selectively placed. However, in the case of block copolymers, as the matrix presents chemically different polymer chains, mechanism can be modified.[37] Based on this theory, lots of works have been published using different NPs,[38,39] and among them, due to the interest that magnetic NPs have attracted, they also have been modified with polymer brushes for being dispersed in block copolymers.

In order to modify magnetic NP surface with polymer brushes, three different methods can be mainly found: *grafting to*, *grafting through*, and *grafting from*.

1.3.1 GRAFTING TO *METHOD*

In this method, polymers with suitable end functional groups react with appropriate surface sites on the inorganic NPs.[40,41] This technique has a limitation on the choice of functional groups available for incorporating into the polymer, because functional groups on the polymer can compete with the anchor moieties for surface sites. This point has important relevance, if the aim is to immobilize functional polymers containing highly polar or charged groups onto polar surfaces, as the adsorption of functional groups into the surface can be strong and competes with chemisorption process.[42] Another limitation is the thickness of the film, and accordingly the number of functional groups per surface area that can be obtained by using this approach. The rate of attachment reaction levels off rather quickly and further polymer is linked to the substrate only at an extremely slow rate due to the kinetic hindrance. Owing to the inability of large polymer chains to diffuse the reactive surface sites, they are sterically hindered by the surrounding bounded chains.[43] However, following this technique different works have been published, demonstrating the usefulness of the method in some cases.

For example, Lauter et al.[44] modified magnetite (Fe_3O_4) NPs with polystyrene brushes to disperse them in poly (styrene-*block-n*-butyl methacrylate) (PS-*b*-PBMA) diblock copolymer. In this case, Fe_3O_4 NPs surface was covered with short PS chains from α-lithium polystyrene sulfonated.[45] They studied the morphology of the block copolymer with and without NPs by neutron reflectometry and atomic force microscopy (AFM). By neutron reflectometry measurements they observed that PS covered magnetic NPs were segregated toward PS domains for 5% of NPs in volume, repeating the same scenario with 10% of NPs. After annealing the samples, due to the NP placement at PS domains, morphology changed from cylindrical to lamellar structure; NPs were incorporated into PS domains swelling them, what provoked morphological change. At higher magnetic NP concentration the same procedure was seen, swelling a larger amount of PS domains.

Aissou et al.[46] also used the *grafting to* method to cover FePt NPs with dopamine-terminated methoxy poly (ethylene oxide) (mPEO-Dopa), to disperse them in poly (styrene-*block*-ethylene oxide) (PS-*b*-PEO) block copolymer. Functionalization of FePt NPs was carried out following the method described by Xu et al.[47] anchoring mPEO-Dopa chains with a molecular weight of 750 g/mol by using a ligand exchange method. The dopamine is a well-established molecular anchor leading to a robust link between functional molecules and iron oxide. Aissou et al. first found that PS-*b*-PEO block copolymer self-assembled into well-ordered hexagonally close-packed (HCP) morphology from AFM and two-dimensional fast Fourier transform (2D-FFT) analysis. This morphology changed after FePt magnetic NP addition. By SAXS measurements they observed two different morphologies for nanocomposites, depending on sample thickness, and, as it can be seen in Figure 1.3, by TEM analysis observed the segregation of FePt NPs coated with mPEO-Dopa chains at PS/ PEO interfaces. They concluded that the location of NP in the interface could be because FePt NP surface was not fully covered by the ligands, what favored the location of NPs at PS/PEO interfaces. The driving force for the localization of NPs at A/B interface of a block copolymer is inherent to their surfactant behavior, which decreases the interfacial tension, γ_{AB}, between the A and B blocks.[48]

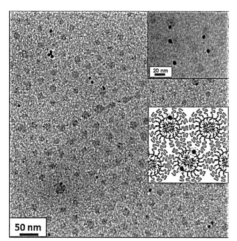

FIGURE 1.3 TEM micrograph of the microphase-separated poly (styrene-block-ethylene oxide) (PS-*b*-PEO)/FePt-m dopamine-terminated -methoxy poly (ethylene oxide) (PEO-Dopa) thin films containing 15 wt.% of FePt particles and annealed under a toluene/water saturated vapor atmosphere. The nanoparticles are located at the PS/PEO interfaces. Inset: magnified TEM image and schematic representation of FePt particle localization.

Source: Adapted with permission from ref 46. Copyright (2011) American Chemical Society.

Basly et al.[49] also modified FePt NPs with PEO brushes, from a thiol-ended poly-(ethylene glycol) methyl ether (mPEO-SH). In addition, they used sacrificial MgO shells as physical barriers during the annealing stage for more efficient anchoring of compatibilizing PEO macromolecules onto NP surface. In this case, $L1_0$-FePt@PEO NPs were selectively located at PEO domains, as it was corroborated by TEM and AFM analysis. Results indicate that at low weight ratio between NPs and block copolymer (0.5 or 1) NPs were isolated from each other (weak aggregation) and mainly positioned close to the center of PEO domains. At higher ratios (≥ 2) large number of aggregates appeared at PEO domains. Those aggregates induced local defects in the hexagonal lattice of PEO cylinders. Finally, 6 % weight ratio the disturbing effect of the NPs led to an inhomogeneity of film thickness.

Following with the *grafting to* method, Barandiaran et al.[50] also modified magnetic NPs by using it. In this case, they functionalized Fe_2O_3 NPs with poly (methyl methacrylate-*block*-ε-caprolactone) brushes with a chlorine terminal group (*Cl*-PMMA-*b*-PCL) synthesized by atom transfer radical polymerization (ATRP) with a number average molecular weight of 21,500 g/mol ($f_{PMMA} = 0.7$ and $f_{PCL} = 0.3$) and a polydispersity of 1.35.[51] Those

NPs were dispersed in poly (styrene-*block*-ε-caprolactone) (PS-*b*-PCL) block copolymer matrix. Fe_2O_3 magnetic NPs were first modified with 3-aminopropyl tryethoxysilane (APTS), and then, *Cl*-PMMA-*b*-PCL polymer brush was anchored by an alkylation reaction of amine groups present at the surface of NPs. They calculated a grafting density of 0.04 chains/nm² by thermogravimetric analysis (TGA). By AFM analysis of nanocomposite surface morphology they observed that, although small NP aggregates were formed during grafting process, functionalized NPs were well dispersed in the block copolymer. The grafting of NPs with PMMA-*b*-PCL copolymer seemed to increase compatibility with matrix, improving their dispersion; they were mainly located at the interface between PCL and PS domains without breaking copolymer nanostructure as it can be seen in Figure 1.4. This fact could be due to the low grafting density (0.04 chain/nm²) of the copolymer on the NP surface, as it was pointed out by other authors.[38]

Those examples demonstrate that *grafting to* is an useful technique to disperse magnetic NPs though block copolymer matrices.

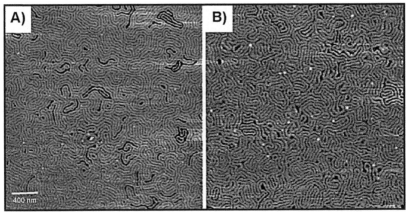

FIGURE 1.4   Atomic force microscopy (AFM) phase images of poly (styrene-block-ε-caprolactone) (PS-*b*-PCL)/Fe_2O_3-*g*-(n-pentyl methacrylate-*b*-methyl methacrylate) (PMMA-*b*-PCL) nanocomposites annealed at 120°C and nanoparticle amounts of (A) 2 wt.% and (B) 5 wt.%.

Source: Adapted with permission from ref 50. Copyright (2014) Elsevier.

1.3.2 GRAFTING THROUGH *METHOD*

It consists on a surface copolymerization through a covalently linked monomer. In this route, the inorganic phase is incorporated inside the polymer chains.[52–55] For a polymer layer generation through this method, the growth of polymer chains is initiated in solution. During propagation, occasionally a surface-bound monomer unit can be integrated into the growing chains, which directly results in a permanent anchoring of the polymer chains. After that, chain growth can proceed and further free or surface attached units can be added to the growing chain. Despite the fact that such processes are widely applied, even in industrial applications mostly for adhesion promotion, from a scientific point of view *grafting through* approach has not been extensively explored and the details of the mechanism are not well understood. However, there are some works that used this method to successfully functionalize NPs to disperse them in block copolymers.

For example, using *grafting through* method, Barandiaran et al. published some works demonstrating the validity of this method to prepare nanocomposites with dispersed and selectively placed magnetic NPs. First, they functionalized Fe_3O_4 magnetic NPs with PMMA polymer brushes following the *grafting through* method.[56] For this purpose, NPmodification was carried out in two steps: first silanization process with 3-methacryloxypropyl trimethoxysilane (MPTS), as this silane contains the vinyl group that would be used for the polymerization of PMMA brushes,[57] and then growth of PMMA brushes by *grafting through*, using 2,2-azobisisobutyronitrile (AIBN) as initiator. They studied the dispersion and selective placement of NPs in PMMA domains of poly (styrene-*b*-methyl methacrylate) (PS-*b*-PMMA) symmetric diblock copolymer by analyzing the surface morphology with AFM. As it can be seen in Figure 1.5, NPs were selectively placed at PMMA domains. In addition, as different NP concentrations were probed, they pointed out that NP addition affected the morphology of block copolymer. At lower concentration lamellar morphology of the block copolymer was maintained, but at higher concentration (5 wt.%) morphology changed to a mixture of lamellae and cylinders perpendicular to the surface. They concluded that this could be due to the selective placement of NPs in one of the blocks, what provoked a swollen of its domain, altering the volume relation between blocks that changed the morphology.

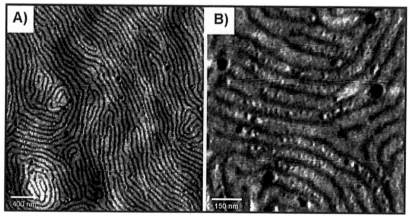

FIGURE 1.5 AFM height images of nanocomposites based on PS-*b*-PMMA block copolymer and 2 wt.% of magnetic nanoparticles functionalized with PMMA polymer brushes following the *grafting through* method.

Barandiaran et al.[58] also modified Fe_2O_3 NPs with PS polymer brushes to prepare nanocomposites based on PS-*b*-P4VP block copolymer. In this case, they similarly modified NP surface with MPTS silane to provide the vinyl group necessary to polymerize PS polymer brushes, and then, they carried out polymerization reaction by *grafting through* technique. Modified NPs were dispersed in PS-*b*-P4VP block copolymer. They studied morphological evolution of block copolymer after solvent vapor annealing (SVA) treatment in dioxane and the effect of NP addition by AFM. Regarding morphological evolution of block copolymer after SVA, they observed that morphology changed from a soft microphase separation without ordered microstructure, to a hexagonal morphology and after longer exposure time, to a stripped morphology. At low NP concentration (1 wt.%), morphology of block copolymer was maintained, but at higher concentration (5 wt.%) morphology changed, probably due to the selective placement of the NPs in the PS domains, what provoked a decrease on domain mobility.[59] In addition, a good dispersion of NPs was corroborated, as big aggregates of NPs were not detected.

In another work, Barandiaran et al.[60] used the same functionalized NPs, both with PMMA brushes[56] and PS brushes,[58] to prepare nanocomposites with poly (styrene-*b*-butadiene-*b*-methyl methacrylate) (SBM) triblock copolymer. They analyzed the morphology of neat block copolymer and nanocomposites with magnetic NPs functionalized with PMMA or PS polymer brushes. SBM triblock copolymer self-assembled into lamellar nanostructure, with a S–B–M–B sequence, as it was seen by AFM. After NP addition, lamellar morphology was maintained and each NP was placed at the corresponding domain, PMMA-modified NPs at PMMA domains and PS-modified NPs at PS ones. This selective placement was corroborated both by AFM and differential scanning calorimetry (DSC) analysis by an increase in glass transition temperature (T_g).

In both cases, after functionalization of NPs with PMMA or PS brushes, they concluded that small aggregates of NPs were created before thin film preparation during NP modification, and not during their dispersion and positioning on the block copolymer matrix. In the *grafting through* functionalization method, polymerization was carried out with silane-modified NPs in the media and, distinctly from *grafting from* method, in which polymer chain grows only from the surface of the NP, the functional group located in the surface of NP could join a growing polymer chain in which there could be previously joined NPs. As the surface of the NPs was multifunctional, with several double bonds, several chains could be bonded to different NPs.[55,61] Owing to these reasons, it was supposed that NPs, instead of being located individually, were bonded together with several polymer chains, creating a kind of network formed by NPs and polymer chains.

1.3.3 GRAFTING FROM *METHOD*

Polymer chains grow in situ from initiator molecules previously pre-grafted onto the surface of NPs.[62,63] There are different polymerization techniques to functionalize surfaces following the *grafting from* method, such as controlled radical polymerization (CRP), ionic living polymerization, or ring-opening polymerization, among others.[64–66] But some issues, such as poor control of molecular weight, polydispersity, end functionality, chain architecture and composition in the case of the CRP[67] or, stringent conditions, and the small number of monomers that can be polymerized in ionic living polymerization method,[68] are disadvantages that should be overcome. Following the aim of improving the CRP technique, controlled living radical polymerization methods have been developed, such as ATRP,[69–72] reversible addition–fragmentation chain transfer (RAFT)[73,74] or nitroxide-mediated polymerization (NMP).[75–77]

ATRP is a metal-catalyzed polymerization involving the reversible activation–deactivation reaction between a growing polymer chain and metal–ligand species. The ATRP initiator is activated in the presence of metals such as Cu, Ru, Fe, and others.[63] In RAFT polymerization, the mechanism involves a sequence of addition fragmentation equilibrium. This technique requires selecting a suitable transfer agent with high transfer constant in radical polymerization of the desired monomer under appropriate reaction conditions.[63] NMP is based on the concept of activation–deactivation equilibrium between dormant species and a small fraction of propagating macroradical. The reactions are carried out at elevated temperatures, where the initiation is rapid and all the chains are formed at the same time.[63]

For dispersing magnetic NPs into self-assembled block copolymers, the most common method for functionalizing NP surface following the *grafting from* technique is ATRP, as many works can be found in the literature. Garcia et al. followed this method to functionalize Fe_3O_4 NPs with PS polymer brushes[78] and then disperse them in SBS triblock copolymer.[79] To functionalize NPs, they used 2-(4-chlorosulfonylphenyl) ethyltrichlorosilane (CTCS) as initiator, which was anchored to NP surface; a polymerization reaction was carried out by ATRP of PS with copper bromide (CuBr)/2,2'-bipyridine (Bip) as a catalyst system and p-toluenesulfonyl chloride (TsCl) as a free initiator. They calculated a grafting density of 0.9 chain/nm². After that, they prepared nanocomposites dispersing those NPs in SBS block copolymer, and measured their T_g by calorimetry and dynamic mechanical analysis, mechanical properties by rheological analysis, and morphology by AFM. They observed that NPs were selectively placed at PS domains, as the T_g of PS domain increased, whereas that of PB ones maintained constant. They corroborated the good dispersion of NPs by AFM analysis. In addition, PS-modified NP addition enhanced mechanical properties of nanocomposites.

Xu et al.[37] investigated on PMMA-modified magnetic NPs, dispersed in PMMA homopolymer and PS-*b*-PMMA block copolymer. Fe_3O_4 NPs were grafted with PMMA brushes with molecular weights from 2.7 to 35.7 kg/mol, synthesized following the method proposed by Ohno et al.[80] After cleavage of polymer brushes from NP surface, they measured molar masses of 2.7, 13.3, and 35.7 kg/mol for different brushes, and calculated a grafting density of around 0.73 chains/nm². They characterized the morphology of nanocomposites by AFM and TEM. Regarding dispersion of NPs in PMMA homopolymer, they found that the longer were polymer brushes attached to NP surface, better was NP dispersion. In as cast nanocomposites, they obtained good dispersions with all NPs, but thermal annealed NPs with shorter polymer brushes (2.7 and 13.3 kg/mol) tended to form aggregates, as was not the case for NPs with the longest polymer brushes. On the contrary, as molecular weight of polymer brushes attached to NPs increased, NP dispersion was reduced, resulting in the formation of aggregates even in the as-cast nanocomposite films. As it can be seen in Figure 1.6, NPs with 2.7 kg/mol polymer brushes were selectively dispersed at 1 wt.% of concentration, but at higher concentration they tended to form aggregates and disrupted the lamellar morphology. In the case of longer polymer brushes, small aggregates tended to locate at defects in the lamellae, whereas larger ones greatly perturbed the morphology and became encapsulated by onion-ring copolymer structures. They concluded that dispersion behavior of NPs in PMMA and PS-*b*-PMMA was controlled by different mechanisms.[37]

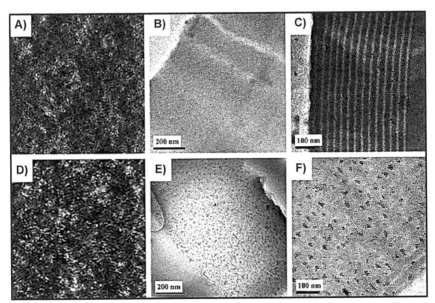

FIGURE 1.6 (A) AFM height image and (B) cross-sectional TEM image of an as-cast PS-*b*-PMMA/Fe$_3$O$_4$-2.7 K (1 wt.%) film. (C) cross-sectional TEM image of PS-*b*-PMMA/Fe$_3$O$_4$-2.7 K (1 wt.%) after annealing at 185°C for 10 days. (D) AFM height image and (E) cross-sectional TEM image of an as-cast PS-*b*-PMMA/Fe$_3$O$_4$-2.7 K (10 wt.%) film. (F) cross-sectional TEM image of PS-*b*-PMMA/Fe$_3$O$_4$-2.7 K (10 wt.%) after annealing at 185°C for 8 days. The size of both AFM images is 2×2 μm^2, and the height scale is $\Delta z = 0$–8 nm.

Source: Adapted with permission from ref 37. Copyright (2014) Elsevier.

Barandiaran et al.[81] also worked with magnetic NPs functionalized by *grafting from*, modifying Fe_2O_3 NP surface with PMMA polymer brushes by ATRP to disperse them in poly (isoprene-*b*-methyl methacrylate) (PI-*b*-PMMA) block copolymer. They functionalized magnetic NPs following the method used by Garcia et al.[82] first modifying NPs with CTCS initiator and then polymerizing PMMA by ATRP with CuBr/Bip as catalyst, obtaining a grafting density of 0.8 chains/nm², and polymer brushes with low molecular weight (1000 g/mol). They analyzed morphology of nanocomposites without any treatment and after SVA in acetone by AFM. They observed that lamellar morphology of PI-*b*-PMMA without any treatment was more disrupted with NP addition, than for solvent vapor annealed nanocomposites. When NPs were added to as-casted block copolymer, orientation of lamellae changed, resulting in a mixture of perpendicular and parallel lamellae. They concluded that NP addition could modify interactions between block copolymer domains and the substrate, evolving an orientation change of the lamellar nanostructure. On the other hand, cylindrical morphology obtained for the neat diblock copolymer after solvent annealing was maintained with NP addition up to 5 wt.% of NP concentration, when cylindrical nanostructure was disrupted and disordered morphology obtained.

Those works demonstrates the usefulness of *grafting from* technique to disperse and selective place magnetic NPs in nanostructured block copolymers.

1.4 MAGNETIC PROPERTIES OF NANOCOMPOSITES

As it has been mentioned before, the aim of dispersing magnetic NPs in block copolymers is the preparation of magnetic nanocomposites, namely, new materials with magnetic properties. The reason to pursue good and selective dispersion of magnetic NPs is that this parameter will condition the correct property transference to the nanocomposite. To analyze magnetic properties of the material, different characterization techniques can be used.

In order to characterize the macroscopic magnetic properties of nanocomposites, superconducting quantum interface device (SQUID) magnetometers have been used. For example, Xu et al.[83] carried out zero-field-cooled (ZFC) and field-cooled (FC) magnetic moment and M-H magnetization curves with a SQUID magnetometer. Nanocomposites were prepared following the procedure used in a previous work.[37] In those magnetic characterizations, they studied nanocomposites with Fe_3O_4 NPs functionalized with PMMA polymer brushes synthesized by *grafting from*. This functionalizing method gave the opportunity to analyze the effect polymer brush length (2.7 and 35.7 kg/mol) in magnetic properties. They observed that, despite creating aggregates with a diameter of about 164 nm, the magnetic properties are quite similar, against what it should be expected. They concluded that the reason could be that, in the case of NPs functionalized with the longest polymer brushes, those polymer brushes were large enough to keep enough distance between NPs, so interactions between aggregated NPs were negligible. In the magnetic characterizations presented in Figure 1.7.

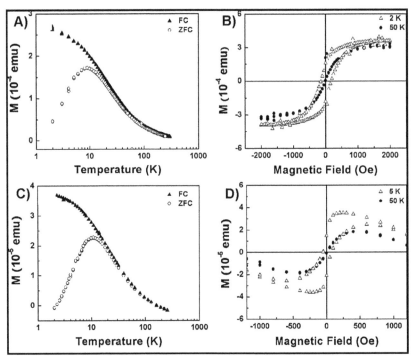

FIGURE 1.7 Magnetic properties of as-cast PS-*b*-PMMA/Fe_3O_4-2.7 K (10 wt.%) films: zero-field-cooled (ZFC) and field-cooled (FC) curves at 100 Oe (A) and *M-H* curves at 2 and 50 K (B). Magnetic properties of PS-*b*-PMMA/Fe_3O_4-35.7 K (16 wt.%) films after annealing at 185°C for 240 h: ZFC and FC curves at 100 Oe (C) and *M-H* curves at 5 and 50 K (D). Note: both (B) and (D) include diamagnetic contributions from the block copolymer matrix and sample holder. In the case of (D), these contributions are sufficient to produce a negative slope at high fields.

Source: Adapted with permission from ref 83. Copyright (2009) American Chemical Society.

Barandiaran et al. also studied the magnetic properties of block copolymer/NP nanocomposites by ZFC/FC measurements and M-H magnetization curves with a SQUID magnetometer. They used magnetic NPs modified with PS or PMMA polymer brushes following the *grafting through* method and dispersed in PS-*b*-P4VP diblock copolymer[58] or SBM triblock copolymer.[60] From the magnetic measurements, they found that the blocking temperature (T_b) was similar in all cases, between 200 and 250 K. This could be due to the functionalization method *grafting through*, where small aggregates are formed during functionalization. As the T_b is related to NP size, being those measured values similar it could be expected that the size of aggregates was similar in all cases.

Yao et al.[84] also studied magnetic properties by SQUID measurements in nanocomposites based on poly (styrene-*b*-*N*-isopropyl acrylamide) (PS-*b*-PNIPAM) block copolymer and γ-Fe_2O_3 magnetic NPs. Those NPs were surface functionalized with PS brushes following the grafting to method,[44] and selectively placed at PS domains of nanostructured PS-*b*-PNIPAM block copolymer. In this case, they studied the effect of NP concentration and temperature in magnetic properties. They observed that remanence (M_r) and saturation magnetization (M_s) showed similar tendency changing the concentration of NPs, both increased with increasing NP concentration, while on the contrary, relative M_r/M_s remained independent to concentration. Regarding temperature, the width of hysteresis loop decreased with increasing temperature. They concluded that, considering all aspects of the magnetic behavior, the investigated hybrid films exhibited the typical superparamagnetic properties.

Aissou et al.[85] also characterized macroscopically magnetic properties of nanocomposites based on PS-*b*-PEO block copolymer and FePt NPs with an SQUID magnetometer, but in addition, they studied microscopic magnetic properties by magnetic force microscopy (MFM). In this work, they dispersed NPs by functionalizing NP surface by *grafting to* with PEO polymer brushes, so NPs were selectively located at PEO cylinders. Magnetic images were obtained with a surface force microscopy equipment in torsional mode by two pass technique (interleave mode). The magnetic probe consists of a silicon cantilever with a pyramidal tip with a magnetic coating (CoCr). From this microscopic study based on magnetic properties of nanocomposite, they detected black signals produced by magnetic NPs in PEO domains. That magnetic phase image was acquired at a 5 nm lift, in the second pass, without an indication of the topography remaining for this lift height. Therefore, distinct dark spots observed in the magnetic phase image can be directly correlated with the $L1_0$-ordered FePt NP.

1.5 CONCLUSIONS

From this review work, it could be concluded that during the last years interesting advances have been done in the synthesis and characterization of nanocomposites based on block copolymers and magnetic NPs. Many works have been published demonstrating the interest of the topic and different ways of obtaining magnetic nanocomposites. The differences mainly come from the functionalization of NPs, as different techniques can be followed. All the methods described in this chapter had its advantages and disadvantage. When using small surfactant molecules, which is a simple and easy method, the selectivity of the same surfactant could be changed depending on the matrix. *Grafting to* method also is a simple an easy method, but obtained low grafting density can lead to nonselectively dispersed NPs. *Grafting through* method is also an easy method, but as it has been demonstrated, small NP aggregates are formed during modification. Finally, *grafting from* is a versatile method, because makes possible to control the length polymer brushes, for a better control of the dispersion and selective placement of NPs. But *grafting from* is the most complex method. Owing to this reason, it should be important to select the adequate method to functionalize NPs depending on the needs.

KEYWORDS

- **magnetic nanoparticle**
- **block copolymer**
- **polymer brushes**
- **dispersion**
- **selectivity**

REFERENCES

1. Gutierrez, J.; Tercjak, A.; Peponi, L.; Mondragon, I. Conductive Properties of Inorganic and Organic TiO_2/Polystyrene-block-Poly (ethylene oxide) Nanocomposites. *J. Phys. Chem. C* **2009**, *113*, 8601–8605.
2. Etxeberria, H.; Tercjak, A.; Mondragon, I.; Eceiza, A.; Kortaberria, G. Electrostatic Force Microscopy Measurements of CdSe-PS Nanoparticles and CdSe-PS/Poly (styrene-*b*-butadiene-*b*-styrene) Nanocomposites. *Collod Polym. Sci.* **2014**, *292*, 229–234.
3. Xia, X.; Metwalli, E.; Ruderer, M. A.; Körstgens, V.; Busch, P.; Böni, P.; Müller-Buschbaum, P. Nanostructured Diblock Copolymer Films with Embedded Magnetic Nanoparticles. *J. Phys. Condens. Matter.* **2011**, *23*(25), 4203 (9pp).
4. Bockstaller, M. R.; Mickiewicz, R. A.; Thomas, E. L. Block Copolymer Nanocomposites: Perspectives for Tailored Functional Materials. *Adv. Mater.* **2005**, *17*, 1331–1349.
5. Darling, S. B.; Yufa, N. A.; Cisse, A. L.; Bader, S. D.; Sibener, S. J. Self-organization of FePt Nanoparticles on Photochemically Modified Diblock Copolymer Templates. *Adv. Mater.* **2005**, *17*, 2446–2450.
6. Park, S.; Lee, D. H.; Xu, J.; Kim, B.; Hong, S. W.; Jeong, U.; Xu, T.; Russell, T. P. Macroscopic 10-Terabit–per–Square-Inch Arrays from Block Copolymers with Lateral Order. *Science* **2009**, *323*, 1030–1033.
7. Jang, Y. H.; Xin, X.; Byun, M.; Jang, Y. J.; Lin, Z.; Kim, D. H. An Unconventional Route to High-Efficiency Dye-Sensitized Solar Cells via Embedding Graphitic Thin Films into TiO_2 Nanoparticle Photoanode. *Nano Lett.* **2012**, *12*, 479–485.
8. Munoz-Menendez, C.; Conde-Leboran, I.; Baldomir, D.; Chubykalo-Fesenkob, O.; Serantes, D. The Role of Size Polydispersity in Magnetic Fluid Hyperthermia: Average vs. Local Infra/Over-heating Effects. *Phys. Chem. Chem. Phys.* **2015**, *17*, 27812–27820.
9. Zhao, X.; Liu, X. A Novel Magnetic $NiFe_2O_4$@graphene–Pd Multifunctional Nanocomposite for Practical Catalytic Application. *RSC Adv.* **2015**, *5*, 79548–79555.
10. Aseri, A.; Garg, S. K.; Nayak, A.; Trivedi, S. K.; Ahsan, J. Magnetic Nanoparticles: Magnetic Nano-technology Using Biomedical Applications and Future Prospects. *Int. J. Pharm. Sci. Rev. Res.* **2015**, *31*, 119–131.
11. Sun, C.; Lee, J. S. H.; Zhang, M. Magnetic Nanoparticles in MR Imaging and Drug Delivery. *Adv. Drug Delivery Rev.* **2008**, *60*, 1252–1265.
12. Galloway, J. M.; Talbot, J. E.; Critchley, K.; Miles, J. J.; Bramble, J. P. Developing Biotemplated Data Storage: Room Temperature Biomineralization of $L1_0$ CoPt Magnetic Nanoparticles. *Adv. Funct. Mater.* **2015**, *25*, 4590–4600.
13. Mirshahghassemi, S.; Lead, J. R. Oil Recovery from Water under Environmentally Relevant Conditions Using Magnetic Nanoparticles. *Environ. Sci. Technol.* 2015, *49*, 11729–11736.
14. Hoheisel, T. N.; Hur, K.; Wiesner, U. B. Block Copolymer-nanoparticle Hybrid Self-assembly. *Prog. Polym. Sci.* **2015**, *40*, 3–32.
15. Leibler L. Theory of Microphase Separation in Block Copolymers. *Macromolecules* **1980**, *13*, 1602–1617.

16. Bates, F. S.; Fredickson, G. H. Block Copolymer Thermodynamic: Theory and Experiment. *Annu. Rev. Phys. Chem.* **1990**, *41,* 525–557.

17. Hadjichristidis, N.; Pispas, S.; Fluodas, G. A. (2003) *Block Copolymers: Synthetic Strategies, Physical Properties, and Applications;* Wiley Interscience: New Jersey.

18. Balsara, N. P. Kinetics of Phase Transitions in Block Copolymers. *Curr. Opin. Solid State Mater. Sci.* **1999**, *4,* 553–5587.

19. Castelletto V.; Hamley I. W. Morphologies of Block Copolymer Melts. *Curr. Opin. Solid State Mater. Sci.* **2004**, *8,* 426–438.

20. Matsen M. W.; Bates F. S. Unifying Weak- and Strong-segregation Block Copolymer Theories. *Macromolecules* **1996**, *29,* 1091–1098.

21. Fasolka M. J.; Mayes A. M. Block Copolymer Thin Films: Physics and Applications. *Annu. Rev. Mater. Res.* **2001**, *31,* 323–355.

22. Nandan, B.; Gowd, E. B.; Bigall, N. C.; Eychmüller, A.; Formanek, P.; Simon, P.; Stamm, M. Arrays of Inorganic Nanodots and Nanowires Using Nanotemplates Based on Switchable Block Copolymer Supramolecular Assemblies. *Adv. Funct. Mater.* **2009**, *19,* 2805–2811.

23. Gowd, E. B.; Nandan, B.; Vyas, M. K.; Bigall, N. C.; Eychmüller, A.; Schlörb, H.; Stamm, M. Highly Ordered Palladium Nanodots and Nanowires from Switchable Block Copolymer Thin Films. *Nanotechnology* **2009**, *20*(41), 5302 (10pp).

24. Lin, Y.; Böker, A.; He, J.; Sill, K.; Xiang, H.; Abetz, C.; Li, X.; Wang, J.; Emrick, T.; Long, S.; Wang, Q.; Balazs, A.; Russell, T. P. Self-directed Self-assembly of Nanoparticle/Copolymer Mixtures. *Nature* **2005**, *434,* 55–59.

25. Horechyy, A.; Zafeiropoulos, N. E.; Nandan, B.; Formanek, P.; Simon, F.; Kiriya, A.; Stamm, M. Highly Ordered Arrays of Magnetic Nanoparticles Prepared via Block Copolymer Assembly. *J. Mater. Chem.* **2010**, *20,* 7734–7741.

26. Park, M. J.; Char, K.; Park, J.; Hyeon, T. Effect of the Casting Solvent on the Morphology of Poly (styrene-*b*-isoprene) Diblock Copolymer/Magnetic Nanoparticle Mixtures. *Langmuir* **2006**, *22,* 1375–1378.

27. Char, K.; Park, M. J. Selective Distribution of Interacting Magnetic Nanoparticles into Block Copolymer Domains Based on the Facile Inversion of Micelles. *React. Funct. Polym.* **2009**, *69,* 546–551.

28. Wu, J.; Li, H.; Wu, S.; Huang, G.; Xing, W.; Tang, M.; Fu, Q. Influence of Magnetic Nanoparticle Size on the Particle Dispersion and Phase Separation in an ABA Triblock Copolymer. *J. Phys. Chem. B* **2014**, *118,* 2186–2193.

29. Horechyy, A.; Nandan, B.; Zafeuropoulos, N. E.; Jehnichen, D.; Göbel, M.; Pospiech, D. Nanoparticle Directed Domain Orientation in Thin Films of Asymmetric Block Copolymers. *Colloid. Polym. Sci.* **2014**, *292,* 2249–2260.

30. Lo, C.-T.; Lin, W.-T. Effect of Rod Length on the Morphology of Block Copolymer/ Magnetic Nanorod Composites. *J. Phys. Chem. B* **2013**, *117,* 5261–5270.

31. Gupta, S.; Zhang, Q.; Emrick, T.; Russell, T. P. "Self-Corralling" Nanorods Under an Applied Electric Field. *Nano Lett.* **2006**, *6,* 2066–2069.

32. Borukhov, I.; Leibler, L. Enthalpic Stabilization of Brush-Coated Particles in a Polymer Melt. *Macromolecules* **2002**, *35,* 5171–5182.

33. Hasegawa, R.; Aoki, Y.; Doi, M. Optimum Graft Density for Dispersing Particles in Polymer Melts. *Macromolecules* **1996**, *29,* 6656–6662.

34. Gay, C. Wetting of a Polymer Brush by a Chemically Identical Polymer Melt. *Macromolecules* **1997,** *30,* 5939–5943.

35. Maas, J. H.; Fleer, G. J.; Leermakers, F. A. M.; Cohen Stuart M. A. Wetting of a Polymer Brush by a Chemically Identical Polymer Melt: Phase Diagram and Film Stability. *Langmuir* **2002,** *18,* 8871–8880.

36. Xu, J.; Qiu, F.; Zhang, H.; Yang, Y. Interactions of Polymer Brush-Coated Spheres in a Polumer Matrix. *J. Polym. Sci. B Polym. Phys.* **2006,** *44,* 2811–2820.

37. Xu, C.; Ohno, K.; Ladmiral, V.; Composto, R. J. Dispersion of Polymer-Grafted Magnetic Nanoparticles in Homopolymers and Block Copolymers. *Polymer* **2008,** *49,* 3568–3577.

38. Kim, B. J.; Bang, J.; Hawker, C. J.; Kramer, E. J. Effect of Areal Chain Density on the Location of Polymer-Modified Gold Nanoparticles in a Block Copolymer Template. *Macromolecules* **2006,** *39,* 4108–4114.

39. Etxeberria, H.; Zalakain, I.; Fernandez, R.; Kortaberria, G.; Mondragon, I. Controlled Placement of Polystyrene-Grafted CdSe Nanoparticles in Self-Assembled Block Copolymers. *Colloid Polym. Sci.* **2013,** *291,* 633–640.

40. Duwez, A.-S.; Guillet, P.; Colard, C.; Gohy, J.-F.; Fustin, C. A. Dithioesters and Trithio-carbonates as Anchoring Groups for the "Grafting-To" Approach. *Macromolecules* **2006,** *39,* 2729–2731.

41. Sawall, D. D.; Villahermosa, R. M.; Lipeles, R. A.; Hopkins, A. R. Interfacial Polym-erization of Polyaniline Nanofibers Grafted to Au Surfaces. *Chem. Mater.* **2004,** *16,* 1606–1608.

42. Minko, S.; Müller, M.; Motornov, M.; Nitschke, M.; Grundke, K.; Stamm, M. Two-Level Structured Self-Adaptive Surfaces with Reversibly Tunable Properties. *J. Am. Chem. Soc.* **2003,** *125,* 3896–3900.

43. Advincula, R.; Ruehe, J.; Brittain, W.; Caster, K. *Polymer Brushes;* Eds. VCH Wiley: Weinheim, 2004.

44. Lauter, V.; Müller-Buschbaum, P.; Lauter, H.; Petry, W. Morphology of Thin Nano-composite Films of Asymmetric Diblock Copolymer and Magnetite Nanoparticles. *J. Phys. Condens. Matter* **2011,** *23*(25), 4215 (6pp).

45. Lauter-Pasyuk, V.; Lauter, H. J.; Gordeev, G. P.; Müller-Buschbaum, P.; Toperverg, B. P.; Jernenkov, M.; Petry, W. Nanoparticles in Block-Copolymer Films Studied by Specular and Off-Specular Neutron Scattering. *Langmuir* **2003,** *19,* 7783–7788.

46. Aissou, K.; Fleury, G.; Pecastaings, G.; Alnasser, T.; Mornet, S.; Goglio, G.; Hadziioannou, G. Hexagonal-to-Cubic Phase Transformation in Composite Thin Films Induced by FePt Nanoparticles Located at PS/PEO Interfaces. *Langmuir* **2011,** *27,* 14481–14488.

47. Xu, C.; Xu, K.; Gu, H.; Zheng, R.; Liu, H.; Zhang, X.; Guo, Z.; Xu, B. Dopamine as A Robust Anchor to Immobilize Functional Molecules on the Iron Oxide Shell of Magnetic Nanoparticles. *J. Am. Chem. Soc.* **2004,** *126,* 9938–9939.

48. Kim, B. J.; Fredrickson, G. H.; Hawker, C. J.; Kramer E. J. Nanoparticle Surfactants as a Route to Bicontinuous Block Copolymer Morphologies. *Langmuir,* **2007,** *23,* 7804–7809.

49. Basly, B.; Alnasser, T.; Aissou, K.; Fleury, G.; Pecastaings, G.; Hadziioannou, G.; Duguet, E.; Goglio, G.; Mornet S. Optimization of Magnetic Inks Made of L1$_0$ Ordered

FePt Nanoparticles and Polystyrene-block-Poly (ethylene oxide) Copolymers. *Langmuir* **2015**, *31,* 6675–6680.

50. Barandiaran, I.; Cappelletti, A.; Strumia, M.; Eceiza, A.; Kortaberria, G. Generation of Nanocomposites Based on (PMMA-*b*-PCL)-Grafted Fe_2O_3 Nanoparticles and PS-*b*-PCL Block Copolymer. *Eur. Polym. J.* **2014**, *58,* 226–232.

51. Zhang, X.; Matyjaszewski, K. Synthesis of Well-Defined Amphiphilic Block Copolymers with 2-(Dimethylamino) Ethyl Methacrylate by Controlled Radical Polymerization. *Macromolecules* **1999**, *32,* 1763–1766.

52. Rozes, L.; Fornasieri, G.; Trabelsi, S.; Creton C.; Zafeiropoulos, N. E.; Stamm, M.; Sanchez, C. Reinforcement of Polystyrene by Covalently Bonded Oxo-Titanium Clusters. *Prog. Solid State Chem.* **2005**, *33,* 127–135.

53. Trabelsi, S.; Janke, A.; Hässler, R.; Zafeiropoulos, N. E.; Fornasieri, G.; Bocchini, S.; Rozes, L.; Stamm, M.; Gérard, J.-F.; Sanchez, C. Novel Organo-Functional Titanium-oxo-cluster-Based Hybrid Materials with Enhanced Thermomechanical and Thermal Properties. *Macromolecules* **2005**, *38,* 6068–6078.

54. Etxeberria, H.; Zalakain, I.; Mondragon, I.; Eceiza, A.; Kortaberria, G. Generation of Nanocomposites Based on Polystyrene-Grafted CdSe Nanoparticles by *Grafting through* and Block Copolymer. *Colloid Polym. Sci.* **2013**, *291,* 1881–1886.

55. Henze, M.; Mädge, D.; Prucker, O.; Rühe, J. "*Grafting Through*": Mechanistic Aspects of Radical Polymerization Reactions with Surface-Attached Monomers. *Macromolecules* **2014**, *47,* 2929–2937.

56. Barandiaran, I.; Kortaberria, G. Selective Placement of Magnetic Fe_3O_4 Nanoparticles into the Lamellar Nanostructure of PS-*b*-PMMA Diblock Copolymer. *Eur. Polym. J.* **2015**, *68,* 57–67.

57. Kim, M.; Hong, C. K.; Choe, S.; Shim, S. E. Synthesis of Polystyrene Brush on Multiwalled Carbon Nanotube Treated with $KMnO_4$ in the Presence of Phase-Transfer Catalyst. *J. Polym. Sci. A Polym. Chem.* **2007**, *45,* 4413–4420.

58. Barandiaran, I.; Kortaberria, G. Synthesis and Characterization of Nanostructured PS-*b*-P4VP/Fe_2O_3 Thin Films with Magnetic Properties Prepared by Solvent Vapor Annealing. *RSC Adv.* **2015**, *5,* 95840–95846.

59. Etxeberria, H.; Fernandez, R.; Zalakain, I.; Mondragon, I.; Eceiza, A.; Kortaberria, G. Effect of CdSe Nanoparticle Addition on Nanostructuring of PS-*b*-P4VP Copolymer via Solvent Vapor Exposure. *J. Colloid Interface Sci.* **2014**, *416,* 25–29.

60. Barandiaran, I.; Kortaberria, G. Magnetic Nanocomposites Based on Poly (styrene-*b*-butadiene-*b*-methyl methacrylate) and Modified Fe_2O_3 Nanoparticles. *Eur. Polym. J.* **2016**, *78,* 340–351.

61. Rahimi-Razin, S.; Haddadi-Asl, V.; Salami-Kalajahi, M.; Gehgoodi-Sadabad, F.; Roghani-Mamaqani H. Matrix-Grafted Multiwalled Carbon Nanotubes/Poly (methyl methacrylate) Nanocomposites Synthesized by in situ RAFT Polymerization: A Kinetic Study. *Int. J. Chem. Kinet.* **2012**, *44,* 555–569.

62. Pyun, J.; Matyjaszewski, K. Synthesis of Nanocomposite Organic/Inorganic Hybrid Materials Using Controlled/"Living" Radical Polymerization. *Chem. Mater.* **2001**, *13,* 3436–3448.

63. Radhakrishnan, B.; Ranjan, R.; Brittain, W. J. Surface Initiated Polymerizations from Silica Nanoparticles. *Soft Matter.* **2006**, *2,* 386–396.

64. Ilgacha, D. M.; Meleshkoa, T. K.; Yakimansky, A. V. Methods of Controlled Radical Polymerization for the Synthesis of Polymer Brushes. *Polym. Sci. Ser. C* **2015**, *57*, 3–19.

65. Zhang, N.; Luxenhofer, R.; Jordan, R. Thermoresponsive Poly (2-oxazoline) Molecular Brushes by Living Ionic Polymerization: Kinetic Investigations of Pendant Chain Grafting and Cloud Point Modulation by Backbone and Side Chain Length Variation. *Macromol. Chem. Phys.* **2012**, *213*, 973–981.

66. Li, W.; Bao, C.; Wright, R. A. E.; Zhao, B. Synthesis of Mixed Poly (ε-Caprolactone)/Polystyrene Brushes from Y-Initiatorfunctionalized Silica Particles by Surface-Initiated Ring-Opening Polymerization and Nitroxidemediated Radical Polymerization. *RSC Adv.* **2014**, *4*, 18772–18781.

67. Matyjaszewski, K.; Davis, T. P. *Handbook of Radical Polymerization;* Eds. VCH Wiley: Weinheim, 2002.

68. Zhou, Q.; Fan, X.; Xia, C.; Mays, J.; Advincula, R. Living Anionic Surface Initiated Polymerization (SIP) of Styrene from Clay Surfaces. *Chem. Mater.* **2001**, *13*, 3057–3057.

69. Kanhakeaw, P.; Rutnakornpituk, B.; Wichai, U.; Rutnakornpituk, M. Surface-Initiated Atom Transfer Radical Polymerization of Magnetite Nanoparticles with Statistical Poly (tert-butyl acrylate)-poly (poly (ethylene glycol) methyl ether methacrylate) Copolymers. *J. Nanomater.* **2015**, *2015*(12), 1369.

70. Mao, X.; Sun, H.; He, X.; Chen, L.; Zhang, Y. Well-Defined Sulfamethazine-Imprinted Magnetic Nanoparticles via Surface-Initiated Atom Transfer Radical Polymerization for Highly Selective Enrichment of Sulfonamides in Food Samples. *Anal. Methods* **2015**, *7*, 4708–4716.

71. Qin, L.; Xu, Y.; Han, H.; Liu, M.; Chen, K.; Wang, S.; Wang, J.; Xu, J.; Li, L.; Guo, X. β-Lactoglobulin (BLG) Binding to Highly Charged Cationic Polymer-Grafted Magnetic Nanoparticles: Effect of Ionic Strength. *J. Colloid Interface Sci.* **2015**, *460*, 221–229.

72. Sun, N.; Meng, X.; Xiao, Z. Functionalized Si_3N_4 Nanoparticles Modified with Hydrophobic Polymer Chains by Surface-Initiated Atom Transfer Radical Polymerization (ATRP). *Ceram. Int.* **2015**, *41*, 13830–13835.

73. Ranka, K.; Brown, P.; Hatton, T. A. Responsive Stabilization of Nanoparticles for Extreme Salinity and High-Temperature Reservoir Applications. *ACS Appl. Mater. Interfaces* **2015**, *7*, 19651–19658.

74. Rossner, C.; Glatter, O.; Saldanha, O.; Köster, S.; Vana, P. The Structure of Gold-Nanoparticle Networks Cross-Linked by Diand Multifunctional RAFT Oligomers. *Langmuir* **2015**, *31*, 10573–10582.

75. Mazurowski, M.; Sondergeld, K.; Elbert, J.; Kim, C. J.; Junyu, Li.; Frielinghaus, H.; Gallei, M.; Stühn, B.; Rehahn, M. Polystyrene Brushes on Fully Deuterated Organic Nanoparticles by Surface-Initiated Nitroxide-Mediated Radical Polymerization. *Macromol. Chem. Phys.* **2013**, *214*, 1094–1106.

76. Robbes, A.-S.; Cousin, F.; Meneau, F.; Chevigny, C.; Gigmes, D.; Fresnais, J.; Schweinse, R.; Jestin, J. Controlled Grafted Brushes of Polystyrene on Magnetic γ-Fe_2O_3 Nanoparticles via Nitroxide-Mediated Polymerization. *Soft Matter.* **2012**, *8*, 3407–3418.

77. Chen, Z.; Yang, Q.; Peng, K.; Guo, Y. Surface-Initiated Nitroxide-Mediated Radical Polymerization of 4-Vinylpyridine on Magnetite Nanoparticles. *J. Appl. Polym. Sci.* **2011**, *119*, 3582–3590.
78. Garcia, I.; Tercjak, A.; Zafeiropoulos, N. E.; Stamm, M.; Mondragon, I. Generation of Core/Shell Iron Oxide Magnetic Nanoparticles with Polystyrene Brushes by Atom Transfer Radical Polymerization. *J. Polym. Sci. A Polym. Chem.* **2007**, *45*, 4744–4750.
79. García, I.; Tercjak, A.; Rueda, L.; Mondragon, I. Self-Assembled Nanomaterials Using Magnetic Nanoparticles Modified with Polystyrene Brushes and Poly (styrene-*b*-butadiene-*b*-styrene). *Macromolecules* **2008**, *41*, 9295–9298.
80. Ohno, K.; Morinaga, T.; Koh, K.; Tsujii, Y.; Fukuda, T. Synthesis of Monodisperse Silica Particles Coated with Well-Defined, High-Density Polymer Brushes by Surface-Initiated Atom Transfer Radical Polymerization. *Macromolecules* **2005**, *38*, 2137–2142.
81. Barandiaran, I.; Grana, E.; Katsigiannopoulos, D.; Avgeropoulos, A.; Kortaberria, G. Nanocomposites Based on Nanostructured PI-*b*-PMMA Copolymer and Selectively Placed PMMA-Modified Magnetic Nanoparticles: Morphological and Magnetic Characterization. *Eur. Polym. J.* **2016**, *75*, 514–524.
82. Garcia, I.; Zafeiropoulos, N. E.; Janke, A.; Tercjak, A.; Eceiza, A.; Stamm, M.; Mondragon, I. Functionalization of Iron Oxide Magnetic Nanoparticles with Poly (methyl methacrylate) Brushes via Grafting-From Atom Transfer Radical Polymerization. *J. Polym. Sci. A Polym. Chem.* **2007**, *45*, 925–932.
83. Xu, C.; Ohno, K.; Ladmiral, V.; Milkie, D. E.; Kikkawa, J. M.; Composto, R. J. Simultaneous Block Copolymer and Magnetic Nanoparticle Assembly in Nanocomposite Films. *Macromolecules* **2009**, *42*, 1219–1228.
84. Yao, Y.; Metwalli, E.; Su, B.; Körstgens, V.; Moseguí González, D.; Miasnikova, A.; Laschewsky, A.; Opel, M.; Santoro, G.; Roth, S. V.; Müller-Buschbaum, P. Arrangement of Maghemite Nanoparticles via Wet Chemical Self-Assembly in PS-*b*-PNIPAM Diblock Copolymer Films. *ACS Appl. Mater. Interfaces* **2015**, *7*, 13080–13091.
85. Aissou, K.; Alnasser, T.; Pecastaings, G.; Goglio, G.; Toulemonde, O.; Mornet, S.; Fleury, G.; Hadziioannou, G. Hierarchical Assembly of Magnetic L1$_0$-Ordered FePt Nanoparticles in Block Copolymer Thin Films. *J. Mater. Chem. C* **2013**, *1*, 1317–1321.

CHAPTER 2

APPLICATIONS OF IONIC LIQUIDS IN POLYMERIC COMPOSITES

ANJALI BISHNOI[1,2,*], SUNIL KUMAR[1,3], SANDEEP RAI[1,4]
and ANKIT MISHRA[1]

[1]*Department of Polymer and Rubber Technology, Shroff S. R. Rotary Institute of Chemical Technology, Block no: 402, Vataria, Ankleshwar-Valia road, Valia, Dist: Bharuch, Gujarat 393003, India*

[2]*Department of Chemistry, Indian Institute of Technology Hauz Khas, New Delhi 110 016, India, Tel.: +919638362368*

[3]*Centre for Polymer Science and Engineering, Indian Institute of Technology Hauz Khas, New Delhi 110016, India*

[4]*GRP Limited, GIDC Estate, Panoli, Bharuch, Gujarat, India*

*Corresponding author. E-mail: anjali.bishnoi@srict.in,
anjali_bishnoi@yahoo.com

CONTENTS

ABSTRACT

Ionic liquids (ILs) as a new kind of media or materials have been widely studied in broad fields such as material science, catalysis, electrochemistry, and analytical chemistry. Many ionic liquids show a unique combination of physical and electrochemical properties such as electric conductivity, electrochemical stability, noninflammability, and very low vapor pressure; as a consequence many of these ILs are interesting materials for different electrochemical applications and have been investigated in the areas such as batteries, electric double layer capacitors, fuel cells, sensors, and actuators. The studies reported in the research article on the utilization of ILs as accelerator of curing of nitrile rubber. The results indicated that IL significantly reduced optimum cure time with a scarifying loss in physical properties.

2.1   INTRODUCTION

Ionic liquids (ILs) have accomplished a rocket-like boost in the past few decades. Ionic liquids are a new kind of materials that comprises ions (anions and cations) and are liquid at room temperatures hence, also known as room temperature ionic liquids (RTILs). The very fascinating feature about ILs is that they are not made up of molecules, that is they comprise ions and that too in equal numbers hence, ILs are electrically neutral. While ordinary liquids such as water and gasoline are predominantly made of electrically neutral molecules. Ionic liquids are largely made of ions and short-lived ion pairs. Mostly, ILs possess melting point below room temperatures and even some have melting point below 0°C. Hence, these are also called ionic melts, fused salts, or salts in the liquid state. They are very much different from inorganic salts. For example, consider the very common salt "sodium chloride" which have melting point around 800°C but unlike these inorganic salts ILs potent fluidity over a wide temperature range (around 300–400°C). Some reports claimed origin of IL (ethyl ammonium nitrate) as old as World War I.[1,2] Later, RTILs have been utilized as nonaqueous polar-like solvents for electrochemical and spectroscopic studies of transition metal complexes.[3,4,5]

If we consider the fact that what is actually inside the nutshell, ILs comprise large, bulky organic cation very efficiently interacting with the anion which is usually inorganic. Hence, one can say that generally ILs offer the unique properties resulting from the hybridization between two chemical entities: organic cation and inorganic cation. Generally, selection of cation affects the properties and stability of the IL. The anion controls the chemistry and functionality of IL. For example, presence of bulky aliphatic side chain on the cation will show a glass transition instead of melting point. The most commonly selected cations are mentioned below in Figure 2.1(a). Anions can be organic or inorganic and some of them are mentioned in Figure 2.1(b):

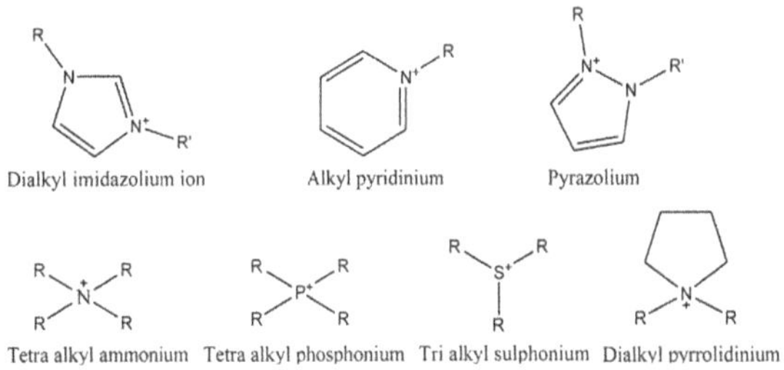

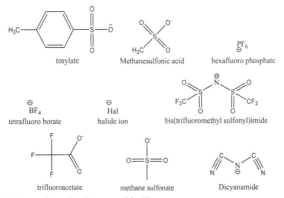

FIGURE 2.1A   Examples of cations used in ionic liquids (ILs).

FIGURE 2.1B   Examples of anions used in IL.

If we refer the above examples of cations and anions various permutations and combinations are possible between them, resulting into different types of ILs. Some of them are mentioned in Table 2.1.

These are the liquids which have various interesting properties such as, negligible vapor pressure, high viscosity, excellent thermal stability, wide liquid range, catalytic properties, and so forth. The reason for very low vapor pressure is the strong Coulomb (ionic) interaction between the cation and anion. Other interesting properties of ILs include nonflammability, thermally, mechanically and electrochemically stable, electrochemical window (2–4.5 V), and tuneable viscosity.

TABLE 2.1   Full Name, Chemical Structure, and General Abbreviation of Common Ionic Liquids (ILs).

ILs	Chemical structure	General abbreviation
1-Ethyl-3-methylimidazolium Chloride		[EMIM] Cl
1-Butyl-3-methylimidazolium Chloride		[BMIM] Cl
1-Butyl-3-methylimidazolium hexafluorophosphate		[BMIM] PF$_6$
1-Ethyl-3-methylimidazolium tetrafluoroborate		[EMIM] BF$_4$
1-Ethyl-3-methylimidazolium dicynamide		[EMIM] DCA
1-Ethyl-3-methylimidazolium dimethyl phosphate		[EMIM] (MeO)$_2$PO$_2$

TABLE 2.1 *(Continued)*

ILs	Chemical structure	General abbreviation
1-Butyl-3-methylpyrrolidinium methanesulfonate		[BMPyrr] (MeSO$_3$)
1-tetradecyl-3-carboxy-methyl imidazolium chloride		[TCMIM] Cl
1-Butyl-3-methylimidazolium bis(trifluoromethylsufonyl) imide		[BMIM] TFSI

Currently major chunk of research in academia and industries is heading to substitute eco-friendlier technologies which involve hazardous and volatile solvents used extensively. ILs possess low vapor pressure, hence are considered as best possible eco-friendly substitutes for volatile solvents, not only this, they are also capable to act as catalysts.[6] ILs have opened up new horizons in the field of chemistry. These are the liquids which have ILs have many applications, such as powerful solvents and electrically conducting fluids. Among other fields of applications; energy, chemical engineering, biotechnology, coating, electrochemistry, and so forth are the promising ones.[7] They can be used as electrolytes, solvents, heat storage, analytics, lubricants, liquid crystals, nanoparticle synthesis, fuel cell, solar energy conversion, and so forth.[8–11] Many classes of chemical reactions such as Diels-Alder reactions and Friedel-Crafts reactions, hydrodimerization, Beckman rearrangement, Fischer Indole synthesis, stereoselective alkylation halogenation, and so forth can be performed using ILs as solvents.[12, 13] Together with these areas, ILs also have been widely investigated for their applications in the field of polymer and rubber composites. ILs have recently evaluated as additives in various ranges of function in polymer compounds.

In the present study, an IL was synthesized and used as an accelerator in medium Acrylonitrile cold acrylonitrile-butadiene rubber (NBR) compounding. The objective of the study is to develop environment-friendly green and low cost IL for rubber compounding to reduce optimum cure time (t_{90}). Due to nonvolatile nature, IL usage in an industry will lead to almost no emission of volatile organic compounds. Additionally, improvement in cure time will have huge bearing on economic of production in rubber industries.

2.1.1 APPLICATIONS OF ILS IN POLYMER AND POLYMERIC COMPOSITES

Recently, exhaustive research has been carried out on the ILs as additives for polymer materials. The use of ILs as performance additives in rubber formulations is getting attention, due to their designable characteristics. Antielectrostatic properties, lubricant properties, and activity in cross-linking reactions of ILs also have been reported. The most important reason for the use of ILs in elastomers was the improvement of filler dispersion (e.g., carbon nanotubes, carbon black, halloysite nanotubes, and silica) in hydrophobic matrices as well as the enhancement of the ionic conductivity, thermal and mechanical properties of polymer composites (Fig. 2.2).

FIGURE 2.2   Applications of IL in rubber technology.

Zhao et al., fabricated poly (methyl methacrylate) (PMMA)/MWCNT nanocomposites by using 1-butyl-3- methylimidazolium hexafluorophosphate (BMIMPF$_6$) IL. They have reported that the use of ILs resulted in positive alterations of more than one properties such as, dispersion of MWCNT in PMMA matrix, enhancement of conductivity, depression in T_g, efficient processing during melt mixing, and so forth.[14]

2.1.1.1   SOLID-STATE POLYMER ELECTROLYTE

Polymer electrolytes also find applications in electrochemical devices, such as rechargeable batteries, fuel cells, super capacitors, electrochromic displays, and so forth. They overcome the problem of corrosion as well. The area of polymer electrolytes has gone through various transformations, that is from dry solid polymer electrolyte (SPE) systems to plasticized, gels, and rubbery to micro/nanocomposite polymer electrolytes. Earlier, polymers were mostly used as structural materials or as insulations in electrical applications. But recently, they have emerged as electron or ion conductors, that is, they possess some ionic conductivity when combined with appropriate salts and can be employed as an electrolyte. Peter V. Wright, a polymer chemist, in 1975 first reported that poly(ethylene oxide) (PEO) can act as a host for sodium and potassium salts, thus producing a solid electrical conductor polymer/salt complex.[15] Further, in this line the polymer electrolytes with high ionic conductivity (1.2×105 S/cm at 30°C) were also prepared by Marwanta et al.[16] by using NBR and N-ethylimidazolium bis(trifluoromethanesulfonyl)imide [EMIM][TFSI] ionic liquid. Further, increase in the ionic conductivity was observed by incorporation of lithium salt to this composite, unaltering other mechanical properties. This research group has also emphasized on the interaction between the anion of imidazolium ionic liquid and cyanide group of NBR. Similarly, Cho et al.[17] have also reported the high conductivity of 2.54×10^{-4} S/cm at 20°C as well as high electrical stability of composite made by using 1-butyl-3-methylimidazolium *bis*(trifluoromethylsulfonyl) imide [BMIM][TFSI] and NBR. It has been reported in many research studies that conductivity of chloroprene rubber (CR)/carbon nanotubes (CNTs) composites has been significantly improved by using imidazolium-based salts.[18–21]

Multiwalled carbon nanotubes (MWCNTs) and 1-butyl-3-methylimidazolium bis(trifluoromethylsulfonyl)imide [BMIM][TFSI] when mixed with CR to make composites, it gives electrical conductivity of 0.1 S/cm with a stretchability more than 500%. This was attributed to the presence of electrons and ions or some kind of network formation with the IL modified tubes. However, it has not been reported that conductivity increases as we increase the quantity of IL.[18]

In another report by Steinhauser et al.,[20] improvement of the conductivity over more than two orders of magnitude was observed for the composite which contains 15 phr of [BMIM][TFSI] IL and 3 phr of CNTs. Which was explained by enhanced CR-CNTs interaction due to the presence of [BMIM][TFSI]. Subramaniam et al. have reported that the electrical conductivity of IL-CR/CNTs composites increased upon increasing thermal ageing, depicting applications of these composites where conductivity at high temperatures is required.[21]

2.1.1.2 LUBRICANT

A lubricant is defined as a substance which is used to reduce friction between the surfaces. They are also known as processing aids. They help in easy processing of polymers. Lubricants in polymers serve to decrease the friction forces between polymer-polymer surface, polymer-metal surface, polymer-filler surface, and so forth. Ding et al. 2011[22] have modified magnesium hydroxide (Mg $(OH)_2$) with 1-tetradecyl-3-carboxymethyl-imidazolium chloride (TcmimCl) IL. This was utilized for making composites with linear low density polyethylene (LLDPE) through melt mixing. According to their findings TcmimCl acted as an efficient lubricant and compatibilizer for Mg $(OH)_2$ and LLDPE and has enhanced processability and mechanical properties of LLDPE/TcmimCl-Mg $(OH)_2$ composites.

2.1.1.3 PLASTICIZER

Plasticizer constitutes a large group of compounding materials, which are incorporated in polymers for smooth processing and shaping of materials for molding. Their presence alters the physical processing of rubber compounds such as hardness, elongation, low temperature performance, antistatic properties, flex life, and so forth. together with this plasticizer are also known to improve the processing of polymers via lowering viscosity, easy filler incorporation, smooth dispersion of the recipe ingredients, and better flow characteristics. They also reduce the glass transitions temperature of rubber compounds. Likozar et al.[23,24] have reported that the glass transition temperature (T_g) of Hydrogenated Nitrile Butadiene Rubber (HNBR)/MWCNT-OH/IL compounds decreased by increasing bis(trifluoromethylsulfonyl) imide anions-based ILs on HNBR matrix. However, hexafluorophosphate and tetrafluoroborate anions-based ILs cause elevation of T_g by 5°C. This may be due to variable interaction between different ions present in ILs and rubber matrix.

2.1.1.4   ANTISTATIC ADDITIVE

Antistatic agents are chemicals that are used to treat the material surfaces in order to eliminate or minimize the buildup of static electricity which is generated due to the triboelectric effect or any noncontact process. The function of antistatic agent is to make the material's surface itself slightly conductive. As we know that polymers are insulators in nature and hence, get static electricity. Ionic liquids serve the purpose effectively on plastics, floors, and wood.[25–27]

A series of bis(trifluoromethanesulfonyl)imide-type ILs having cations as 1-butyl-3-methylimidazolium, 1-(2-hydroxyethyl)-3-methylimidazolium, tris(2-hydroxyethyl)methylammonium, 1-ethylpyridinium, *N*, *N*-diethyl-*N*-methyl-*N*-(2-methoxyethyl)ammonium,   1-butyl-1-methylpyrrolidinium, tributyl-N-octylphosphonium, and so forth. with bromide anion have been used as antistatic additives for polyurethanes.[28] Similarly, several reports have been published for polypropylene (PP). Wang et al. have reported IL as an antistatic additive for PP.[29] 1-*n*-tetradecyl-3-methylimidazolium bromide ([C14mim]Br) IL was also utilized by Ding et al., for PP film.[22]

2.1.1.5   ANTIMICROBIAL ADDITIVE

Antimicrobial additives are the chemicals that generate an antimicrobial surface that prevents or minimizes the ability of microorganisms to grow on the surface of material. Generally, a coating of these compounds is applied or they are incorporated within the material which has an inherent toxicity toward microorganisms. Imidazole-based ILs exhibits antimicrobial properties so are used as additives for multiple functions in various compounds of polymers. Docherty et al. have examined the antimicrobial effects of butyl-, hexyl-and octyl-imidazolium and pyridinium bromide ILs. They reported butyl-imidazolium and pyridinium bromides as less antimicrobial than ILs with longer alkyl chain lengths to microorganisms examined (*Escherichia coli*, *Staphylococcus aureus*, *Bacillus subtilis*, *Pseudomonas fluorescens*, and *Saccharomyces cerevisiae*). They have reported significant antimicrobial activity of hexyl-and octyl-imidazolium and pyridinium bromides-based ILs.[30]

2.1.1.6 FILLER DISPERSION IMPROVEMENT AGENT

ILs are found to facilitate the dispersion of different ingredients in rubber compound. Dispersion is considered to be the integral phenomenon for achieving optimum properties of cured rubber products. To ensure even distribution of filler dispersion in rubber matrix a series of ILs were used ([AMIM][Cl], [EMIM][SCN], [MOIM][Cl], 3-(triphenylphosphonic)-1-sulfonic acid tosylate, and trihexyl(tetradecyl)phosphonium decanoate) for the blend of Solution Styrene-Butadiene Rubber (SSBR)/polybutadiene rubber (BR) and MWCNTs. Results showed that the enhancement in the mechanical properties of blend of SSBR/BR with only at ~3 wt.% of MWCNTs loading was observed, and which may be through the chemical coupling between MWCNTs and rubber chains by 1-allyl-3-methylimidazolium chloride [AMIM][Cl], as proved by Raman spectra, dynamic mechanical analysis (DMA), and electrical properties.[31] IL-coated HNTs (1-butyl-3-methylimidazolium hexafluorophosphate [BMIM][PF6]) when used for SBR compounds showed accelerated curing in comparison with pristine HNTs. This is again attributed to the interaction between IL and HNTs and improvements in the state of filler dispersion.[32] Similarly, physical (cation–π/π–π) interaction between [BMIM][TFSI] and MWCNTs promotes thermal stability of composites and homogenous dispersion of nanotubes in elastomer matrix.[33]

2.1.1.7 UV STABILIZER

Polymers generally get weathered due to direct or indirect effects of heat and light. The degradation generally occurs due to undesirable oxidation, chain scission due to breakdown of crosslinks.

The main role of UV stabilizers is to dissipate the energy absorbed from UV radiations as heat and hence, the UV absorption reduces and subsequently reduces the rate of weathering.

The ILs also possess UV resistant properties and are found to be the effective agents for different polymer applications. Application of alkylimidazolium-based ILs as potential UV stabilizing agents for wood[34,35] and cellulose materials[36] is very well documented in several reports. As per Subramaniam et al. MWCNTs modified with 1-butyl-3-methylimidazolium bis(trifluoromethylsulfonyl)imide IL exhibited higher mechanical properties (tensile modulus, hardness) and thermal stability than the chloroprene rubber composites with unmodified MWCNTs.[37,38]

2.1.1.8 CURE CHARACTERISTIC IMPROVEMENT

Vulcanization is essential to make rubber usable. There are various agents that can be utilized for this purpose, for example, sulfur, metal oxides, resins, peroxides, and so forth. However, sulfur vulcanization is the most common and the economically viable process. ILs catalyze the formation of interface crosslinks by facilitating the dispersion of curing agents (like ZnO) resulting in the minimization of curing time and subsequently increasing the crosslink density of vulcanizates. As it is known from the literature, imidazolium ILs may affect the crosslinking behavior of rubber composites.[39,40]

Various ILs have been synthesized for different applications, for example, imidazolium, benzalkonium, and phosphonium ILs combined with the 2-mercaptobenzothiazolate in NBR compounds where sulfur was used as cure system. IL synthesiszed by Pernak et al. was also found to reduce the amount of 2-mercaptobenzothiazolate. It has been assumed that the rate of vulcanization varies with nature of anion. It was reported that higher cure rate of NBR vulcanized with sulfur was exhibited by rubber compounds containing alkylimidazolium chlorides and bromides, whereas the longest vulcanization times were observed for rubber compounds with alkylimidazolium hexafluorophosphates.[41] Similarly, the applications of hydrophilic and hydrophobic ILs and their mixtures for making advanced polymer blends and composites have been extensively reviewed by Shamsuri, and Daik.[42] In the present research article, we report synthesis of a low cost IL. Synthesized IL was mixed in NBR compounds and its effect on rheological, mechanical, and physical properties was evaluated.

2.2 EXPERIMENTAL SECTION

2.2.1 MATERIALS

Formic acid (Rankem, India) and 2-amino ethanol (Rankem, India) used without further purifications. The NBR medium ACN (33%) was procured from Omnova Solutions Pvt Ltd, Valia, India. The curing agents for NBR were sulfur (Rankem, India), zinc oxide (Qualigens, India), mercaptobenzothiazole (NOCIL, India), and stearic acid (Qualigens, India). Silica (Madhu silica, India) was used as reinforcement fillers for NBR.

2.2.2 SYNTHESIS OF IONIC LIQUID (IL)

The IL was synthesized by using equimolar ratio of formic acid and 2-amino ethanol according to the procedure described in literature.[43] The entire addition process was carried out with continuous stirring for 1 h in an ice bath. Subsequently, this mixture was continuously stirred for another 24 h at room temperature. The resulting IL was colorless viscous liquid with no crystallization.

2.2.3 COMPOUNDING AND TESTING METHODS OF NBR COMPOSITES

2.2.3.1 PREPARATION OF COMPOUNDS OF NBR

Composition and samples designation are depicted in Table 2.2. The compounds of NBR were prepared with the loading of 2.5, 5.0, 7.5, and 10 phr of IL. The fillers, curing agent, and IL were gradually mixed on a lab two roll mill. The total mixing time was kept at 15 min. The raw NBR rubber was masticated prior to the addition of compounding ingredients. The mixture of silica and IL was homogenized with NBR in two roll mill (Future Foundation, Delhi, India) at a rotor speed of 40 rpm and temperature of 40°C. The rubber compounds were processed at a rotor speed of 40 rpm, and the initial temperature was set to 40°C. The compounds were kept for 16 h for conditioning prior to curing.

TABLE 2.2   Compositions and Nomenclature of NBR Compounds.

S. no.	Ingredients	IL-0	IL-2.5	IL-5	IL-7.5	IL-10
1	NBR	100	100	100	100	100
2	Sulfur	2	2	2	2	2
3	MBTB/MBTS	2	2	2	2	2
4	Zinc oxide	5	5	5	5	5
5	Stearic acid	1	1	1	1	1
6	ppt silica/silica	30	30	30	30	30
7	Ionic liquid	0	2.5	5	7.5	10

The conditioned compounds were cured on Curing Press (Future Foundation, Delhi, India) at 160°C for 30 min.

2.2.3.2   RHEOLOGICAL BEHAVIOR

The curing characteristics of the NBR uncured compounds were determined with a moving die rheometer (Future Foundation, Delhi, India) at 160°C for 30 min. The optimum cure time (t_{90}), scorch time (t_{s2}), minimum torque (M_L), and maximum torque (MH) were determined from the curing curves. The time required to reach 60 and 90% of curing was termed t_{60} and t_{90}, respectively.

2.2.3.3   MECHANICAL AND PHYSICAL PROPERTIES

Stress-strain tests were performed with a universal material testing machine (Future Foundation, Delhi, India) with a crosshead speed of 500 mm·min^{-1}. To measure mechanical properties, three different dumbbell-shaped specimens were punched from each rubber sample. The tensile strength (TS) at break was measured at room temperature. Hardness was measured with the help of Hardness Tester (Zwick, India). Average of five samples was considered for final readings.

2.2.3.4   DETERMINATION OF CROSSLINK DENSITY

The crosslink density of vulcanized NBR and NBR compounds were determined by equilibrium swelling method. The weighed samples were kept in toluene for 48 h at room temperature. The swollen samples were weighed on an analytical balance and then dried in a vacuum oven for 24 h at 80°C to remove all the solvent and reweighed. The crosslink density (V_e) was determined by using the Flory-Rehner equation:[44]

$$V_e = \frac{-\ln\left(1 - V_r\right) + V_r + \chi V_r^2}{V_s\left(V_r^{1/3} - V_r/2\right)}$$

Where, V_r is the volume fraction of the rubber in the swollen equilibrium samples, V_s is the solvent molar volume (107 cm^3/mol for toluene), χ is the

NBR-toluene interaction parameter and is taken as 0.435 calculated according to the reference.[45] The volume fraction (V_r) of swollen investigated NBR samples was calculated by using equation:[46]

$$V_r = \frac{m_o \times \phi \times (1-\alpha)/\rho_r}{m_o \times \phi \times (1-\alpha)/\rho_r + (m_1 - m_2)/\rho_s}$$

Where, α is the mass loss of the NBR vulcanizate during swelling, m_0 is the sample mass before swelling, m_1 and m_2 are sample masses before and after drying, ρ_r and ρ_s are the density of rubber and solvent, respectively, and ϕ is the mass fraction of vulcanizate during swelling.

2.3 RESULTS AND DISCUSSION

2.3.1 RHEOLOGICAL BEHAVIOR OF NBR VULCANIZATES

The influence of IL was estimated based on the rheometer measurements. The curing behaviors of NBR and NBR compound containing ILs at a range of 2.5–10 phr loadings are shown in Figure 2.3. The curing process of the nitrile rubber compounds was carried out at 160°C for 30 min. Cross-linking by sulfur utilizes double bonds in diene-containing elastomers, for example, NBR.

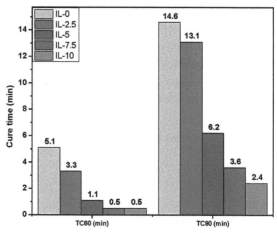

FIGURE 2.3   Effect of IL on cure time (NBR compound).

The incorporation of IL at entire concentrations as in rubber compounds leads to a systematic decrease in the cure time. Significant changes with varying levels of IL can be observed from the rheometric data obtained. Variation in minimum torque (M_L) directly affects the viscosity of the rubber compound. An increase in the values of torque is an indication of the formation of crosslinks. The main reason for the changes in these values can be attributed to the plasticizing effect of the IL employed. Moreover, IL may act as a secondary accelerator by the interface crosslink formation and improve dispersion of ZnO. It can be concluded that higher IL concentrations have a detrimental effect on the formation of covalent bonds in NBR compounds. It is visible from the Figure 2.3, that IL has significantly increased the cure rate of the nitrile rubber.

2.3.2 MECHANICAL AND PHYSICAL PROPERTIES OF NBR VULCANIZATES

The influence of IL on the tensile properties of the nitrile composites was also investigated (Fig. 2.4). Tensile strength of the NBR composites clearly displays a decreasing trend when 10 phr of IL, as the IL can also acts as a plasticizer and affects the physical properties. In general, we may observe that an increase in IL content yields poorer mechanical properties due to the softening of the polymer but it increases the processing characteristics of rubber. Compared with the virgin nitrile rubber (without IL), the incorporation of IL leads to decrease in elongation-at-break parameter, this is related to the elasticity.

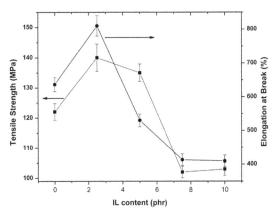

FIGURE 2.4   Effect of IL on mechanical properties (NBR compound).

The TS and elongation-at-break as a function ionic liquid (IL) contents diagrams corresponding to the investigated NBR compounds are shown in Figure 2.4. An increase in the TS (~16% at 2.5 phr and ~11% at 5 phr of IL) is registered with IL content till 5 phr followed by a sharp decrease with further addition. In general, with an increase in IL content, mechanical properties of rubber became poor due to the softening of the rubber. Further, ~27% increase in elongation-at-break is observed in the composition of 2.5 phr of IL content followed by a systematic decrease in the elongation-at-break with the addition of IL.

Effect of IL on the hardness shore A was studied at different load levels. It was observed that, hardness increased with lower level (2.5 phr) of IL. However, the subsequent increase in IL concentration (i.e., 5, 7.5, and 10 phr), hardness decreased and attained a constant value (Fig. 2.5).

Hence, it can be summarized that the higher loading of IL decrease the TS and increases the elongation-at-break. This phenomenon can be explained by the lowering of crosslink density. Theoretically, it can be predicted that higher loading of IL can lead to the formation of a thin layer onto the filler particle surface and in turn may lower the rubber filler interactions and also the formation of crosslinks.

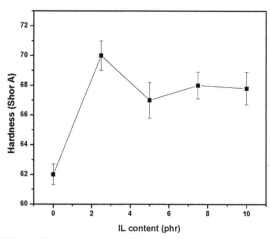

FIGURE 2.5   Effect of IL on hardness (NBR compound).

2.3.3   CROSSLINK DENSITY OF NBR VULCANIZATES

The crosslink densities of NBR compounds as function of IL contents are shown in Figure 2.6. An increase in crosslink density is observed in the composition range 2.5–5 phr of IL content followed by a systematic decrease in the crosslink density is observed with the addition of IL. Similar trend was found in TS and elongation-at-break revealed that at the lower content of IL a 3-dimensional crosslink network formed. Therefore, at lower IL content, the mechanical properties of NBR compounds are improved.

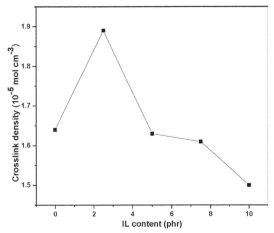

FIGURE 2.6   Effect of IL on crosslink density (NBR compound).

2.4   CONCLUSION AND FUTURE SCOPE

The application of ILs predominantly in the field of polymer composites have been reviewed. A low-cost IL was synthesized, characterized, and tried in NBR compounding as an accelerator in rubber curing. The effect of IL was evaluated at varying levels (2.5–10 phr) in NBR compounding. It was concluded that at a lower level (2.5 phr) IL, exhibited an increase in mechanical properties including TS. However, on a higher level (5.0–10 phr levels) of IL, TS, and other physical properties dropped. Optimum cure time (t_{90}) also drastically reduced with the increase in IL concentration. The IL synthesized, therefore, has potential to be utilized as cure time enhancer and in turn improving the productivity of rubber processing industries with the scarifying loss of physical properties like TS and elongation-at-break (IL level >2.5 phr). Based on the preliminary results of our lab studies, we propose an expansion of this as our future plan of research.

KEYWORDS

- ionic liquid
- composites
- rubber
- physical properties
- cure characteristics
- NBR

REFERENCES

1. Walden, P. Molecular Weights and Electrical Conductivity of Several Fused Salts. *Bull. Acad. Imp. Sci. (Saint Petersburg)*, **1914,** *1800,* 405–422.
2. Wasserscheid, P.; Keim, W. Ionic Liquids—New "Solutions" for Transition Metal Catalysis, *Angew. Chem. Int. Ed.* **2000,** *39,* 3772–3789.
3. Appleby, D.; Hussey, C. L.; Seddon, K. R.; Turp, J. E. Room-Temperature Ionic Liquids as Solvents for Electronic Absorption Spectroscopy of Halide Complexes. *Nature* **1996,** *323,* 614–616.
4. Seddon, K. R. Room-Temperature Ionic Liquids: Neoteric Solvents for Clean Catalysis. *Kinet. Catal.* **1996,** *37,* 693–697.
5. Earle, N. J.; Seddon, K. R. Ionic Liquids: Green Solvents for the Future. *Pure Appl. Chem.* **2000,** *72,* 1391–1398.
6. Rogers, R. D.; Seddon, K. R. Ionic Liquids–Solvents of the Future. *Science* **2003,** *302,* 792–793.
7. Natalia, V. P.; Kenneth, R. S. Applications of Ionic Liquids in the Chemical Industry. *Chem. Soc. Rev.* **2008,** *37,* 123–150.
8. Abbott, A. P.; Ryder, K. S. Electroplating using Ionic Liquids. *Annu. Rev. Mater. Res.* **2013,** *43,* 335–358.
9. Armand, M.; Tarascon, J. M. Building Better Batteries. *Nature* **2008,** *451,* 652–657.
10. Scrosati, B.; Garche, J. Lithium Batteries: Status, Prospects and Future. *J. Power Sources* **2010,** *195,* 2419–2430.
11. Pringle, J. M.; Golding, J.; Forsyth, C. M.; Deacon, G. B.; Forsyth, M.; MacFarlane, D. R. Physical Trends and Structural Features in Organic Salts of the Thiocyanate Anion. *Mater. Chem.* **2002,** *12,* 3475–3480.

12. Earle, M. J.; Seddon, K. R.; Adams, J. A.; Roberts, G. Friedel-Crafts Reactions in Room Temperature Ionic Liquids. *Chem. Comm.* **1998**, *19*, 2097–2098.
13. Zhao, H.; Malhotra, S. V. Applications of Ionic Liquids in Organic Synthesis. *Aldrichimica Acta.* **2002**, *35*, 14–22.
14. Zhao, L.; Li, Y.; Cao, X.; You, J.; Dong, W. Multifunctional Role of an Ionic Liquid in Melt-Blended Poly (methyl methacrylate)/multi-walled Carbon Nanotube Nanocomposites. *Nanotechnology* **2012**, *23*, 1–8.
15. Wright, P. V. Electrical Conductivity in Ionic Complexes of Poly (ethylene oxide). *Polym. Int.* **1975**, *7*, 319–327.
16. Marwanta, E.; Mizumo, T.; Nakamura, N.; Ohno, H. Improved Ionic Conductivity of Nitrile Rubber/Ionic Liquid Composites. *Polymer* **2005**, *46*, 3795–3800.
17. Cho, M.; Seo, H.; Nam, J.; Choi, H.; Koo, J.; Lee, Y. High Ionic Conductivity and Mechanical Strength of Solid Polymer Electrolytes Based on NBR/Ionic Liquid and its Application to an Electrochemical Actuator. *Sens. Actuators B.* **2007**, *128*, 70–74.
18. Subramaniam, K.; Das, A.; Heinrich, G. Development of Conducting Polychloroprene Rubber using Imidazolium Based Ionic Liquid Modi_Edmulti-Walled Carbon Nanotubes. *Comp. Sci. Technol.* **2011**, *71*, 1441–1449.
19. Subramaniam, K.; Das, A.; Steinhauser, D.; Kluppel, M.; Heinrich, G. Influence of Ionic Liquids on the Dielectric Relaxation Behavior of CNT Based Elastomer Nanocomposites. *Eur. Polym. J.* **2011**, *47*, 2234–2243.
20. Steinhauser, D.; Subramaniam, K.; Das, A.; Heinrich, G.; Kluppel, M. Influence of Ionic Liquids on the Dielectric Relaxation Behavior of CNT Based Elastomer Nanocomposites. *Express Polym. Lett.* **2012**, *11*, 927–936.
21. Subramaniam, K.; Das, A.; Heinrich, G. Improved Oxidation Resistance of Conducting Polychloroprene Composites. *Compos. Sci. Technol.* **2013**, *74*, 14–19.
22. Ding, Y.; Wang, P.; Wang, Z.; Chen, L.; Xu, H.; Chen, S. Magnesium Hydroxide Modified by 1-n-tetradecyl-3-carboxymethyl Imidazolium Chloride and its Effects on the Properties of LLDPE. *Polym. Eng. Sci.* **2011**, *51*, 1519–1524.
23. Likozar, B. The Effect of Ionic Liquid Type on the Properties of Hydrogenated Nitrile Elastomer/Hydroxy-Functionalized Multi-Walled Carbon Nanotube/Ionic Liquid. *Soft. Matter.* **2011**, *7*, 970–977.
24. Likozar, B. Diffusion of Ionic Liquids into Elastomer/Carbon Nanotubes Composites and Tensile Mechanical Properties of Resulting Materials. *Nanotechnology* **2010**, *17*, 35–42.
25. Gibon, C.; Herbst, H.; Minder, E. Antistatic Thermoplastic Compositions. W.O. Patent 2,011,069,960, 2011.
26. Krausche, C.; Wong, W. M.; Stefanie, S.; Hiller, M.; Pedro C. Thick Floor Coating having Antistatic Properties. U.S. Patent 20,090,068,435 A1, 2007.
27. Roessler, A.; Schottenberger, H. Antistatic Coatings for Wood-Floorings by Imidazolium Salt-Based Ionic Liquids. *Prog. Org. Coat.* **2014**, *77*, 579–582.
28. Iwata, T.; Tsurumaki, A.; Tajima, S.; Ohno, H. Bis (trifluoromethanesulfonyl) Imide-Type Ionic Liquids Asexcellent Antistatic Agents for Polyurethanes. *Macromol. Mater. Eng.* **2013**, *299*, 794–798.
29. Wang, X.; Liu, L.; Tan, J. Preparation of an Ionic-Liquid Antistatic/Photostabilization Additive and its Effectson Polypropylene. *J. Vinyl. Addit. Techn.* **2010**, *16*, 58–63.
30. Docherty, K. M.; Kulpa, C. F. Toxicity and Antimicrobial Activity of Imidazolium and Pyridinium Ionic Liquids. *Green Chem.* **2005**, *7*, 185–189.

31. Das, A.; Stoclekhuber, K. W.; Jurk R.; Fritzsche, J.; Kluppel M.; Heinrich G. Coupling Activity of Ionic Liquids Between Diene Elastomers and Multi-Walled Carbon Nanotubes. *Carbon* **2009,** *47,* 3313–3321.

32. Guo, B.; Liu, X.; Zhou, W. Y.; Lei, Y.; Jia, D. J. Adsorption of Ionic Liquid on to Halloysite Nanotubes: Mechanism and Reinforcement of the Modified Clay to Rubber. *Macromol. Sci. Part B: Phys.* **2010,** 49, 1029–1043.

33. Yin, B.; Zhang, X.; Zhang, X.; Wang, J.; Wen, Y.; Jia, H.; Ji, Q.; Ding, L. Ionic Liquid Functionalized Graphene Oxide for Enhancement of Styrene-Butadiene Rubber Nanocomposites. *Polym. Adv. Technol.* **2017,** *28,* 293–302.

34. Patachia, S.; Croitoru, C.; Friedrich, Ch. Effect of UV Exposure on the Surface Chemistry of Wood Veneers Treated with Ionic Liquids. *Appl. Surf. Sci.* **2012,** *258,* 6723–6729.

35. Yuan, T. Q.; Zhang, L. M; Xu, F.; Sun, R. C. Enhanced Photostability and Thermal Stability of Wood by Benzoylation in an Ionic Liquid System. *Ind. Crop. Prod.* **2013,** *45,* 36–43.

36. Croitoru, C.; Patachia, S.; Porzsolt, A.; Friedrich, C. Effect of Alkylimidazolium Based Ionic Liquids on the Structure of UV-Irradiated Cellulose. *Cellulose* **2011,** *18,* 1469–1479.

37. Subramaniam, K.; Das, A.; Häußler, L.; Harnisch, C.; Stöckelhuber, K. W.; Heinrich, G. Enhanced Thermal Stability of Polychloroprene Rubber Composites with Ionic Liquid Modified MWCNTs. *Polym. Degrad. Stab.* **2012,** *97,* 776–785.

38. Subramaniam, K.; Das, A.; Heinrich, G. Improved Oxidation Resistance of Conducting Polychloroprene Composites. *Compos. Sci. Technol.* **2013,** *74,* 14–19.

39. Wang, S.; Hou, L. Application of Four Ionic Liquids as Plasticizers for PVC Paste Resin. *Iran. Polym. J.* **2011,** *20,* 989–997.

40. Przybyszewska, M.; Zaborski, M. Effect of Ionic Liquids and Surfactants on Zinc Oxide Nanoparticle Activity in Crosslinking of Acrylonitrile Butadiene Elastomer. *J. Appl. Polym. Sci.* **2010,** *116,* 155–164.

41. Pernak, J.; Walkiewicz, F.; Maciejewska, M. Zaborski, M. Ionic Liquids as Vulcani-zation Accelerators, *Ind. Eng. Chem. Res.* **2010,** *49,* 5012–5017.

42. Shamsuri, A. A.; Daik, R. Applications of Ionic Liquids and their Mixtures for Prep-aration of Advanced Polymer Blends and Composites: A Short Review. *Rev. Adv. Mater. Sci.* **2015,** *40,* 45–59.

43. Niyazi, B. A New Ionic Liquid: 2-hydroxy Ethylammonium Formate. *J. Mol. Liq.* **2005,** *116,* 15–18.

44. Flory, P. J.; Rehner, J. Jr. Statistical Mechanics of Cross-Linked Polymer Networks II. *Swelling Chem. Phys.* **1943,** *2,* 521–526.

45. Hwang, W; Wei, K. Mechanical, Thermal, and Barrier Properties of NBR/Organosili-cate Nanocomposites. *Polym. Eng. Sci.* **2004,** *44,* 2117–2124.

46. Flory, P. J. Statistical Mechanics of Swelling of Network Structures. *Chem. Phys.* **1950,** *18,* 108–111.

CHAPTER 3

ROLE OF EQUINE ORDURE IN ENHANCING PHYSICAL AND MECHANICAL PROPERTIES OF NATURAL BIO-ACTIVE COMPOSITES

RAJ KUMAR SATANKAR[1], AMRITA KAURWAR[1], SANDEEP GUPTA[1], K. USHA[2], SALIFU T. AZEKO[3], WINSTON O. SOBOYEJO[4], ALFRED B. O. SOBOYEJO[5], and ANAND PLAPPALLY[1]

[1]*Indian Institute of Technology, Jodhpur, Rajasthan 342011, India*

[2]*ACESSD, M G University, Kottayam, Kerala 686560, India*

[3]*Nigerian Turkish Nile University, Abuja, Nigeria*

[4]*Princeton University, NJ 08544, USA*

[5]*The Ohio State University, Columbus, OH 43210, USA*

*Corresponding author. E-mail: anandk@iitj.ac.in

CONTENTS

ABSTRACT

An empirical study validating the timely increment of flexural strength in green bio-composite beams, containing dispersed equine ordure (EO) in naturally occurring clayey soil matrix is performed. These materials containing 50% by volume of EO mimicking thermal properties of the clayey matrix has sustained importance in rural households of Western Rajasthan, India, for the manufacture of cantilever slabs. High gradient improvement in flexural strength in the composite with equal volume fraction of the components was observed after 21 days of ambient curing. Density development in these biocompatible MM-EO composites is stochastic in nature. Comparatively high SiO_2 content is observed in the composites manufactured from equal volume of MM and EO compared to other MM-EO composites with distinct constituent volume fractions. The fracture toughness of these materials follows a bimodal dissimilitude with time. The implication of this work is towards a new set of bioactive materials that can be locally manufactured to substitute cement-based structures.

3.1   INTRODUCTION

The equine ordure (EO) dispersed in an aged slurry of Meth Mitti (MM) has been used traditionally to manufacture grain storage composite structures (used to subdue microbial activity) and load-bearing cantilever shelves.[36] The recent discovery of microbe-inhibiting nature of the derivatives from EO supports this traditional usage.[21,24] In this article, composites manufactured with MM-EO in a volumetric ratio of 1:1 used in the construction of light cantilever shelves is discussed.[36] The article also attests that knowledge of this technology rests with the women folk of the region and has been inherited from their ancestors. The use of locally available cellulosic waste by rural people may show their inherent inertia in adopting urban construction technology as well as materials such as cement.[43,52] This inertia helps them to follow traditional practices thus preserving their culture knowledge of this construction technology.[53] Some of the MM-EO structures (as illustrated in Fig. 3.1) within rural indoor settings in different arid locations of Barmer, Rajasthan, India, are over 20 years old.[36] During this long period of use, these structures are checked for cracks and imminent climate change effects.[36]

     The structural use and microbial inhibition discovery come at a time when India is experiencing a dwindling equine population.[59] Rajasthan is a state in India with a large fraction of equine population compared to its other states.[59] This work points towards recycling equine ordure in a meaningful way, as a new bio-composite with a sustainability quotient. Lignocellulosic content similar to that in equine ordure has been used as reinforcement in construction materials such as concrete.[4,8] The work also implies at finding local waste with cellulosic material, which when adopted, will effectively reduce material transportation energy cost and manufacturing energy cost in building construction.[46] This composite aims to utilize the biological as well as mechanical properties of EO in improving the MM matrix strength.[71] The image of the load-bearing aesthetic cantilever shelf structure manufactured from this cellulose-rich composite is shown in Figure 3.1 below.

FIGURE 3.1   The MM-EO composite cantilever shelves at Arna Jharna Museum.
Image Courtesy: Rupayan Institute, Arna Jharna Museum, Jodhpur, India.

Equine ordure (EO) particulates may increase damping, reduce dilation properties and help in surface texture improvement in MM-EO composites.[25,54] Variability in specific heat capacity and improvement in heat conduction can also help increasing speed and simplicity in composite production and working with it manually.[25,54,67] Additives like EO are often introduced into the MM matrices to evolve novel functional characteristics and durability in the MM composites.[8,9,32,58,64] For example, cellulosic fiber within EO when mixed with MM may cause reduction in gross weight and handling cost.[10] This reduction is due to its superior compressive properties when compared with composite solely made out of soil or MM. Similarly, polymeric fiber-reinforced mud helps maintain uniformity of indoor temperature both in winter and summer climatic regimes.[11] Maintaining thermal equilibrium indoors will keep the household energy expenses at bay.[1]

Fracture toughness of these EO reinforced composites discussed here may be affected by the size of particles.[38] Therefore, functionally cow dung-based composites are used for flooring while equine ordure-based composites are used for making aesthetic yet load-bearing cantilever shelf structures.[36] Cow dung has smaller cellulosic fibers as compared to those of EO. Toughness of similar composites varied depending on the constituents used and their volume fractions.[38]

In addition to the above discussions, this article concentrates on bending, densification, and thermal properties. It sheds light on modeling strategies of these properties for this particular cellulose-rich composite. Stochastic aspects of property development in MM-EO composite are discussed and multi-parameter modeling strategies are adopted for these heterogeneous particulate mixtures.[34] An empirical structural analysis and microstructural study are also carried out in support of the traditional use of 50% by volume EO-based composites as a lightweight construction material.

3.2   EXPERIMENTAL METHOD

3.2.1   RAW MATERIAL CHARACTERIZATION AND COMPOSITE MANUFACTURING

A day-aged slurry of MM in water and sieved EO (35 × 1000 mesh) are the raw materials used here. The surface morphology of the raw materials EO and MM were studied using scanning electron microscopy (SEM) [Carl Zeiss EVO 18 SEM IIT Jodhpur, India, Nova Nano FE-SEM 450 (FEI) MRC, MNIT, Jaipur, India]. Chemical microanalysis was performed using energy dispersive spectroscopy (EDS) in conjunction with above described SEM [Carl Zeiss EVO 18 SEM IIT Jodhpur, India, Nova Nano FE-SEM 450 (FEI) MRC, MNIT, Jaipur, India].

A mixture of volume fractions of MM and EO, viz. 100:0, 70:30, 60:40, 50:50, and 40:60 were used to make test samples. The nomenclature of these particulate composites is defined based on its constituents as M10E0, M7E3, M6E4, M5E5, and M4E6 respectively. Particulate composites made out of MM and EO are said to have isotropic properties according to local women folk who manufacture cantilever shelves.[36]

Day-aged MM and EO in equal volumes are mixed uniformly and made into a wet mass. This wet mass is spread on a flat wooden panel as illustrated in Figure 3.2 and pressed manually. Further, similar lumps are added and cumulative pressure is applied signifying a stochastic material design process. Thickness control and geometrical alignment is an observable feature while manufacturing. The manually pressed composite samples of size $275 \times 105 \times 15$ mm^3 were cured at room temperature for 28 days. Weights of the samples were measured at small intervals of seven days each for all the distinct volume fractions defined above. Weighing was performed on an electronic weighing machine (Model Virgo, Danwer Scales, India) with an accuracy of 0.01 g. Weight loss measurement was performed since strength of any material thought to substitute concrete is dependent on the specific fraction of water used for preparation of the composite.[66]

FIGURE 3.2   The procedure of making the MM-EO composite.

Weight reduction of all samples is illustrated in Figure 3.3 for a curing time span of 28 days. This elaborates a negligible change in weight of samples with approximately 45–55% by volume of EO. The variation of weights is in the range 0.50–0.54 kg for 50:50 composition, from almost the 5th day of curing. This non-variable weight of the M5E5 composite supports its use for construction by rural people.[36] It is important to understand that volume fraction of the constituent, the microclimate (humidity and temperature variation in the specific geographic location of composite manufacture or immediate environment), and time play an important role in defining the weight of the sample.[8] The above discussion strengthens the concept of density variation in MM-EO composites due to shrinkage or volume change.[66]

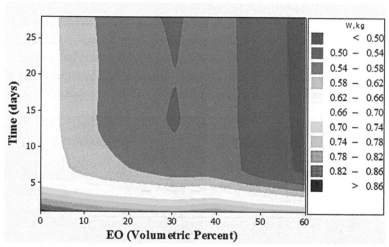

FIGURE 3.3   The variation of weight of MM-EO samples with age and composition.
Source: Mintab 16 software, IIT Jodhpur license.

3.2.2   COMPOSITE CHARACTERIZATION METHODS

The brackish groundwater at Jodhpur is blended with MM for a day.[30] This blend is mixed with distinct volume fractions of EO. The description of salinity of groundwater is beyond the purview of this chapter, hence the readers are encouraged to read about the brackishness of ground-water in Rajasthan.[30] The availability of brackish water across the state can be correlated with the design and local development of this MM-EO composite mix technology for construction.

To identify the various compounds present in these MM-EO compos-ites, the 28 day samples were analyzed using X-ray Fluorescence (XRF) spectroscopy (Bruker S4 Pioneer facility at AIRF, Jawaharlal Nehru University, New Delhi, India). Further, composite density and thermal conductivity property variation for distinct specimens aged for 28 days was also performed.[23] Interested readers are requested to refer Carlson et al., 2008[13] for detailed experimental framework. The guarded hot plate method[33] for thermal conductivity measurement was used in this case.

3.2.3   FLEXURAL STRENGTH AND FRACTURE TOUGHNESS

The samples of size $100 \times 25 \times 12.5$ mm^3 were tested for their flexure strength and fracture toughness following ASTM C99/C99M standard on three point bend universal platform, (Model EZ50 Lloyd, Germany; IIT Jodhpur) under a loading rate of 1 mm/min.[47] The flexural tests with a 1 mm/min loading rate were performed on unnotched 80 mm long MM-EO composite specimens. The fracture toughness tests on the notched 80 mm long speci-mens with a pre-crack to width ratio of 0.4 was performed following the ASTM E399 standard.[60] The test samples are subjected to Mode 1 fracture using a monotonic load of 3N/s using the three point bend platform.[52]

3.3   RESULTS AND DISCUSSIONS

3.3.1   CHARACTERIZATION OF MM, EO, AND MM-EO COMPOSITE

The matrix part of the composite is made using MM. The surface morphology and EDS spectrum of the MM resembles the phyllosilicates group.[65] Figure 3.4 illustrates distinct, well-developed, and separate flakes kept one over the other indicating a sheet or layered structure.

FIGURE 3.4   SEM micro graph of *Meth Mitti* (MM).
Source: Carl Zeiss EVO 18 SEM IIT Jodhpur, India.

Figure 3.5 confirms the presence of interconnected six membered rings of silicate sheets bonded with cations (Al, Mg, and Fe), sharing the hydroxyl ions situated at the core of the rings as illustrated in Figure 3.4.[48] The reinforcement material used for composite manufacture was EO. Figures 3.6 and 3.7 elaborate the specific results of its surface and chemical microanalysis.

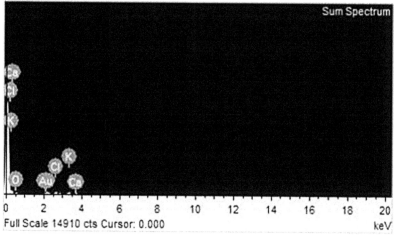

FIGURE 3.5   EDS spectra of MM.
Source: Carl Zeiss EVO 18 SEM, IIT Jodhpur, India.

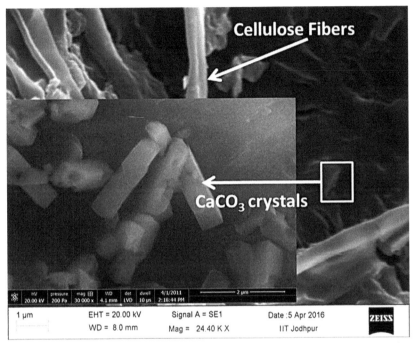

FIGURE 3.6   FESEM micro graph of EO at a 2 μm showing CaCO$_3$ crystals (Welton, 2003) inset on its cellulose network.

Source: Nova Nano FE-SEM 450 (FEI) MRC, MNIT, Jaipur, India.

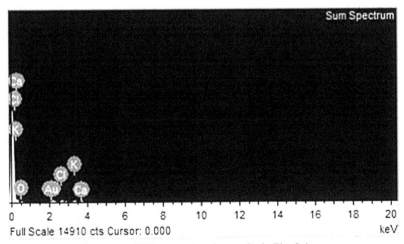

FIGURE 3.7   EDS spectra of EO supporting the results in Fig. 3.4.

Source: Nova Nano FE-SEM 450 (FEI) MRC, MNIT, Jaipur, India.

The presence of calcium carbonates in EO, along with the cellulose content influences biocementation, and makes it a good reinforcing agent, thus producing a strong fibrous network between the matrix and the reinforcement.[57] This would mean an increase in $CaCO_3$, which in turn would prevent failures due to bending fractures.[37] The presence of calcium carbonate will help in improving surface finish and decrease the surface energy of the material.[37] The composites with cellulose interacting with calcium carbonate exhibit adsorption of calcium carbonate onto cellulosic fibers in aqueous media.[20,26] Thus, this composite finds potential application in the paper manufacturing industry.[14] This would also mean that animal dung such as equine ordure will be an optimal material for use in paper manufacturing.[19,40,50]

Figure 3.8 shows comparatively higher molecular SiO_2 present in M5E5, which may provide an edge to improving its physical properties, compared to other composites. M5E5 contains the maximum SiO_2 content making it the least bioactive compared to other distinct samples. The MM-EO composites can be considered as bioactive material because of the high ratio of CaO to P_2O_5 content and low (60%) SiO_2.[41] Further presence of nontoxic cations present in the physiological environment may also render the composite bio-compatible due to which its manual manufacture is feasible.[63]

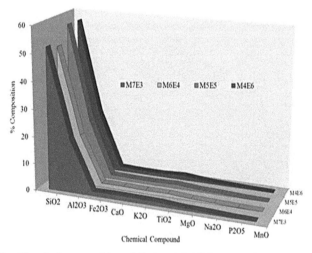

FIGURE 3.8   Chemical composition of distinct MM-EO composites.

Source: Bruker S4 Pioneer facility at AIRF; Jawaharlal Nehru University, New Delhi, India.

Presence of alkaline elements in groundwater used in the preparation of the blend leads to permeation of sodium oxides into the composites.[30] Therefore, the composites show (because of High K_2O/Na_2O ratio) clayey character.[41] Highly-reactive oxides of Ca, K, and Na oxide show a fractional decrease in volume with increase in EO in the samples.

3.3.2 DENSITY AND THERMAL CONDUCTIVITY OF MM-EO COMPOSITES

It is very significant to note (as shown in Figure 3.9) that thermal conductivity of pure MM is almost mirrored again after an addition of 50% volume fraction of EO. This also means that preservation of thermal property of a specific volume of MM is performed by addition of the same volume of EO. This attests the fact that selection of specific volume fraction MM and EO for manufacturing the composite can help design controlled thermal environments.[68] It can also be noted that thermal conductivity dips with increase in EO from 0 to 40% and then regains again with further increase in volume of EO to 60%.

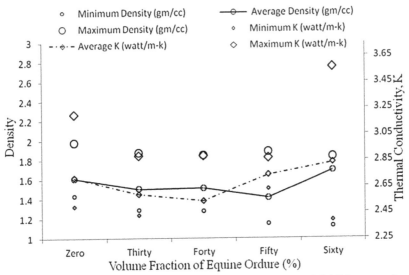

FIGURE 3.9   Variation of density and thermal conductivity in the MM-EO composites.

From the framework of quotient response modeling developed,[52] for better prediction of thermal conductivity, mass (from Fig. 3.3) is assumed to be an observable variable in the development of thermal conductivity. Therefore a new variable is defined as,[53]

$$G = X_i / Y \tag{3.1}$$

Here X_i is the mass of the EO-MM composite whose thermal conductivity K (Y) is to be calculated. Then G can be predicted by an expression,

$$G = 0.0519 + 0.0115 \ln X_i \tag{3.2}$$

Equation 3.2 has a coefficient of determination value of 80.9% and a model error of 0.00088 [Minitab v. 16 IIT Jodhpur License]. The distribution defined for G is acceptablesince the Anderson-Darling test statistic, 0.282, is less than the c_α of 1.321 for exponential distribution of G at a level of significance of 0.05.[2] Since mass is an observable quantity for these composites manufactured from powdery materials like sieved EO, the density is plotted along with thermal conductivity. It is to be understood that densification of such materials is a stochastic process.[34] Density will be discussed in relation with dilation aspects, hence is treated under transient flexural properties in the succeeding section.

3.3.3 FLEXURAL CHARACTER

It is known that incorporation of cellulosic fibers in load bearing composites increases their tensile response.[42] As seen in Figure 3.10 MM-EO composite samples are found to gain flexural strength after two weeks of curing at room temperature. It is found that the M5E5 attains better flexural strength than concrete when taken as a function of a very low maximum service temperature.[5] Further, at the end of 28 days, the strength was at par with those of rigid polymer foams.[5] In this flexural test, when the specimens are loaded centrally, the upper part of the specimen experiences compressive stress, while the lower layers experience tensile stress.[27,17] Figure 3.11 provides a snapshot of loading style during a three point band test for flexure or fracture toughness.

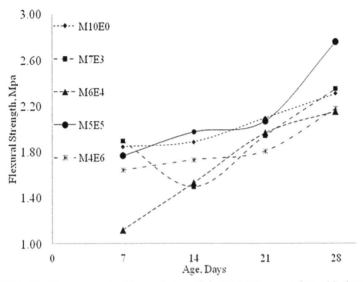

FIGURE 3.10   Development of flexural strength in MM-EO composites with time.

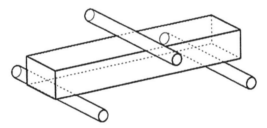

FIGURE 3.11   Representation of three point loading framework for rectangular specimens.

Overall increase in strength may be due to frictional interaction between the fibrous elements of the EO and particles of MM.[12] The inclusion of EO into MM matrix may also affect the dilative behavior of the composites.[12] This can be easily correlated with a small positive discontinuity in density and related improvement in flexural strength with age, as elaborated in Figures 3.8 and 3.9. The variation in density (after 2 weeks of curing) is to be critically analyzed for understanding this aspect as shown in Figure 3.9. This is in contention of the fact that calcium carbonate content will help improve shrinkage properties, and with time, of post-molding interaction with organic molecules of cellulose, thus improving properties of the composite.[3]

In randomly reinforced fiber composites, the strength and thermal properties are said to be dependent on its constituents and their volume fractions respectively.[15] This would mean flexural strength can be expressed in a way similar to the framework derived in Eqs. 3.1 and 3.2. Here, it is important to understand that the most observable variable for development of flexural strength is time and hence the following new variable Q can be defined as,[62]

$$Q = T_i / F_s \qquad\qquad (3.3)$$

Here, F_s is the measured value of flexural strength enumerated in Figure 3.10 and T_i is the age of curing and consequent testing of the MM-EO composite samples. Then, Q can be predicted as,

$$Q = -6.39 + 5.55 \, lnT_i \qquad\qquad (3.4)$$

The variance in the prediction of nonlinear continous Q using Equation 3.4 is $R^2 = 91.4\%$ and with a model error of 0.932 [Minitab v. 16 IIT Jodhpur License]. The distribution defined for Q is acceptable since the Anderson-Darling test statistic is 0.559, which is less than the c_α of 1.321 for exponential distribution at α-0.05.[2] It is important to note from Figure 3.12 that density for MM-EO composites decreases with a steep slope within a wek and then follows an asymptote.[35] This transient fall in density D_i can be easily related to its prior development D_{i-1} as a random function of T_i for $i=1, 2...n$.[34] This can expressed as,

$$\frac{D_i}{D_{i-1}} = T_i^{d_i} \qquad\qquad (3.5)$$

Here, d_i is a constant and $T_i^{d_i}$ becomes a transfer function.[51] Due to the heterogeneity of the MM-EO composite, the density decreases independently from time to time. Thus, the curves in Figure 3.12 represent a discrete time birth-death process. If density D_i is transformed using T_i in the following manner, then quotient characterizes both density D_i and T_i, respectively.[62] The transformation is,

$$\delta \to \frac{T_i}{D_i} \qquad\qquad (3.6)$$

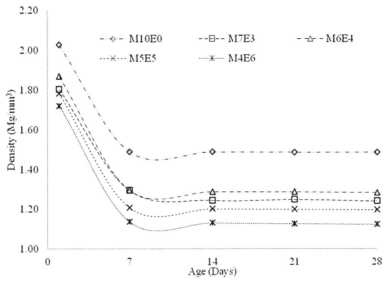

FIGURE 3.12 Density developments in MM-EO composites with time indicating possible dilatation.

The new variable δ will characterize nonlinearity of density of MM-EO composites. This would mean that linearization is required to define this new variable.[52] This linearization is written as,

$$\delta = m_i + d_i T_i \qquad (3.7)$$

Here, m_i and d_i are coefficients representative of a MM-EO composite with a distinct composition and time respectively. The linearization developed for the different MM-EO composites are tabulated in Table 3.1 below.

TABLE 3.1   Modeling Density Variation as a Function of Time.

MM-EO Samples/Coefficients	m_i	d_i	R^2	Error, S
M7E3	−0.0488	0.764	99.9	0.249
M6E4	−0.306	0.813	99.9	0.244
M5E5	−0.408	0.878	99.3	0.729
M4E6	−0.392	0.881	99.2	0.792

It is important to characterize the stochastic evolution of density of all different MM-EO composites using a generalized equation. In order to perform, the framework developed by Plappally et al in 2010 is utilized. Here, the coefficient representative of a MM-EO composite with a distinct composition, m_i, is considered as a new parameter and Equation 3.7 is rewritten as,[51]

$$\delta = A_i + d_i^* T_i + e_i m_i \tag{3.8}$$

Here, A_i, d_i^* and e_i represent the coefficient of model, time, and the newly derived parameter characterizing distinct MM-EO composites. Table 3.2 illustrates the coefficient of determination of the model elaborated using Equation 3.8.

TABLE 3.2 Modeling Density Variation in MM-EO Composites as a Function of Independent Variables Time and Parameter Characterizing Distinct MM-EO Composites, Respectively.

Variables/Coefficients	A_i	d_i^*	e_i	R^2	Error, S
T_i	−0.289	0.834		98.7	0.91
m_i	−1.31	0.834	−3.51	99.1	0.75

The model for density variation as elaborated in Equation 3.8 and Table 3.2 follows the variance test,[52]

$$Var\,\delta = \left(d_i^*\right)^2 Var\,T_i + \left(e_i\right)^2 Var\,m_i \tag{3.9}$$

$$Var\,\delta = 65.74 \approx \left(0.834\right)^2 \times 93.33 + \left(-3.51\right)^2 \times 0.021 \tag{3.10}$$

It is important to note that the variability in density due to time $\left(d_i^*\right)^2 Var\,T_i$ is large compared to that of the individual MM-EO composition effects, $\left(e_i\right)^2 Var\,m_i$.

3.3.4 FRACTURE TOUGHNESS

It is to be noted that smaller the particles, better the fracture toughness in particulate composites.[38] This is the main reason why women sieve the horse dung using a 35–1000 mesh before mixing it with water saturated MM dough. Figure 3.13 visually illustrates the bimodal fracture toughness distribution for similar volume fraction MM-EO samples.[49] A distinct yet increasing trend in fracture toughness is observed for the first 3 weeks and then a sudden decline is noted for the families of M5E5–M4E6 and M7E3–M6E4, respectively. Similar variation in fracture toughness was observed[39] in particulate polypropylene and $CaCO_3$ composite,[67] enumerating a visual illustration of fracture toughness for wood flake-reinforced polyester, and sand particle-reinforced polyesters.

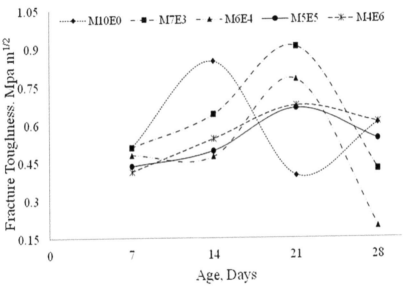

FIGURE 3.13   Bimodal distribution of fracture toughness with age.

The two basic toughening mechanisms observed in these particulate composites are increase of roughness of the fractured surface and the crack being trapped by the fibers.[56] It can be observed from Figure 3.14, that the fractured surface M7E3 sample contains numerous small nonuniform pits from which the cellulosic fibers might have sheared or slid away during the single edge notch bend test.[28] In the 60:40 sample of the MM-EO composite, an increase in the size of pits is observed or in other words, creation of small number of large valleys due to cellulosic fiber pull-out is observed. Figure 3.15 illustrates the extended cellulosic fiber protruding from the cracked surface of the M5E5 specimen which are thought to have created numerous small pits on the specimen's cracked surface due to the fiber pull out.[47,61,70,69]

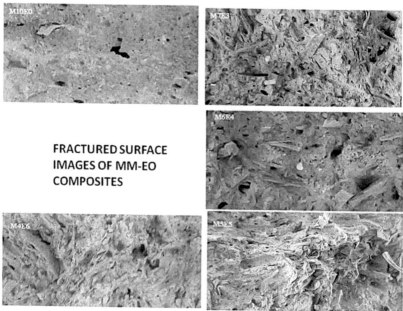

FRACTURED SURFACE IMAGES OF MM-EO COMPOSITES

FIGURE 3.14 Fractured surfaces of MM composites (100:0, 70:30, 60:40, 50:50 and 40:60) observed using a 4X optical zoom camera.

FIGURE 3.15 The fractured M5E5 specimen and the crack surfaces with protruding cellulosic fibers.

But if the fractured surface of composite samples with equal volume of MM and EO are observed, uniform spread of MM and EO interface has resulted in uniform sliding or de-bonding of cellulosic fiber from the MM matrix.[60] This uniform narrower de-bonding suggests that more work was performed to fracture the M5E5 specimen. The surface roughness illustration in Figure 3.14 helps to explain the high value of 28 day fracture toughness of 50:50 MM-EO sample compared to 70:30 and 60:40 samples shown in Figure 3.13.[28,55,56,72]

The above discussion supports the fact that composites with the same volume fraction constituents exhibit higher fracture toughness compared to MM-EO composites containing lesser EO content.[22] This would also mean that interfacial shearing between MM and cellulosic fiber in EO is pertinent and supports the 28 day data of fracture toughness variation enumerated in Figure 3.13.[15,31,61,70,69] It is difficult to model the fracture toughness of these particulate composite.[29] The difficulty may be due to heterogeneous make up, and feeble MM-EO interface.[29]

3.4 CONCLUSIONS

The sustainable use of equal volume fractions of MM and EO in the construction of cantilever shelves in rural households is justified and supported. The major findings in this empirical study and its statistical analysis are:

1. The thermal conductivity of pure MM mirrors again in the composite with 50% volume fraction of EO and MM. A novel multi-parameter model for thermal conductivity of MM-EO composites as a function of mass is proposed.
2. The cellulose fiber-$CaCO_3$ content within EO makes it an important commodity in construction materials as well as paper manufacturing industry.
3. Composite samples with approximately 50% volume of EO showed negligible weight loss with time after a week's curing. This stability in weight encourages densification due to dilation or feature of shrinkage.
4. Brackish water was used during the manufacture of MM-EO composites. In other words, a novel application of brackish water in design of fiber reinforced cementitious materials is enumerated.

5. With increase in volume fraction of EO in the MM-EO composite, the content of $CaCO_3$ in association with cellulose also improves. This may help reduce surface energy, improve surface finish of products, and increase resistance to bending fracture in MM-EO composites.

6. Resistance to flexural forces increases with increase in age for different composites with distinct compositions. A new quotient-response multi-parameter model for flexural strength as a function of time.[53]

7. M5E5 attains maximum flexural strength compared to composites with other distinct compositions.

8. Density development in composites with distinct constituent volume fractions followed a birth-death process as a function of age. A new stochastic multi-parameter model of density variation as a function of time and distinct composition constant is promulgated.

9. Shearing or sliding of cellulosic fiber is observed in mode 1 fracture experiments. Roughness of the fracture surface is observed to indicate the toughness of the composites. Roughness of the fractured surface may be a new predictor to model toughness of such fiber-reinforced composites.

10. Models derived for flexure and thermal conductivity of MM-EO composites are mathematically similar in formulation.

11. It is important to note the transient nature properties of toughness of the MM-EO composite. Indoor MM-EO composite cantilever shelves within different rural residential buildings are more than two decades old.[36] They have a decadal maintenance cycle which probably means that they require two services during a span of 20 years.[36] This may point towards a requirement of long term property characterization rather than a 28 day study cycle.

3.5 ACKNOWLEDGMENT

The research team is very much indebted to Rupayan Insititute which provided access to Arna-Jharna: The Desert Museum of Rajasthan, MNIT Jaipur, DRL, Jodhpur and AIRF, JNU, New Delhi for providing its research facilities and also to IIT, Jodhpur for providing the seed grant support for this research.

KEYWORDS

- **strength**
- **stochastic**
- **toughness**
- **regression**
- **composite**

REFERENCES

1. Al-Khawaja, M. J. Determination and Selecting the Optimum Thickness of Insulation for Buildings in Hot Countries by Accounting for Solar Radiation. *Appl. Therm. Eng.* **2004,** *24*(17–18), 2601–2610.
2. Ang, A. H. S.; Tang, W. H. *Probability Concepts in Engineering, Planning and Design, Volume 1;* John Wiley and Sons: New York, NY. 2007.
3. Arencón, D.; Velasco J. I. Fracture Toughness of Polypropylene-Based Particulate Composites. *Materials* **2009,** *2,* 2046–2094.
4. Arsene M-A.; Bilba, K.; Soboyejo, A. B. O.; Soboyejo, W. O. Influence of Chemical and Thermal Treatments on Tensile Strength of Fibers from Sugar Cane Bagasse and Banana Tree Trunk. Inter American Conference on Non-Conventional Materials and Technologies in Ecological and Sustainable Construction IAC-NOCMAT 2005—Rio, Rio de Janeiro, Brazil, Nov 11–15, 2005.
5. Ashby, M. F. *Materials Selection in Mechanical Design,* 3rd ed.; Elsevier Butterworth Heinemann: Boston, 2005.
6. ASTM C99/C99M-15 Standard Test Method for Modulus of Rupture of Dimension Stone, ASTM International, West Conshohocken, PA, 2015, http://dx.doi.org/10.1520/C0099_C0099M-15.
7. ASTM E399-12e3 Standard Test Method for Linear-Elastic Plane-Strain Fracture Toughness KIc of Metallic Materials, ASTM International, West Conshohocken, PA, 2012, http://dx.doi.org/10.1520.
8. Bentur, A.; Mindess, S. *Fibre Reinforced Cementitious Composites, Modern Concrete Technology Series;* Taylor and Francis: NY, 2007; p 595.
9. Bertelli, G.; Camino, G.; Marchetti, E.; Costa, L.; Casorati, E.; Locatelli, R. Parameters Affecting Fire Retardant Effectiveness in Intumescent Systems. *Polym. Degrad. Stabil.* **1989,** *25,* 277–292.
10. Binici, H.; Aksogan, O.; Shah, T. Investigation of Fibre Reinforced Mud Brick as a Building Material. *Constr. Build. Mater.* **2005,** *19,* 313–318.
11. Binici, H.; Aksogan, Orhan.; Bodur, M. N.; Akca, Erhan.; Kapur, S. Thermal Isolation and Mechanical Properties of Fibre Reinforced Mud Bricks as Wall Materials. *Constr. Build. Mater.* **2007,** *21*(4), 901–906.

12. Bower, T.; Jefferson, A.; Cleall, P.; Lyons, P. A Micro-Mechanics Based Soil-Fibre Composite Model for use with Finite Element Analysis, Proceedings of the 24th UK Conference of the Association for Computational Mechanics in Engineering, 31 March–01 April, 2016; Cardiff University: Cardiff.

13. Carlson, K.; Yakub, I.; Plappally, A.; Soboyejo, W. O. Low Cost Ceramics: Applications in Water Filtration and Building Insulation, Summer 2008 REU program, Princeton University.

14. Cheng X-F.; Qian, H.; Zhang S-W.; Zhang, Z-S.; He, Y.; Ma, M-G. Preparation and Characterization of Cellulose—$CaCO_3$ Composites by an Eco-Friendly Microwave-assisted Route in a Mixed Solution of Ionic Liquid and Ethylene Glycol. *BioResources* **2016,** *11*(2), 4392–4401.

15. Chamis, C. C. Design Properties of Randomly Reinforced Fiber Composites, NASA TN D-6696, Report No. E-6569, March 1972, 27.

16. Chamis, C. C. Mechanics of Load Transfer at the Fiber/Matrix Interface, NASA TN D-6588, Report No. E-6447, February 1972, 63.

17. Chanap, R. Study of Mechanical and Flexural Properties of Coconut Shell Ash Reinforced Epoxy Composites, B. Tech Thesis, 2012, Mechanical Engineering, NIT Rourkela, India.

18. Cox, D. R.; Miller, H. D. The Theory of Stochastic Processes, Methuen and Co Ltd, 1965 London E C 4.

19. Cox H. L. The Elasticity and Strength of Paper and Other Fibrous Materials. *Br. J. Appl. Phys.* **1952,** *3*(3), 72.

20. Dalas E.; Klepetsanis P. G.; Koutsoukos P. G. Calcium Carbonate Deposition on Cellulose. *J. Colloid Interface Sci.* **2000,** *224*(1), 56–62.

21. Drury, J. Horse Dung has Scientists on Scent of Antibiotic Success, Reuters, Technology [Online], http://www.reuters.com/article/us-switzerland-horse-dung-mushrooms-idUSKBN0MC12920150316) (accessed Mar 16, 2016).

22. Du, Y.; Jain, N.; Shukla, A. Effect of Particle Size on Fracture Behavior of Polyester/Al_2O_3 Composites, Proceedings of the 2006 SEM Annual Conference, Society of Experimental Mechanics, Saint Louis, Missouri USA.

23. Eckert, E. R. G.; Goldstein, R. J. Measurements in Heat Transfer, 2nd Ed.; Hemisphere Publishing Corporation 1976; p 642.

24. Essig, A.; Hofmann, D.; Münch, D.; Gayathri, S.; Künzler, M; Kallio, P. T.; Sahl, H. G; Wider, G.; Schneider, T.; Aebi, M Copsin, a Novel Peptide-Based Fungal Antibiotic Interfering with the Peptidoglycan Synthesis. *J. Biol. Chem.* **2014,** *289*(5), 34953–34964.

25. Ferrigno, T. H. Principles of Filler Selection and Use. In *Handbook of Fillers and Reinforcements for Plastics;* Katz, H. S., Milewski, J. V., Eds.; Van Nostrand Reinhold: New York, NY, USA, 1987; pp 8–62.

26. Fimbel P.; Siffert B. Interaction of Calcium Carbonate (calcite) with Cellulose Fibres in Aqueous Medium. *Colloids Surf.* **1986,** *20*(1–2), 1–16.

27. Flinn, R. A.; Trojan, P. A. *Engineering Materials and Their Applications;* Houghton Mifflin Company: Boston, 1975; p 551.

28. Gahr, K-H. Z. Wear by Hard Particles. *Tribol. Int.* **1998,** *31*(10), 587–596.

29. German, R. M. Particulate Composite: Fundamentals and Application, Springer; 1st ed. 2016, 436.

30. Gopal, R.; Bhargava, T. N.; Appraisal of the Quality of Ground Waters in the Arid Zone of Rajasthan and Kutch. *Def. Sci. J.* **1981,** *31*(1), 73–86.

31. Haspel, B.; Hoffmann, C.; Elsner, P.; Weidenmann, K. A. Characterization of the Interfacial Shear Strength of Glass-Fiber Reinforced Polymers Made from Novel RTM Processes. *Int. J. Plast. Technol.* **2015,** *19*, 333–346.

32. Heathcote, K. A. Durability of Earth Wall Building. *Constr. Build. Mater.* **1995,** *9*(3),185–189.

33. IS 3346 Method of the Determination of Thermal Conductivity of Thermal Insulation Materials (two slab guarded hot plate method), Thermal insulation materials sectional committee, Bureau of Indian Standards, 1980.

34. Kaminskii, V. M.; Nikolenko, A. N.; Sidorenko, I. Y. A Two-Dimensional Stochastic Model of the Densification of Powdered Materials. *Soviet Powder Metallurgy and Metal Ceramics* (Poroshkovaya Metallurgiya, No. 2(230)), **1982,** *21*(2), 104–106.

35. Kou, S. C.; Kou, S. G. Modeling Growth Stocks via Birth–Death Processes. *Adv. Appl. Prob.* **2003,** *35*, 641–664.

36. Kothari, K. Personal Communication, Director, Rupayan Sansthan and Arna-Jharna Desert Museum, Jodhpur, Rajasthan, India, 2013.

37. K-tron, Calcium Carbonate in Plastics Compounding, [Online], Coperion K-Tron Woodbury-Glassboro Road Sewell, NJ http://www.ktron.com/industries_served/Plastics/Calcium_Carbonate_in_Plastics_Compounding.cfm (accessed Jan 2016).

38. Lauke, B. Effect of Particle Size on Fracture Toughness of Polymer Composites, ICF12, Ottawa 2009, 8962.

39. Li, D.; Zheng, W.; Qi, Z. The J-Integral Fracture Toughness of PP/CaCO$_3$ Composites. *J. Mater. Sci.* **1994,** *29*, 3754–3758.

40. Mai, C. Marisa, http://chiangmaiwithmarisa.com/portfolio/animal-dung-paper/ (accessed March 13, 2016).

41. Marikani, A.; Maheswaran, A.; Premanathan, M.; Amalraj, L. Synthesis and Characterization of Calcium Phosphate Based Bioactive Quaternary P_2O_5–CaO–Na_2O–K_2O Glasses. *J. Non-Cryst. Solids.* **2008,** *354*(33), 3929–3934.

42. Markovich, I.; van Meir, J. G. M.; Walraven J. C. Single Fiber Pullout from Hybrid Fiber Reinforced Concrete. *HERON* **2001,** *46*(3), 191–200.

43. Mathur, V. K. Composite Materials from Local Resources. *Constr. Build. Mater.* **2006,** *20*(7), 470–477.

44. Minitab, Minitab v. 16 software, License purchased by IIT Jodhpur, 2016.

45. Moore, D. S.; McCabe, G. P.; Craig, B. A. *Introduction to the Practice of Statistics;* W. H. Freeman and Company, 6th Ed.; 2011; pp 1010.

46. Morel, J. C.; Mesbah, A.; Oggerob, M.; Walker, P. Building Houses with Local Materials: Means to Drastically Reduce the Environmental Impact of Construction. *Build. Environ.* **2001,** *36*, 1119–1126.

47. Mustapha, K.; Azeko, S. T.; Annan, E.; Kana, M. G. Z.; Daniel, L.; Soboyejo, W. O. Pull-Out Behavior of Natural Fiber from Earth-Based Matrix, *J. Compos. Mater.* 0021998315622247, first published on December 17, 2015 DOI: 10.1177/0021998315622247.

48. Nelson, S. A. Phyllosilicates (Sheet Silicates) in Phyllosilicates (Micas, Chlorite, Talc, and Serpentine), EENS 2110, [Online], Aug 19, 2015 Mineralogy, Tulane University. http://www.tulane.edu/~sanelson/eens211/phyllosilicates.htm (accessed March, 2016).

49. Nevasmaa, P.; Laukkanen, A.; Planman, T.; Wallin, K. A Novel Method for Fracture Toughness Assessment of Inhomogeneous Ferritic Steel Weldments using Bimodal Master Curve Analysis. *International Conference on Fracture,* ICF11, Italy 2005.

50. Njeru, G. Don't Pooh-Pooh It: Making Paper from Elephant Dung, 5 May 2016, http://www.bbc.com/news/business-36162953 (accessed May 16, 2016).

51. Plappally, A.; Soboyejo, A.; Fausey, N.; Soboyejo, W.; Brown, L. Stochastic Modeling of Filtrate Alkalinity in Water Filtration Devices: Transport through Micro/Nano Porous Clay Based Ceramic Materials. *J. Nat. Env. Sci.* **2010,** *1*(2), 96–105.

52. Plappally, A. K.; Yakub, I.; Brown, L. C.; Soboyejo, W. O.; Soboyejo, A. B. O. Physical Properties of Porous Clay Ceramic-Ware. *J. Eng. Mater. Technol.* **2011,** *113*(3), 311004-1–311004-9.

53. Plappally, A.; Chen, H.; Ayinde, W.; Alayande, S.; Usoro, A.; Friedman, K. C.; Dare, E.; Ogunyale, T.; Yakub, I.; Leftwich, M.; Malatesta, K.; Rivera, R.; Brown, L.; Soboyejo, A.; Soboyejo, W. A Field Study on the Use of Clay Ceramic Water Filters and Influences on the General Health in Nigeria. *J. Health Behav. Public Health* **2011,** *1*(1), 1–14.

54. Pukánszky, B. Particulate Filled Polypropylene: Structure and Properties. In *Polypropylene, Structure, Blends and Composites;* Karger-Kocsis, J., Ed.; Chapman and Hall: London, UK, 1995; Volume 3, pp 1–70.

55. Pukanszky, B.; Voros, G. Mechanism of Interfacial Interactions in Particulate Filled Composites. *Compos. Interf.* **1993,** *1*, 411–27.

56. Qiao, Y. Fracture Toughness of Composite Materials Reinforced by Debondable Particulates. *Scripta Materialia* **2003,** *49*, 491–496.

57. Rae Cho; Kang. Direct Observation of Mineral–organic Composite Formation Reveals Occlusion Mechanism. *Nat. Commun.* **2016,** *7*, 10187. PMC. Web. 6 July 2016.

58. Ren, K. B.; Kagi, D. A. Upgrading the Durability of Mud Bricks by Impregnation. *Build. Environ.* **1995,** *30*(3), 433–440.

59. Sheik, Z. Shaktiman is a rare horse, and there is more than one reason.[online] 2016, The Indian Express. http://indianexpress.com/article/explained/shaktiman-is-a-rare-horse-and-there-is-more-than-one-reason-for-it/ (accessed March, 2016).

60. Soboyejo, W. *Mechanical Properties of Engineered Materials;* Marcel Dekker: New York, 2003; p 583.

61. Sockalingam, S.; Nilakantan, G. Fiber-Matrix Interface Characterization through the Microbond Test. *Int'l J. Aeronaut. Space Sci.* **2012,** *13*(3), 282–295.

62. Sussman, H. Existence and Uniqueness of Minimal Realizations of Nonlinear Systems. *Math. Syst. Theory* **1976,** *10*, 263–284.

63. Wang, M. Developing Bioactive Composite Materials for Tissue Replacement. *Biomaterials* **2003,** *24*, 2133–2151.

64. Weidenfeller, B.; Höfer, M.; Schillling, F. Cooling Behaviour of Particle Filled Polypropylene During Injection Moulding Process. *Composites: Part A* **2005,** *36*, 345–351.

65. Welton, J. E. SEM Petrology Atlas, Methods in Exploration Series No. 4, Chevron Oil Field Research Company, AAPG publications, Tulsa, OK., 2003.

66. Wilby, C. B. *Concrete for Structural Engineers: A Text to CP 110;* Newnes-Butterworths: Boston, 1977; p 212.

67. Wong, K. J.; Yousif, B. F.; Low, K. O.; Ng, Y.; Tan, S. L. Effects of Fillers on the Fracture Behavior of Particulate Polyester Composites. *J. Strain Anal.* **2009,** *45,* 67–78.

68. Xu, Y.; Kinugawa, J.; Yagi K. Development of Thermal Conductivity Prediction System for Composites. *Mater. Trans.* **2003,** *44*(4), 629–632.

69. Zhandarov, S.; Pisanova, E.; Mäder, E.; Nairn, J. A. Investigation of Load Transfer Between the Fiber and the Matrix in Pull-Out Tests with Fibers having Different Diameters. *J. Adhesion Sci. Technol.* **2001,** *15*(2).

70. Zhandarov, S.; Mader, E. Characterization of Fiber/Matrix Interface Strength: Applicability of Different Tests, Approaches and Parameters. *Compos. Sci. Technol.* **2005,** *65,* 149–160.

71. Zenkert, D. *An Introduction to Sandwich Construction;* Chameleon Press Ltd: London, 1995.

72. Zhu, Z. K.; Yang, Y.; Yin, J.; Qi, Z. N. Preparation and Properties of Organo-Soluble Polyimide/Silica Hybrid Materials by Sol–Gel Process. *J. Appl. Polym. Sci.* **1999,** *73,* 2977–2984.

CHAPTER 4

PREPARATION AND CHARACTERIZATION OF POLYBENZOXAZINE/PLASTICIZED PVC-BASED FUMED SILICA COMPOSITES

HUSSEIN ALI, S. RADHAKRISHNAN, and M. B. KULKARNI*

Department of Polymer Engineering, Maharashtra Institute of Technology, Kothrud, Pune 411038, India

Corresponding author. E-mail: malhari.kulkarni@mitpune.edu.in

CONTENTS

ABSTRACT

Polybenzoxazine/plasticized polyvinyl chloride (PPVC)-fumed silica based flexible composites have been prepared by blending benzoxazine (BZ-a) monomer with PVC along with various other additives using a melt blending method. The fumed silica content varied from 0 to 20 wt%. The mechanical and thermal properties of the composite material were evaluated, and the microstructure was investigated through scanning electron microscopy. Addition of 3 wt% fumed silica in the composition showed higher mechanical properties as compared with other filled composites. It has been observed that the thermal stability of the composites increased with increasing amount of BZ-a and fumed silica in the compositions. The composite prepared with 20 wt% PBZ and 3 wt% fumed silica possesses very superior mechanical properties, good thermal stability, and easy processability. These results have been explained on the basis of formation of interpenetrating cross-linked structure of polybenzoxazine and partial reduction of plasticizing effect.

4.1 INTRODUCTION

Polyvinyl chloride (PVC) is a rigid polymer and difficult to melt process due to degradation at comparatively lower temperature than other common commodity plastics. Hence, it is mixed with plasticizers, heat stabilizers, and so forth.[1–4] before it is processed. It is most widely used because of its advantages such as low cost, good mechanical properties, low flammability, and so forth. The additives in PVC have been determined for toxicity and many of these are now being faced out.[5–8] Hence, the green additives, especially the plasticizers such as epoxidized vegetable oils, have been studied for PVC. In recent years, there has been a high demand for plant oils as an alternative resource to produce additives for various applications such as polymer coating, adhesive, and nanocomposite.[9–13] The major sources of vegetable oils are annual plants such as soybean, corn, linseed, cottonseed, or peanuts. However, other sources are oil-bearing perennials such as the palm, olive, or coconut.[14] Naturally occurring plant oils and fatty acids derived thereof are the most important renewable feedstock processed in the chemical industry and in the preparation of bio-based functional polymers and polymeric materials.[15–17] Nowadays, there is an increase in the use of natural-based plasticizers that are characterized by low toxicity and low migration and have good compatibility with several plastics, resins, rubbers, and elastomers in substitution of conventional plasticizers. This group includes epoxidized triglyceride vegetable oils from soybean oil, linseed oil, castor oil, sunflower oil, and fatty acid esters

(FAEs). Although it has many advantages, PVC has two shortcomings in commercial applications: poor impact strength and poor processability. To improve the thermal stability, it was thought that the blend of PVC with thermally stable polymer would lead to a better product. Most of the thermally stable polymers that are melt-processable belong to engineering plastics with high processing temperatures $> 250°C$ and can lead to higher degradation of PVC component. On the other hand, polybenzoxazine polymers have good thermal stability and get cured at 180°C, which can lead to interpenetrating network formation during processing. Polybenzoxazine exhibits superior properties than those of traditional phenolic resins, such as (i) near-zero volumetric change upon curing, (ii) very low moisture absorption, (iii) for some polybenzoxazines, glass transition (T_g) is much higher than cure temperature, (iv) high char yield, (v) no strong acid catalysts are required for curing, (vi) release of no by-product during curing, (vii) good thermal and flame-retarding properties, and (vii) zero shrinkage during curing.[18–26] However, pure polybenzoxazine-based polymers have number of disadvantages such as brittleness and difficulty in processing; thus, polybenzoxazine can be coupled with suitable substances that will overcome these problems.[27–31] In this chapter, we report the effect of blending PVC plasticized with epoxidized soy oil with polybenzoxazine. These studies report the compositional dependence of performance of unfilled polybenzoxazine-PVC blends and also the effect of addition of fumed silica on the properties of the material.

4.2 EXPERIMENTAL

4.2.1 MATERIALS

The chemicals used initially were bisphenol-A, paraformaldehyde, and aniline (99% Merck, India); chloroform (99.7%, Qualigens Fine Chemicals, India); PVC (67GER01) (G1311096) (having a density of 0.55 g/ml), with dibutyl octoate (DOP) (Idikha Impex, Chennai, India); epoxidized soybean oil (ESBO) (Bebb India Private Limited, Haryana, India) as plasticizer and calcium stearate stabilizer (Salasar Oxides Pvt. Ltd., Kota, India); and fumed silica from Henan Minmetals East Industrial Co, Ltd. (HJSIL 200 is a hydrophilic fumed silica with a specific surface area of 200 m^2/g) and average grain size (20–30 nm). All chemicals were used as received.

4.2.2   SYNTHESIS OF BENZOXAZINE MONOMER

Benzoxazine (BZ-a) monomer (aniline base) was synthesized using a solventless method by reacting bisphenol-A, aniline, and paraformaldehyde. In a typical synthesis, bisphenol-A (0.02 mol, 4.48 g), aniline (0.04 mol, 3.68 ml), and paraformaldehyde (0.08 mol, 2.4 g) were mixed in a round bottom flask and heated slowly at 90°C in an oil bath for 90 min. After cooling, BZ-a monomer was extracted from the reaction mixture by dissolving in $CHCl_3$ followed by filtration. Pure BZ-a monomer was finally obtained by evaporating $CHCl_3$. BZ-a monomer was then dried in a vacuum oven for 24 h at 55°C to remove traces of chloroform.

4.2.3   PREPARATION OF POLYVINYL CHLORIDE/ POLYBENZOXAZINES BLENDS AND NANOCOMPOSITE

Blends of BZ-a and PVC with several compositions were prepared by mixing BZ-a monomer, PVC, calcium stearate stabilizer, dioctyl phthalate (DOP), and epoxidized soybean oil (ESBO) with fumed silica and without fumed silica by overhead stirrer for 10 min, after that processed in the Brabender mixing chamber (Plasti-corder) to the equilibrium state of torque at a temperature range of 165–180°C. The compound was homogenized using mixer, type 50 EHT Brabender at mixing speed of 40 rpm. The sample was degassed under vacuum for few minutes to remove the air bubbles. All the samples for property measurements were prepared by casting their blends into a metal mold. Samples for PBZ-PVC were prepared in the same way for comparison. Various compositions of blends are listed in Table 4.1 (without fumed silica) and 4.2 (with content of fumed silica 3%).

Similar compositions were made for increasing silica content of 5, 7.5, 10, and 20% with same levels of ESBO and DOP.

The composites containing PVC and PBZ with fumed silica will have interaction between silica particles with PVC at the interface and cross-linked polybenzoxazine network, which is depicted in Scheme 4.2. It should be remembered that this will be interpenetrating network with silica particles embedded inside.

TABLE 4.1   Preparation of Blends PBZ-PVC (Without Fumed Silica).

Sr. no.	PVC (g)	ESBO (g)	Heat stabilizer (g)	DOP (g)	BZ (g)
1	100	25	1	25	0
2	95	25	1	25	5
3	90	25	1	25	10
4	85	25	1	25	15
5	80	25	1	25	20
6	75	25	1	25	25
7	70	2	1	25	30

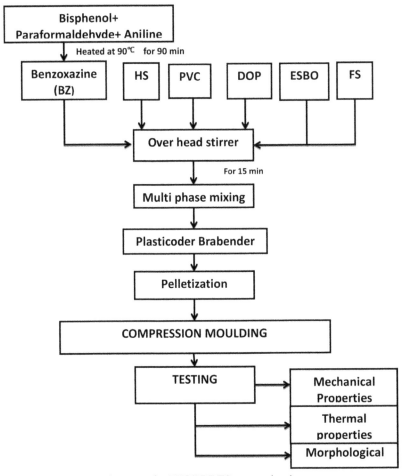

SCHEME 4.1   Preparative route for PVC/PBZ/FS composite sheets.

TABLE 4.2　Preparation of Blends PBZ-PVC (with Contents of Fumed Silica 3%).

Sr. no.	PVC (g)	ESBO (g)	Heat stabilizer (g)	DOP (g)	BZ (g)	Fumed silica
1	97	25	1	25	0	3
2	92.15	25	1	25	4.85	3
3	87.3	25	1	25	9.7	3
4	82.45	25	1	25	14.55	3
5	77.6	25	1	25	19.4	3
6	72.75	25	1	25	24.25	3
7	67.9	25	1	25	29.1	3

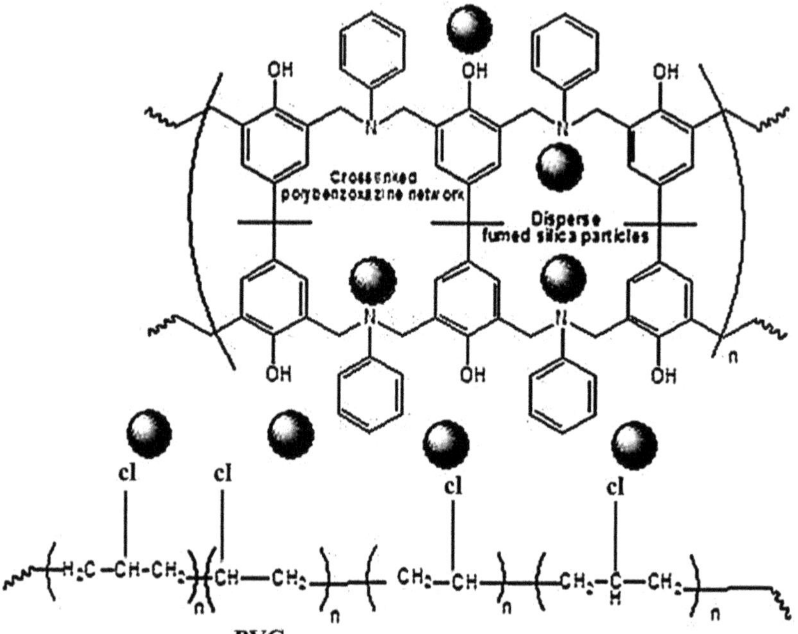

SCHEME 4.2　Possible interaction mechanism of fumed silica with polybenzoxazine and polyvinyl chloride at the interface and cross-linked polybenzoxazine network.

4.2.4　SAMPLE CHARACTERIZATION

Fourier-transform infrared (FTIR) spectra of the samples were acquired using a functional group analysis (Shimadzu, Japan), equipped with a potassium bromide (KBr) beam splitter. All spectra were recorded with 50 scans at resolution of 4 cm^{-1} and spectral range between 4000 and 400 cm^{-1}. Thermogravimetric analysis (TGA) and differential scanning calorimetric (DSC) analysis were carried out for the neat polymers and composites by using TGA-50 and DSC-60 (Shimadzu, Japan), respectively. Thermal analysis was performed from 5 to 350°C at a constant heating rate 10°C/min in air atmosphere for DSC and TGA from 0 to 900°C.

Tensile properties were measured using dumbbell-shaped specimens on a Universal Testing Machine (computerized, software based), Model No. STS-248, India, (Praj Laboratory, Kothrud, Pune) according to ASTM D638M-91 test procedure at 100% strain rate; the crosshead speed of 50 mm/min was maintained for testing. The values of tensile modulus were also determined at low strains. Room temperature measurements were carried out at a constant crosshead speed of 5 mm/min. The flexural properties of the neat polymers and blend were determined in accordance with ASTM D 790 using a Universal Testing Machine (computerized, software based), Model No. STS-248, India, (Praj Laboratory, Kothrud, Pune) according to ASTM D638M-91 with 10 kN load cell. Specimens were tested in a three-point loading with 50 mm support span at a crosshead speed of 5 mm/min at room temperature.

Izod impact strength values were evaluated on a Zwick Izod impact tester (digital), Model No. S102, Germany, (Praj Laboratory, Kothrud, Pune) according to ASTM D256 test procedure using notch samples. Flexural strength of all compositions was measured according to ASTM D-790. The average values of the mechanical properties and their standard deviations have been reported. All mechanical tests were performed at room temperature. Shore A hardness tester Model No. 026243, ID No. PRAJ/INST/08, Kori, Japan (Praj Laboratory, Kothrud, Pune) and Shore D hardness tester, Blue Steel Engineering, India.

The surface morphology of the composites was studied using a scanning electron microscope (SEM) (EVO 18, ZEISS, Germany). In pellet holder, tensile-fractured sample were pasted and fixed to sample holder. A 6380 SEM was used to evaluate the microparticle dispersion in the polymer matrix. The cryogenic fracture surface was used to take SEM micrograph.

4.3   RESULTS AND DISCUSSION

4.3.1   FTIR ANALYSIS

The FTIR spectra for Bz-a (monomer), PVC and PVC-PBO-FS are indicated in Figure 4.1. In the case of Bz-O monomer characteristics peaks at 954 cm^{-1} and 1496 cm^{-1} were observed which correspond to C-H out plane deformation of 1,2,4-trisubstituted phenyl ring present in the monomer. On the other hand these reduce considerably in intensity (almost disappear) in PVC-PBO-FS composite since the polymerization of Bz-a to PBO takes place during processing and compression molding. The FTIR of the composite indicates the major peaks of PVC (2910, 1240 and cm^{-1}) as well as those of PBO and silica.

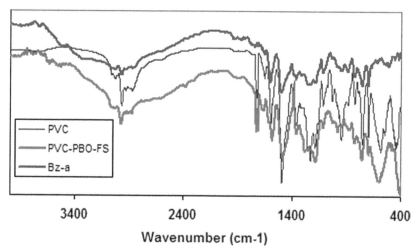

FIGURE 4.1   FTIR of pure polybenzoxazine and pure polyvinyl chloride with their blends.

4.3.2   THERMAL ANALYSIS

The DSC of the PVC-PBZ blends indicated in Figure 4.2 shows only an exotherm above 250°C, indicating the degradation of the blend possibly and leaching out of the plasticizer. With the addition of PBZ, higher heat is needed for degradation of the PVC since the network formation gives tight binding of the additive and plasticizer. This leads to shift of the peak to higher temperature.

DSC PVC-PBZ

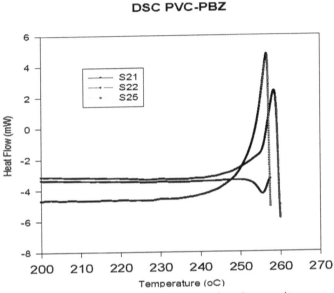

FIGURE 4.2   DSC thermogram of PVC/PBZ/FS blends and composite.

The limiting oxygen index (LOI) is the minimum concentration of oxygen, expressed as a percentage that will support combustion of a polymer. LOI values were determined by standardized tests (ASTM D2863) (Table 4.3). The LOI of 77.6PVC/19.4PBZ/3FS provided a value as high as 27, which is much greater than that of the PPVC, 23 %. Therefore, these composites can be considered as good prospects with flame retardancy which can be useful in several applications such as electronic industries, automotive, as well as in aerospace industries.

TABLE 4.3   Limiting Oxygen Index (LOI) Test for Various Composites of PVC/PBZ/FS.

Sample code	LOI
100PVC/0PBZ	23
95PVC/5PBZ	23.5
80PVC/20PBZ	23.3
70PVC/30PBZ	22.8

TABLE 4.3   *(Continued)*

Sample code	LOI
77.6PVC/19.4PBZ/3FS	27
90.25PVC/4.75PBZ/5FS	23.9
76PVC/4PBZ/20FS	24
64PVC/16PBZ/20FS	23.8

Thermal stability of the pure PVC and composites was investigated by TGA. Comparative TGA thermograms of the pure PVC and its composites are shown in Figure 4.3. The degradation of all the composites was found in the temperature range of 350–550°C. From the graphs, we can observe that the degradation temperature at 10% weight loss of the composite was shifted to a higher side at high content of fumed silica as compared to the PVC. The degradation temperature for 10% weight loss of 64PVC/14PBZ/20FS composite was 285°C. However, the degradation temperature for 10% weight loss of 77.6PVC/19.4PBZ/3FS composite was 230°C, while the degradation temperature of the neat PVC was measured to be about 260°C. In another way, the degradation temperature for 50% weight loss of 64PVC/14PBZ/20FS composite was 475°C. However, the degradation temperature for 50% weight loss of 77.6PVC/19.4PBZ/3FS composite was 450°C, while the degradation temperature of the neat PVC was measured to be about 340°C. Thus, the dispersion of fumed silica into the PVC/PBZ polymeric metrics was found to effectively improve the degradation temperature shifts to higher level. Another interesting feature in the TGA thermograms is an amount of residue at 900°C for all samples. The high char yield of these composites may be due to the cross-linked structure found in PBZ and higher thermally stable inert fumed silica particles embedded in the same. The char yield at 750°C of the composite was found to increase with increasing fumed silica content. However, at low content of fumed silica, no significant increase was observed. These thermograms revealed that the presence of more thermally stable PBZ and FS in the composites enhances the overall thermal stability of the composites.

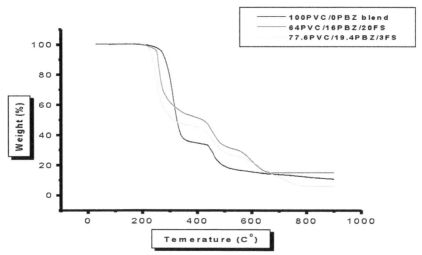

FIGURE 4.3   TGA thermograms of PVC/PBZ/FS composite.

4.3.3 MECHANICAL PROPERTIES

The mechanical properties of the prepared composites and blends were determined using standard methods. The results of mechanical tests are shown in Table 4.4. The synergistic effect of PBZ on mechanical and physical properties may help increase these properties. However, addition of ESBO and DOP at the same time as plasticizer may decrease the mechanical properties. For 80PVC/20PBZ, the mechanical properties were found to be lower than the pure PVC that may be because of ESBO effect dominated in the blend. After addition of fumed silica to the PVC-PBZ, there was significant improvement in tensile strength and impact strength at certain composition (3% FS) while at higher loading there was gradual decrease seen in the mechanical properties. This could be due to agglomeration of the FS at higher concentration which behaved like macro particles. The improvement at lower concentration of FS can be attributed to nano-particulate reinforcement arising from large surface area for interaction with the matrix as indicated in Scheme 4.2.

TABLE 4.4   Mechanical Properties of Neat PPVC/PBZ and Filled with Fumed Silica (wt%).

Compositions	Tensile strength (MPa)	Elongation (%)	Flexural strength (MPa)	Hardness	Impact strength (J/m)
80PVC/20PBZ/0FS	10.6	197	0.87	81–82	350
77.6PVC/19.4PBZ/3FS	17.46	34	3.55	94–95	461
76PVC/19PBZ/5FS	10.66	197	1.34	92–93	255
74PVC/18.5-7PBZ/7.5FS	9.21	174	0.96	91–92	210
72PVC/18PBZ/10FS	8.68	158	1.33	88–89	186
64PVC/16PBZ/20FS	7.88	90	2.17	82–83	381

4.3.4   TENSILE STRENGTH

The effects of nanoparticle fumed silica on the tensile strength of filled PVC/PBZ are presented in Table 4.5. The tensile strength of PVC/PBZ nanocomposites increased with increase in fumed silica contents, up to 3 wt% fumed silica was used. The strength of the 77.6PVC/19.4PBZ/3FS nanocomposites was increased from 10.6 to 17 MPa. Furthermore, the tensile strength of the filled PVC/PBZ/FS nanocomposites was found to be decreased when the filler content was increased beyond 3 wt%. The increase in tensile could be attributed to the reinforcing effect, that is, the addition of rigid inorganic filler into the polymer matrix was capable to improve the stiffness of the polymer composite. It was assumed that at higher content of fumed silica, the phenomenon of agglomeration and surface area of filler lead to formation of stress concentration in the nanocomposites. The variation in tensile strength of fumed silica filled PVC/PBZ composites is presented in Figure 4.4. The experimental results envisaged better dispersion of the smaller sized filler in the PVC/PBZ matrix, and good nanofiller–matrix interaction may be a factor responsible for the trend observed in this study.

TABLE 4.5   Tensile Strength (MPa) of the Composites with the Variation of Fumed Silica in the Composite.

Sample code	Content of fumed silica					
PVC/PBZ	0%	3%	5%	7.5%	10%	20%
	(MPa)	(MPa)	(MPa)	(MPa)	(MPa)	(MPa)
100PVC/0PBZ	14.12	14.645	12.565	12.00	9.14	8.89
95PVC/5PBZ	13.275	12.65	12.185	11.9	11.87	11.235
90PVC/10PBZ	13.82	13.03	10.34	12.82	11.32	9.61
85PVC/15PBZ	11.465	11.8	11.6	11.59	11.18	9.56
80PVC/20PBZ	10.6	17.46	10.655	9.21	8.68	7.875
75PVC/25PBZ	9.245	10.12	11.86	10.59	8.62	7.785
70PVC/30PBZ	9.3	9.405	9.035	6.705	6.6	6.44

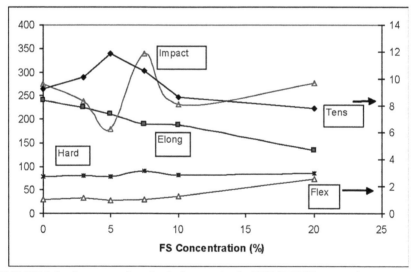

FIGURE 4.4   Summary of mechanical properties for PVC/PBZ/FS composite.

4.3.5   ELONGATION AT BREAK (%)

Elongation at break (%EB) is a measure of the ductility of a material. The results of %EB are shown in Table 4.6. The experimental results showed the reduction in the ductility of the nanocomposite with an increase in the fumed silica contents which support the formation of stiff material due to addition of nano-fumed silica and formation of a rigid interface of interaction between the fillers and matrix. The trend in variation of %EB is represented in Figure 4.4. The variation also supports the effect of significant amounts of PBZ in the blend, which helped to enhance the cross-linking density of PVC beyond the certain level and lead to enhance intermolecular stress and to decrease the values of elongation at break of the filled blend compositions gradually.

TABLE 4.6 Elongation at Break (%EB) of the Composites with the Variation of Fumed Silica in the Composite.

Sample code	Content of fumed silica					
PVC/PBZ%	0%	3%	5%	7.5%	10%	20%
100PVC/0PBZ	196.865	186.605	163.995	176.22	161.355	90.44
95PVC/5PBZ	188.805	136.72	168.185	148.22	266.14	178.5
90PVC/10PBZ	217.46	193.495	159.885	204.11	182.68	118.77
85PVC/15PBZ	199.05	184	144.72	216.825	203.66	145.995
80PVC/20PBZ	197.075	34.885	197.00	174.995	158.33	90.55
75PVC/25PBZ	239.685	224.16	210.72	189.94	187.775	133.05
70PVC/30PBZ	201.265	200.02	160.16	158	147.22	117.44

4.3.6   FLEXURAL AND IMPACT STRENGTH

The effects of nanoparticle sizes on the flexural and impact strength of filled PVC/PBZ are presented in Tables 4.7 and 4.8, respectively. The experimental data on the flexural strength of PVC/PBA nanocomposites showed enhancement in values of the flexural strength of the nanocomposites at initial nanofiller contents and then decrease in flexural strength of the nanocomposites as the nanofillers increase beyond 3 wt% of fumed silica. The strength of the 77.6PVC/19.4PBZ/3FS nanocomposite was increased from 0.865 to 3.55 MPa that supports the improvement in the dispersion at lower levels of fumed silica and at higher loading effect of poor stress distribution lead to decrease in the values beyond 3 wt%. The impact strength of the nanocomposites increased marginally at low concentration of fumed silica content and decreased with further addition of nanofillers. The trend in variation in flexural and impact strength of PVC/PBZ filled with fumed silica against content of fumed silica (wt%) is presented in Figure 4.4.

TABLE 4.7 Flexural Strength (MPa) of the Composites with the Variation of Fumed Silica in the Composite.

Samples code	FS					
PVC/PBZ%	0%	3%	5%	7.5%	10%	20%
	(MPa)	(MPa)	(MPa)	(MPa)	(MPa)	(MPa)
100PVC/0PBZ	1.395	1.55	1.62	1.49	2.41	2.41
95PVC/5PBZ	1.21	1.27	1.43	1.16	2.47	1.95
90PVC/10PBZ	1.13	1.15	1.04	1.38	1.43	2.15
85PVC/15PBZ	1.37	1.28	1.05	1.16	1.23	1.67
80PVC/20PBZ	0.865	3.55	1.34	0.96	1.33	2.17
75PVC/25PBZ	1.06	1.14	0.99	1.04	1.29	2.57
70PVC/30PBZ	1.005	0.97	0.98	1.37	1.44	1.52

TABLE 4.8 Impact Strength (MPa) of the Composites with the Variation of Fumed Silica in the Composite.

Sample code	FS					
PVC/PBZ%	0%	3%	5%	7.5%	10%	20%
	(MPa)	(MPa)	(MPa)	(MPa)	(MPa)	(MPa)
0	400	608.75	316.25	430	426.25	357.5
5	480	430	467.5	320	335	422.5
10	455	281.25	367.5	378.75	322.5	397.5
15	415	257.5	378.75	257.5	222.5	373.75
20	350	461.25	255	210	186.25	381.25
25	275	238.75	178.75	340	231.25	276.25
30	275	212.5	158.75	190	228.75	341.25

4.3.7 HARDNESS

The effect of composition of blends on the hardness of PBZ/PVC/FS composites was determined by Shore (A, D) hardness tester. The variation in hardness of PVC/PBZ composition as the function of content of fumed silica is shown in Figure 4.4 and Table 4.9. The hardness values of the composites increase with an increasing amount of fumed silica content in the blend. When the fumed silica nanoparticles were applied, the hardness value of composition PBZ/PVC/FS was raised from 81 MPa (at 0 wt% FS) to 94 MPa (at 3 wt% FS). Hence, the addition of fumed silica nanoparticles to PBZ/PVC was found to enhance the resistance of the blend deformation. Therefore, the behaviors of the PVC/polybenzoxazine blend nanocomposites are under investigation, which can be used as a high wear-resistant coating material.

TABLE 4.9 Hardness of the Composites with the Variation of Fumed Silica in the Composite.

Sample code	FS					
PVC/PBZ%	0%	3%	5%	7.5%	10%	20%
	(MPa)	(MPa)	(MPa)	(MPa)	(MPa)	(MPa)
0	85	84	87	89	85	87
5	84	84	81	89	86	87
10	82	80	80	87	83	89
15	82	85	76	85	78	87
20	81	94	92	91	88	83
25	78	80	79	91	81	86
30	83	81	77	88	79	85

4.3.8 SEM OBSERVATIONS

The SEM micrographs of a fractured surface of PPVC/PBZ blends and their filled composites obtained at various magnification ranges which were used to study the dispersion and adhesion between filler and polymer are presented in Figure 4.5. The silica particles are very well dispersed as seen in Figure 4.5(b). At higher concentrations of FS, the particles may form the agglomeration in the PVC/PBZ/%FS polymeric matrices, which cause void defects in the composites. The presence of voids and denuded particles on the fracture surface was increased with the increasing content of fumed silica as shown in Figure 4.6, which supports decrease in mechanical properties.

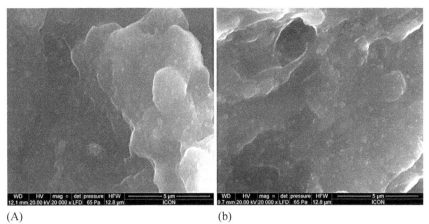

(A)   (b)

FIGURE 4.5   SEM of (a) PVC-PBZ blend and (b) PVC-PBZ+3% FS. The scale bar is 5 μm, the silica particles are much less than 1.0 μm.

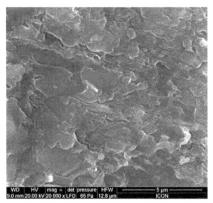

FIGURE 4.6   SEM of PVC-PBZ+10% FS. Particles are large agglomerates greater than 1 μm.

4.4   CONCLUSIONS

Polybenzoxazine-PPVC-based nanocomposites were prepared with and without fumed silica by melt compounding. A range of compositions were made containing different concentrations of PBZ and FS, while maintaining the same plasticizer content. These compounds were characterized by FTIR, DSC, and TGA. The blend was well formed and homogenous. The mechanical properties such as tensile and flexural strength, elongation, and impact strength were studied with composition. There was improvement in the mechanical properties at certain concentration of PBZ and FS. The blend consisting of 77.6PVC/19.4PBZ/3FS exhibited the best mechanical properties along with good thermal stability. Above the concentration of 10% FS, there was a drop in the mechanical properties in all respect. This can be attributed to good dispersion of FS to nanolevels at low concentration but formation of agglomerates at higher levels of FS. The morphological studies with SEM confirm these findings. Thermal properties revealed that the presence of more thermally stable PBZ and FS in the composites enhances the overall thermal stability of the composites. These blends/nanocomposites have the potential to be used in various industrial applications.

KEYWORDS

- **nanocomposite**
- **polybenzoxazine**
- **PPVC**
- **fumed silica**
- **mechanical property**
- **thermal and morphological properties**

REFERENCES

1. Nicholas, P.; Cheremisinoff, P. H. D. *Handbook of Engineering Polymeric Materials*; Plasticizer PVC Blend: New York, 1997; pp 137–145.
2. Robeson, L. M. Miscible Polymer Blends Containing Poly(vinyl Chloride). *J. Vinyl Technol.* **1990**, *12*(2), 89–94.
3. Emmett, R. A. Acrylonitrile-Butadiene Copolymers-Mixtures with Plasticized Polyvinyl Chloride. *Ind. Eng. Chem.* **1944**, *36*, 730–734.
4. Wilkes, C. E.; Daniels, C. A.; Summers, J. W. *PVC Handbook*; Hanser Publications, Hanser Verlag, Munich, 2005, 316–335.
5. Riaz, U.; Vashist, A.; Ahmad, S. A.; Ahmad, S.; Ashraf S. M. Compatibility and Biodegradability Studies of Linseed Oil Epoxy and PVC Blends. *Biomass Bioenergy* **2010**, *34*, 396–401.

6. Nass, L. I. *Encyclopedia Of Poly(vinyl Chloride)*; Mercel-Dekker, Inc.: New York, 1976, Vol. 1; pp 2–5.

7. Fenollar, O.; García, D.; Sánchez, L.; López, J.; Balart, R. Optimization of the Curing Conditions of PVC Plastisols Based on the Use of an Epoxidized Fatty Acid Ester Plasticizer. *Eur. Polym. J.* **2009**, *45*, 2674–2684.

8. Silva, M. A.; et al. Polyvinyl chloride (PVC) and Natural Rubber Films Plasticized with a Natural Polymeric Plasticizer Obtained Through Polyesterification of Rice Fatty Acid. *Polym. Test.* **2011**, *30*, 478–484.

9. Wool R. P.; Sun X. S. *Polymers and Composite Resins from Plant Oils in Bio-based Polymers and Composites;* Elsevier Academic Press: Burlington (USA), 2005; pp 56–113.

10. Belgacem M. N.; Gandini A.; Belgacem M. N. *Materials from Vegetable Oils: Major Sources, Properties and Applications in Monomers, Polymers and Composites from Renewable Resources;* Elsevier: Oxford (UK), 2008; pp 39–66.

11. Lu Y.; Larock R. C. Novel Polymeric Materials from Vegetable Oils and Vinyl Monomers: Preparation, Properties, and Applications. *Chem. Sus. Chem.* **2009**, *2*(2), 136–147.

12. Xia Y.; Larock R. C. Vegetable Oil-Based Polymeric Materials: Synthesis, Properties, and Applications. *Green Chem.* **2010**, *12*(11), 1893–909.

13. Khot S. N.; Lascala J. J.; Can E.; Morye S. S.; Williams G. I.; Palmese G. R. Development and Application of Triglyceride-Based Polymers and Composites. *J. Appl. Polym. Sci.* **2001**, *82*, 703–23.

14. Hui, Y. H. *Baileycs Industrial Oil and Fats Products, Edible Oil and Fat Products: General Application,* 5th ed.; 2000, Vol. 1; pp 19–44.

15. Baltacoglu H.; Balkose D. Effect of Zinc Stearate and/or Epoxidized Soybean Oil on Gelation and Thermal Stability of PVC-DOP Plastigels. *J. Appl. Polym. Sci.* **1999**, *74*(10), 2488–2498.

16. Krauskopf L. G.; Godwin, A. Plasticizers. In *PVC Handbook;* Wilkes, C. E., Summers, J. W., Daniels, C. A., Eds.; Carl Hanser Verlag: Munich, 2005; pp 173–193.

17. Formo, M. W. In *Industrial Use of Soybean Oil*, Proceedings of the 21st World Congress of the International Society of Fat Research (ISF), The Hague; P. J. Barnes & Associates, Eds.; 1995, pp 519–527.

18. Ghosh, N. N.; Kiskan, B.; Yagci, Y. Polybenzoxazines—New High Performance Thermosetting Resins: Synthesis and Properties. *Prog. Polym. Sci.* **2007**, *32*, 1344–1391.

19. Yagci, Y.; Kiskan, B.; Ghosh, N. N. Recent Advancement on Polybenzoxazine—A Newly Developed High Performance Thermoset. *J. Polym. Sci. Pol. Chem.* **2009**, *47*, 5565–5576.

20. Kiskan, B.; Dogan, F.; Durmaz, Y. Y.; Yagci, Y. Synthesis, Characterization and Thermally Activated Curing of Azobenzene-Containing Benzoxazines. *Des. Monomers Polym.* **2008**, *11*, 473–482.

21. Ghetiya, R. M.; Kundariya, D. S.; Parsania, P. H.; Patel, V. A. Synthesis and Characterization of Cardo Bisbenzoxazines and Their Thermal Polymerization. *Polym. Plast. Technol.* **2008**, *47*, 836–841.

22. Jang, J.; Seo, D. Performance Improvement of Rubber Modified Polybenzoxazine. *J. Appl. Polym. Sci.* **1998**, *67*, 1–10.

23. Kiskan, B.; Ghosh, N. N.; Yagci, Y. Polybenzoxazine-Based Composites as High-Performance Materials. *Polym. Int.* **2011**, *60*, 167–177.
24. Ishida, H. Overview and Historical Background of Polybenzoxazine Research, Chapter 1. In *Handbook of Benzoxazine Resins;* Ishida, H., Agag, T., Eds., Elsevier: Amsterdam, Netherlands, 2011; pp 3–81.
25. Rajput, A. B.; Seikh, J. R.; Sarkhel, G.; Ghosh, N. N. Preparation and Characterization of Flexible Polybenzoxazine-LLDPE Composites. *Des. Monomers Polym.* **2012**, *16*, 177–184.
26. Rajput, A. B.; Ghosh, N. N. Preparation and Characterization of Novel Polybenzoxazine Polyester Resin Blends. *Int. J. Polym. Mater.* **2010**, *59*, 1–13.
27. Rimdusit, S.; Punson, K.; Dueramae, I.; Somwangthanaroj, A.; Tiptipakorn, S. Rheological and Thermomechanical Characterizations of Fumed Silica-Filled Polybenzoxazine, Nanocomposites. *Eng. J.* **2011**, *15*, 27–38.
28. Santhosh Kumar, K. S.; Reghunadhan Nair, C. P. *Polybenzoxazines: Chemistry and Properties;* iSmithers, 2010. ISBN: 978-1-84735-500-3.
29. Rimdusit, S., Kunopast P., Dueramae, I. Thermomechanical Properties of Arylamine-Based Benzoxazine Resins Alloyed with Epoxy Resin. *Polym. Eng. Sci.* **2011**, *51*, 1797–1807.
30. Rimdusit, S.; Ishida H. Development of New Class of Electronic Packaging Materials Based on Ternary Systems of Benzoxazine, Epoxy, and Phenolic Resins. *Polymer* **2000**, *41*, 7941–7949.
31. Takeichi, T.; Guo, Y.; Agag, T. Synthesis and Characterization of Poly(urethane-Benzoxazine) Films as Novel Type of Polyurethane/Phenolic Resin Composites. *J. Polym. Sci. Part A: Polym. Chem.* **2000**, *38*, 4165–4176.

CHAPTER 5

INVESTIGATION ON THE THERMAL CHARACTERISTICS OF DRY BONDING SYSTEM FOR TIRE APPLICATIONS

D. JEYABALA[1], J. DHANALAKSHMI[2], and C. T. VIJAYAKUMAR[1,*]

[1]*Department of Polymer Technology, Kamaraj College of Engineering and Technology, K. Vellakulam, Tamil Nadu 625701, India*

[2]*Department of Chemistry, Kamaraj College of Engineering and Technology, K. Vellakulam, Tamil Nadu 625701, India*

Corresponding author. E-mail: ctvijay22@yahoo.com

CONTENTS

ABSTRACT

Resorcinol-formaldehyde resin (R) and hexamethoxymethylmelamine (HMMM) are extensively used as methylene acceptor and donor, respectively, in rubber compounding. This dry bonding system is used by most of the tire manufacturers. Detailed investigations on the chemistry of the dry bonding systems are scarce. Differential scanning calorimetric (DSC) investigations have been carried out for R, HMMM, and R (60%)/HMMM (40%) blend (B) and found that the thermal curing is taking place within a temperature span of 75°C and the curing is starting at around 135°C. The study of the thermal stability of the materials using thermogravimetric analyzer (TGA) indicates that the polymer from the blend is thermally more stable compared to the neat resin R. The Fourier-transform infrared (FTIR) spectroscopic studies of these materials provide ample evidence for the formation of quinone methide structures during the reaction of R and HMMM.

5.1 INTRODUCTION

5.1.1 TIRE

Geometrically, a tire is a torus. Mechanically, a tire is a flexible membrane pressure container or a loop of air. Chemically, a tire consists of materials from long chain macromolecules. Structurally, a tire is a high-performance composite. Robert William Thomson was the inventor of the pneumatic tire. He was granted a patent in France in 1846 and in the USA in 1847. Thomson's "Aerial Wheels" were considered as the forerunner to the modern pneumatic tires. John Boyd Dunlop reinvented the pneumatic tire in 1888 and began to achieve wide applications.[1]

5.1.2 TIRE CHARACTERISTICS

A tire must have following characteristics:

- Provide load carrying capacity
- Provide cushioning and enveloping
- Transmit driving and braking torque
- Produce cornering force

- Provide dimensional stability
- Resist abrasion
- Provide steering response
- Have low rolling resistance
- Provide minimum noise and permit minimum road vibrations
- Be durable and safe

Due to the unique deformability and dampening properties of the pneumatic tire, it is the only product that satisfies all these characteristics.

5.1.3   TIRE REQUIREMENTS

Basically, a tire's functions can be considered in relation to the following three areas:

a)   Vehicle mobility
b)   Performance and integrity
c)   Esthetics and comfort

Vehicle mobility includes cornering, steering response, and abrasion that act in the lateral direction.[5] Performance includes driving and braking torque and rolling resistance, which exert or transfer forces or moments (tangentially) forward. The forces involved with esthetics and comfort act vertically.

5.1.4   TIRE TYPES

Tire can be classified into following types:

- Bias or cross-ply tire
- Bias belted tire
- Radial tire

5.1.4.1   BIAS OR CROSS-PLY TIRE

The cross-ply tire is the older form. It is also called a bias ply or conventional tire. It is constructed of two or more plies or layers of textile casing cords positioned diagonally from bead-to-bead. The rubber-encased cords run at an angle between 30 and 40° to the center line, with each cord wrapped around the beads. A latticed crisscrossed structure is formed with alternate layers crossing over each other and laid with the cord angles in opposite directions.

This provides a strong, stable casing, with relatively stiff sidewalls. However, during cornering, stiff sidewalls can distort the tread and partially lifts it off the road surface. This reduces the friction between the road and the tire. Stiff sidewalls can also make tires run at a high tempera-ture. This is because, as the tire rotates, the cords in the plies flex over each other, causes friction and heat. A tire that overheats can wear prematurely.[6]

5.1.4.2   BIAS BELTED TIRE

A belted bias tire starts with two or more bias plies to which stabilizer belts are bond directly beneath the tread. This provides a smoother ride that is similar to the bias tire, but there is a decrease in rolling resistance because the belts increase tread stiffness. The carcass angle is generally 25–40° and that of the belt is between 25 and 30°. In addition, the angle of the belt is at least 5° lower than that of the carcass. This type of tires was commercially introduced in 1976.

5.1.4.3   RADIAL TIRE

The radial tire was introduced by Michelin Tire Company in 1913. Based on the construction, radial ply tires have much more flexible sidewalls. They use two or more layers of casing plies with the cord loops running radially from bead-to-bead.

The sidewalls are more flexible because the casing cords do not cross over each other. However, a belt of two or more bracing layers must be placed under the tread. The cords of the bracing layers may be of fabric or of steel and are placed at 12–15° to the circumference line. This forms trian-gles where the belt cords cross over the radial cords. The stiff bracing layer links the cord loops together to give fore and aft stability while accelerating or braking and it prevents movement of the cords during cornering. The cord plies flex and deform only in the area above the road contact patch.

There are no heavy plies to distort, and flexing of the thin casing gener-ates little heat that is easily dispersed. A radial ply tire runs cooler than a comparable cross-ply tire and this increase tread life. A radial tire has less rolling resistance as it moves over the road surface.

5.1.5   TIRE STRUCTURAL COMPONENTS

A tire comprises a number of parts and subassemblies. Each part has a specific and unique function. Tire components are tread, sipes, grooves, blocks, rips, dimples, undertread, tread cushion, shoulder, belt/breaker, sidewall, plies, carcass, inner liner, bead, chafers, and rim strip.[7]

5.1.5.1   TREAD

Tread is the wearing surface of the tire, which comes in contact with the road surface. The tread has to provide the necessary traction, sufficient grip, low rolling resistance, less noise, low heat buildup, good abrasion resistance, resistance to chipping, cut growth resistance, flex resistance, and so forth The tread compound is hence designed to meet these requirements and also other processing requirements such as scorch safety, good flow properties, good green tackiness, good dimensional stability, and so forth The elastomers used for tire tread are natural rubber (NR), styrene butadiene rubber (SBR), their oil extended varieties and butadiene rubber (BR). Sometimes cis-poly isoprene, ethylene propylene monomer (EPM) rubber, and ethylene propylene diene monomer (EPDM) rubber are also used. Among the polymers NR, BR, and SBR, the lowest heat buildup is offered by NR and highest road holding capacity by SBR. There is a good difference in the loading of front and back tires. Generally, front tire will be of rib design and rear tire will be of the lug design.

5.1.5.1.1   Tread Pattern

The tread pattern has two basic functions: it must provide drainage and bite. The tread needs a quantity of reasonably sharp well-defined edges that will engage the road surface to provide grip and to transmit driving and breaking torques. These edges need to be transverse for good traction and breaking, and longitudinal for steering and cornering. The properties and applications of the basic thread patterns are tabulated in Table 5.1.

TABLE 5.1 Basic Tire Tread Patterns.

Types	Rib	Lug	Rib—Lug	Block
Profile	Grooves flow along the circumference of a tire.	Grooves flow in a lateral direction.	Combination of rib and lug patterns.	Patterns in which individual blocks are arranged.
Feature	Superb driving stability Fewer rolling resistance Less noise generation Superior water drainage performance Less skidding	Excellent drive and braking forces Stronger traction force Better resistance to cut	Rib type pattern acts to increase driving stability and to prevent skidding. Lug type pattern acts to transfer traction drive and braking forces effectively.	Excellent drive and braking forces. Superb driving stability on the normal paved road.
Applications	Best for driving on cemented roads and highways	Suitable for driving on normal cemented roads, and uncemented ones, suitable for work sites such as a building site	Suitable for driving on normal cemented roads and uncemented ones at a middle/low speed	Good performance when driving on normal cemented roads

5.1.5.2 SIPES

Sipes are the small, slit-like grooves in the tread blocks that allow the blocks to flex. This added flexibility increases traction by creating an additional biting edge. Sipes are especially helpful on the ice, light snow, and loose dirt.

5.1.5.3 GROOVES

Grooves are the continuous circumferential channels between the tread ribs. Grooves create voids for better water channeling on wet road surfaces (like the aqua channel tires below).

Grooves are the most efficient way of channeling water from the tires, front side to the backside. By designing grooves circumferentially, water has less distance to be channeled. Tread wear indicators are molded into the bottom of the tread grooves to indicate when the tire should be replaced.

5.1.5.4 BLOCKS

Blocks are the segments that make up the majority of a tire's tread. Their primary function is to provide traction.

5.1.5.5 RIPS

Rips are continuous circumferential rows of tread rubber, which are in direct contact with the road surface. Designs are molded into the rips for traction, noise suppression, and so forth.

5.1.5.6 DIMPLES

Dimples are the indentations in the tread, normally toward the outer edge of the tire. They improve cooling.

5.1.5.7 UNDERTREAD

It is compounded for high heat resistance to increase high-speed durability and sometimes fuel economy. Undertread rubber thickness is very necessary for tread life as well as for the safety of carcass. Normally, 30–50% of tread pattern depth is kept as undertread rubber depending upon the size and application of the tire.

5.1.5.8 TREAD CUSHION

The primary requirement of cushion compound is adhesion between the tread and outer plies. Normally, NR is used for cushion compound due to its high green tackiness. But in passenger car compounds NR/SBR blend may be used for getting good compatibility if tread compound is oil-extended SBR (OESBR). To improve adhesion property suitable tackifiers are also added.

5.1.5.9   SHOULDER

The shoulders are the upper portion of the sidewall just below the tread edge. Shoulder design affects tire heat behavior and cornering characteristics.

5.1.5.10   BELT/BREAKER

The belt is a layer or layers of material underneath the tread of a radial ply tire to restrict the carcass against extending itself in a circumferential direction when the tire is inflated. The belts have low angles between 20 and 35°. This gives directional stability, steering response, reduce growth, and stiffens the tread portion.

The breaker is the intermediate ply or plies, fitted between the carcass and tread. Bias ply tire constructions use components known as breaker or cap plies. These are similar to belts in the way that they are positioned over the carcass plies in the tread area. They differ from belts in the manner that their cord angles are the same or nearly the same as the carcass plies. Breakers have minimal effect on tire's shape because they do not restrict circumferential growth but are used primarily to add strength in the tread and shoulder regions.

Belts in radial ply tires are made of glass fibers. Hence, special bonding systems are used in their compounding to improve adhesion. Breaker compounds are designed for a gradual reduction in modulus from plies to tread. It should also have good compatibility and adhesion with the cushion compound and the inner plies.

5.1.5.11   SIDEWALL

The sidewalls are the portion of the tire contour between the beads and the tread that primarily control ride. The main requirements of sidewall compound are weather resistance, ozone resistance, low heat buildup, good green and cured adhesion, and so forth.

Passenger car sidewalls are usually extruded as a single unit along with the tread. Tread compound with proper adjustments can also be used for sidewalls. Radial ply sidewalls should have more flex resistance. Some tires with white sidewall or white lettering are available, which have good aesthetic appeal.

5.1.5.12   PLIES

The plies are layers of rubber-coated cord fabric extending from bead-to-bead and the reinforcing members of the tire. The plies consist of cords of rayon, nylon, or polyester woven as the warp of the fabric with very light yarns and widely spaced as the weft. These weft strands serve to maintain the uniformity of the cord during handling but play no part in the performance of the product.

5.1.5.13   CARCASS

The plies are turned up around the bead, thereby locking the bead into the assembly. This assembly is often called the tire body or carcass. Carcass compound should provide good tackiness, high fatigue life, and resistance to reversion.

5.1.5.14   INNER LINER

Inner liner is a thin layer of rubber compound on the inside of the tire. This will insulate the carcass and thus prevent the cords from possible degradation due to the atmospheric moisture absorption. It also prevents the chafing action between the inner tube and tire cords, thus, protecting the tube from the damage.

For tube tire, inner liner may be calendered directly on the underside of the first ply, or sometimes directly applied on the drum as the first stage operation in tire building; hence, it is also called as drum squeegee.

In a tubeless tire, the inner liner is the air retaining member and is usually calendered as a two-layer laminate having stepped edges. The overall gauge may be as high as 2.5 mm and the width may ensure that the edges are overlapped by the inner edges of the chafer; thus, providing a low permeability layer from bead-to-bead.

5.1.5.15   BEAD

The beads are composed of multi-strand bronze-coated high-tensile steel wires coated with a suitable rubber compound and formed into a stiff, inextensible ring. They have the function of providing rigid, practically inextensible units, which retain the inflated tire on the rim under all conditions of loading.

The bead insulation compound should provide good green and cured adhesion to the bead wires, good heat resistance, and high modulus. High loaded SBR with some tackifiers to improve adhesion is usually used.

5.1.5.16   BEAD APEX

The top face of the bead is a triangular rubber compound strip called bead apex. Its primary function is to pack out the area of the structure immediately above the bead coil and provide a steady gradation of thickness between the bead coil and the sidewall zone, the thickness of which is same as that of the casing plies and sidewall rubber. The component which is triangular is formed by extruding from a die.

The apex compound should have high modulus and heat resistance. SBR or a blend of NR/SBR is usually used. Modulus is improved by increasing the amount of sulfur.

5.1.5.17   FLIPPER

Flipper is a bias-cut fabric wrapping the bead as well as the apex into one unit. Flipper compound should give good compatibility to the plies as well as to the apex and the bead is usually a blend of NR/SBR.

5.1.5.18   CHAFER

Chafers are narrow circumferential strips, which enclose the complete bead area. Chafer has to protect the bead portion from chafing with the rim to distribute the flexing above the rim and to prevent penetration of moisture and dirt into the tires. Chafer compound requires high modulus. The nature of the outer ply compound influences the polymer section for chafer. The aim is to have good compatibility and adhesion. Suitable tackifiers can be incorporated to have excellent adhesion with the plies.

5.1.5.19   RIM STRIP

Rim strip or clinch strip may be considered as an extension to the lower edge of the sidewall, introduced as a further anti-abrasion measure on radial ply tires, which are more subjected to rim chafing than the cross-ply tire. It is a narrow extruded strip tapered at both edges.

5.1.6   TIRE RAW MATERIALS

Every tire may look similar but tires are so diverse in terms of their internal structure, rubber constituents, shape, and design that it would even be possible to say that every tire is different. Each tire is made up of more than 100 kinds of materials and can be regarded as a combination of high technologies or as a product of the latest chemical and engineering research.

5.1.6.1   RUBBER

In general, there are four major rubbers used: NR, SBR, BR, and butyl rubber (along with halogenated butyl rubber). The first three are primarily used as tread and sidewall compounds, while butyl rubber and halogenated butyl rubber are primarily used for the inner liner or the inside portion that holds the compressed air inside the tire.

The most popular fillers are carbon black and silica and there are several types of each. The selection depends on the performance requirements as they are different for the tread, sidewall, and apex. Other ingredients also come into play to aid in the processing of the tire or to function as antioxidants, antiozonants, and antiaging agents. In addition, the "cure package" a combination of curatives and accelerators is used to form the tire and give it its elasticity.

5.1.6.2   REINFORCEMENT

Carbon black forms a high percentage of the rubber compound. This gives reinforcement and abrasion resistance. Silica is used together with carbon black in high-performance tires as low heat buildup reinforcement. Sulfur cross-links the rubber molecules in the vulcanization process. Vulcanizing accelerators are complex organic compounds that speed up the vulcanization. Activators assist the vulcanization. The main one is zinc oxide. Antioxidants prevent sidewall cracking due to the action of sunlight and ozone. Textile fabric (mainly Kevlar and carbon fiber treads) reinforces the carcass of the tire. To promote the tack of rubber compounds, resins such as coumarone-indene resins (CI resin), petroleum resins, phenolic resins, and wood resins are added.

Steel cord has been the main reinforcing material for tires. Steel cords are incorporated into the belts and carcass to improve the mechanical stability. The performance of tires depends on the strength and durability of cord to rubber bonds that comprise these materials. A common mode of failure is the separation between the steel belt components. In tires, the belt area has the highest amount of stress and weakest adhesion potential because of the difficulty in bonding rubber to metal.[3] Due to a problem with the bonding of bright steel to rubber compounds, steel is often coated with brass to improve the adhesion. The adhesion between the brass coated steel and rubber is important for the durability of tires. In this adhesion, the structure and composition of adhesion interface determine the strength and stability of brass-plated steel adhesion with rubber compounds.

5.1.6.3   COMPOSITION OF TIRE

The main component of a tire is rubber compound and the compositions of the tires produced by different manufacturers are very similar (Table 5.2).[2]

TABLE 5.2   Material Composition of Passenger Car and Truck Tires.

Materials	Passenger car (%)	Truck (%)
Rubbers/elastomers	47	45
Carbon black	21.5	22
Metal	16.5	25
Textile	5.5	0
Zinc oxide	1	2
Sulfur	1	2
Additives	7.5	5

5.1.7   DRY BONDING SYSTEM FOR TIRE

Dry bonding agents are used as a direct bonding system, which leads to higher adhesion value between rubber and coated steel cords for the manufacturing of tires.[4]

Dry bonding system consists of two components. They are methylene acceptor and methylene donor.

5.1.7.1 METHYLENE ACCEPTOR

The two well-known methylene acceptors are resorcinol and resorcinolic resin. Resorcinol and resorcinolic resins are considered as high polar molecules in rubber compound (Fig. 5.1). Therefore, due to incompatibility, they are basically insoluble in rubber. During vulcanization, the polar molecules migrate out of the rubber and move toward the adhesion interface between the brass and rubber surface. At this interface the resorcinolic resin forms tightly cross-linked network structures from the reaction with a methylene donor. Some of the common methylene acceptor resorcinolic resins are shown below.

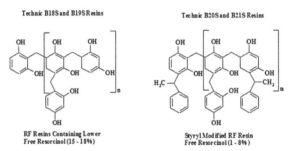

FIGURE 5.1 Methylene acceptors.

5.1.7.1.1 Preparation of Phenol Formaldehyde Resin

Phenolic resins are prepared by a step-growth polymerization of formaldehyde and phenol or phenol derivative using an acid or a base catalyst (Fig. 5.2).

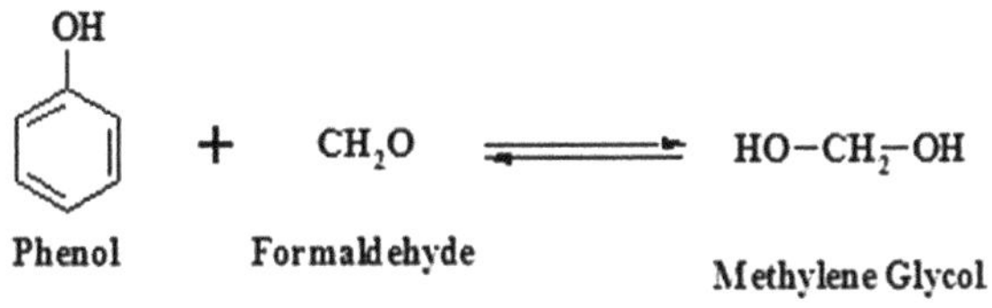

FIGURE 5.2 Formaldehyde in equilibrium with methylene glycol.

Depending on the pH of the catalyst, these monomers react to form one of two general resin types: Novolac resins and resole resins.[8]

5.1.7.1.2 Novolac Resin

An acidic catalyst and a molar excess of phenol to formaldehyde are conditions used to make novolac resins. The following simplified chemistry illustrates the wide range of polymers possible. The initial reaction is between methylene glycol and phenol. The reaction continues with additional phenol and splitting of water.

The reaction creates a methylene bridge at either the ortho position or at the para position of the phenolic aromatic rings. The "rule of thumb" is that the para position is approximately twice as reactive as the ortho position, but there are twice as many ortho sites (two per phenol molecule), so the fractions of ortho–ortho, para–para, and ortho–para linkages are same.

Branching occurs because the reaction can occur at any of three sites on each ring. As the reaction continues, the random orientations and branching quickly result in an extremely complex mixture of polymers of different sizes and structures. The reaction stops when the formaldehyde reactant is exhausted, often leaving up to 10% of unreacted phenol. Distillation of the molten resin during manufacturing removes the excess phenol and water. The final novolac resin is unable to react further without the addition of a cross-linking agent (Fig. 5.3).

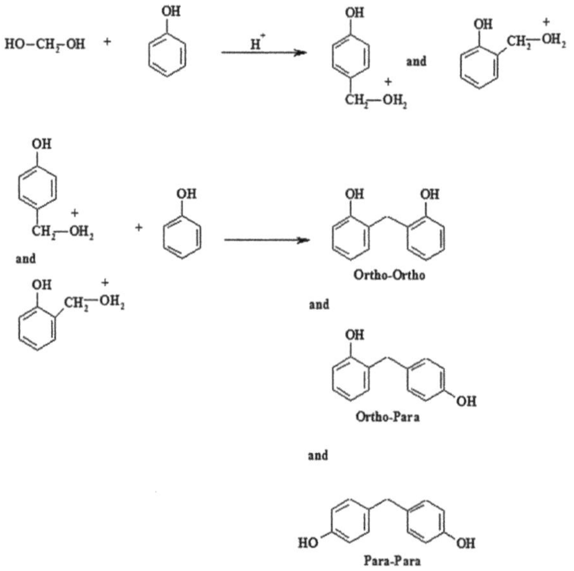

FIGURE 5.3   Formation of methylene bridge.

The most common phenolic resin cross-linking agent is hexamethy-lenetetramine, also known as hexa (hexamine or HMT). Ground and blended with the resin, hexa serves as a convenient source of formalde-hyde, when heated to molding and curing temperatures. A special attribute of hexa is that it reacts directly with resin and phenol without producing appreciable amounts of free formaldehyde. Hexa cures the resin by further linking and polymerizing the molecules to an infusible state. Due to the bond angles and multiple reaction sites involved in the reaction chemistry, the resulting polymer is not a long straight chain but rather a complex three-dimensional polymer network of extreme molecular weight. This tightly cured bonding network of aromatic phenolic accounts for the cured materials hardness and heat and solvent resistant properties.

Novolac resins made with these catalysts tend to cure more rapidly than the standard randomly linked resins. Novolac resins are amorphous (not crystalline) thermoplastics. As they are most typically used, they are solid at room temperature and will soften and flow between 65 and 105°C. The number average molecular weight (Mn) of a standard phenol novolac resin is between 250 and 900. As the molecular weight of phenol is 94 g/mol, a Mn of 500 corresponds to a resin where the average polymer size in the entire distribution of polymers is five-linked phenol rings. Novolac resins are soluble in many polar organic solvents (e.g., alcohols and acetone) but not in water.

5.1.7.1.3   Resole Resin

Resole is produced by the reaction of a phenol or a phenol derivative with an excess amount of formaldehyde in the presence of a base catalyst.

Methylol phenol can react with itself to form a longer chain methylol phenolic or form dibenzyl ether or react with phenol to form a methylene bridge.

The most important point in resole resin chemistry is that, when an excess of formaldehyde is used, a sufficient number of methylol and dibenzyl ether groups remain reactive to complete the polymerization and cure the resin without incorporation of a curing agent such as hexa.[9] For this reason, the industry commonly refers to resole resins as "single-stage" or "one-step" type products. Resole resin manufacture includes polym-erizing to the desired extent, distilling off excess water, and quenching or tempering the polymerization reaction by rapid cooling (Fig. 5.4).

As resole resins continue the polymerization reaction at even ambient temperatures, albeit at much slower rates than during manufacturing, they demonstrate limited shelf lives dependent on the resin character, storage conditions, and applications.

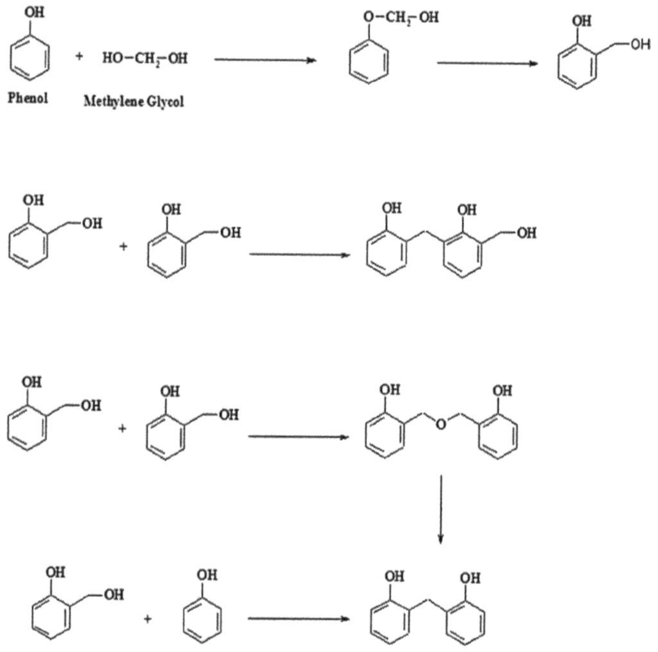

FIGURE 5.4   Synthesis of resole.

5.1.7.2   METHYLENE DONOR

The two well-known methylene donors are hexamethylenetetramine (Hexa or HMT) and hexamethoxymethylmelamine (HMMM) (Fig. 5.5). The materials HMT and HMMM are very effective in their reaction toward resorcinolic compound under rubber curing conditions to produce high cross-link network structures. The presence of resorcinolic networks throughout the rubber matrix provides good physical and mechanical properties for cured rubber. Though HMT has been used as the methylene donor in the NR skim compound formulation, the release of ammonia from the HMT was found to affect the steel cord adhesion due to corrosion.

Therefore, HMT is no longer in use as compound formulations used for steel wire bonding.

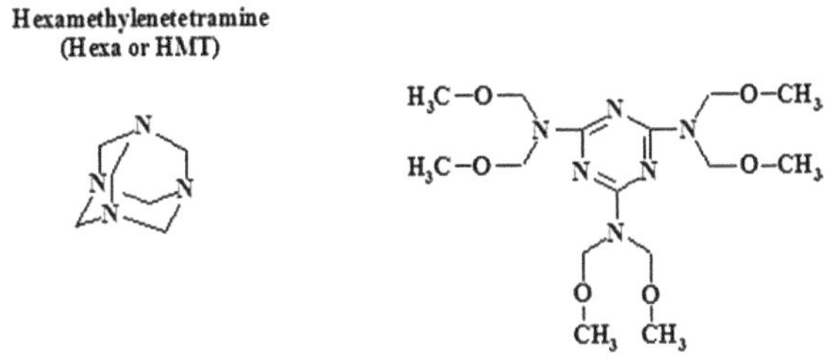

FIGURE 5.5   Methylene donors.

On the other hand, HMMM methylene donor reacts with resorcinolic methylene acceptors more effectively under rubber curing conditions and provides higher unaged adhesion and maintains this adhesion even in the heat, humidity, and steam-aged conditions. Now, HMMM has been widely used as a methylene donor in steel cord skim compounds.

5.1.7.2.1 Preparation of HMT

Hexamethylenetetramine is prepared by the condensation of formaldehyde and ammonia (Fig. 5.6).

$$6\ CH_2O\ +\ 6\ NH_3\ \longrightarrow\ \ +\ 6H_2O$$

FIGURE 5.6   Preparation of hexamethylenetetramine (HMT).

A slurry of 3 mol of formaldehyde, as paraformaldehyde, is formed in about 400 ml of 85 vis. neutral oil using mechanical agitation. Anhydrous ammonia is bubbled through the suspension and the mixture is stirred for 2 h. The amount of ammonia bubbled through the suspension is several times the stoichiometric amount required for reaction with the paraformaldehyde. An exothermic reaction takes place during this period, which is sufficient to raise the temperature of the mixture to about 80°C early in the reaction period. Additional heat is supplied to raise the temperature to over 100°C to remove the by-product water. All of the paraformaldehyde is consumed during the reaction and hexamethylenetetramine is formed as fine white crystals, which settle at the bottom of the reactor. Using this procedure, the paraformaldehyde is converted quantitatively to hexamethylenetetramine and is recovered by filtration as the anhydrous pure compound.[10]

5.1.7.2.2 Preparation of HMMM

Method A
Around 3.1 g (0.01 mol) of hexamethylolmelamine, 20.0 g (0.065 mol) methanol, and 1.0 ml concentrated hydrochloric acid are added in the flask. The reaction mixture is stirred for 8 min at 25°C, then neutralized and concentrated under reduced pressure to afford 2.2 g (56.4%) of product, m.p. 30–35°C (Fig. 5.7).

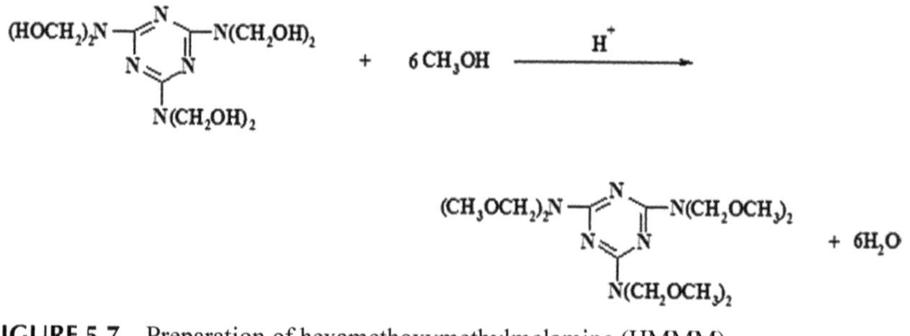

FIGURE 5.7   Preparation of hexamethoxymethylmelamine (HMMM).

Method B
   To a resin flask equipped with a mechanical stirrer, condenser, thermometer, and gas sparge 70 g of 20–50 mesh dehydrated (16 h at 80°C in a vacuum oven) sulfonated polystyrene and cation exchange resin (Dowex 50W-2) are added. Methanol is added to wash the resin and then the methanol is removed through the sparger with stirring for 10 min at reflux.[11] A slurry of 153 g (0.5 mol) of hexamethylolmelamine in 500 g (15.6 mol) of methanol is then added and the mixture is heated while stirring to about 50°C under a nitrogen atmosphere. The temperature is maintained at 45–50°C for 2 h and 15 min, at which time it appears that all the hexamethylolmelamine has dissolved. The solution is then drawn off

and evaporated under reduced pressure at 45°C over a 3 h period to afford 116 g (82%) of product, m.p. 79°C.

5.2   EXPERIMENTAL

5.2.1   MATERIALS

Resorcinol-formaldehyde resin (R) and HMMM were obtained as gift samples from Techno Waxchem Pvt. Ltd., Kolkata—700046, India.

5.2.1.1   PREPARATION OF BLEND (B)

R was dried at 60°C for 24 h in a hot air oven. The blend B was prepared by grinding 60 wt.% of dry R and 40 wt.% of HMMM in a porcelain mortar with minimum quantities of acetone. The blended material was again dried in an air oven so that it is free from acetone and then stored in a desiccator for further analysis.

5.2.1.2   THERMAL CURING

The pure resin R and the blend B were taken in separate micro test tubes and flushed with dry oxygen-free nitrogen and then heated to 150°C and kept at this temperature for 3 h. After the stipulated time, the samples were allowed to cool at room temperature and then removed from the micro test tubes, ground to coarse powder, packed and stored in a desiccator for further analysis.

5.2.2   METHODS

The DSC curves for the pure resin R, HMMM, and the blend B were recorded in a TA Instruments DSC Q20. The materials were hermetically sealed in the aluminum pans and heated at a rate of $10°C \cdot min^{-1}$ in dry nitrogen (Flow rate = $50 \ ml \cdot min^{-1}$) atmosphere. The TG curves for the thermally treated materials were recorded in a TA Instrument TG Q50 system using $10°C \cdot min^{-1}$ heating rate in nitrogen atmosphere. Generally,

4–5 mg of the sample was used with a balance purge of 40 ml min^{-1} and a sample purge of 60 ml·min^{-1}. FTIR spectra of the materials were recorded in a Shimadzu-8400S infrared spectrophotometer using KBr pellet technique. The absorption bands in IR spectra were used to identify the functional groups in the materials investigated.

5.3 RESULTS AND DISCUSSION

5.3.1 THERMAL STUDIES

The DSC curves recorded at a heating rate (β) of 10°C min^{-1} in an inert atmosphere (oxygen-free dry nitrogen; flow rate = 50 ml min^{-1}) for pure R and HMMM are shown in Figure 5.8. From these studies, it is possible to understand their thermal characteristics. Similarly, the thermogravimetry (TG) and differential thermogravimetry (DTG) curves recorded for these materials are presented in Figure 5.9. The DSC study of the R (60%)/ HMMM (40%) blend (B) will provide sufficient information regarding the interaction existing between R and HMMM. The thermal parameters such as endotherm temperature (T_E), enthalpy of the endotherm (ΔH), the onset of curing (T_s), curing maximum (T_{max}), endset of curing (T_e), and enthalpy of cure reaction (ΔH_c) obtained from the DSC curves are discussed further.

Pure R shows the broad endotherm in the temperature region 79–96°C and the endotherm peak was at 88°C, it may be due to the evaporation of the absorbed and adsorbed water molecules in the material. The softening of the material starts around 105°C. The material shows multiple endotherms in the temperature region 136–190°C. The endotherm peaks are found to be at 149 and 164°C with the total enthalpy of (ΔH) 53 Jg^{-1}. Since the resin R contains 15–18% free resorcinol the sublimation of resorcinol from the resin and probably the melting of the resin may be taking place during this temperature region. From Figure 5.9, it is obvious that the material (R) starts to lose weight slowly from around 100°C and this weight loss proceeds till around 262°C. The amount of weight lost during this temperature region is around 14%. From 262°C till 385°C major weight loss from R is taking place and the amount of weight loss occurring at this point is 29%. Another very slow thermal degradation is proceeding from 450 to 715°C and the amount of weight lost during this temperature region is 17%. From these data, one can explicitly state that the endothermic behavior noted in the DSC curve of R (105–190°C) is due to the combined melting and weight loss behavior.

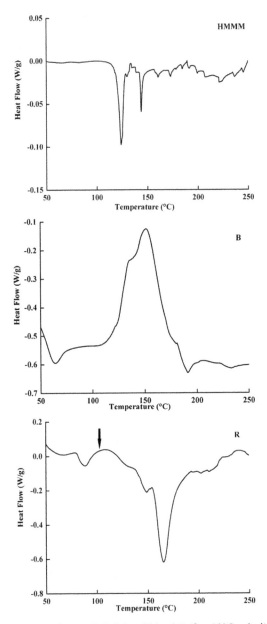

FIGURE 5.8 DSC curves for R, HMMM and blend B ($\beta = 10°C \cdot min^{-1}$).

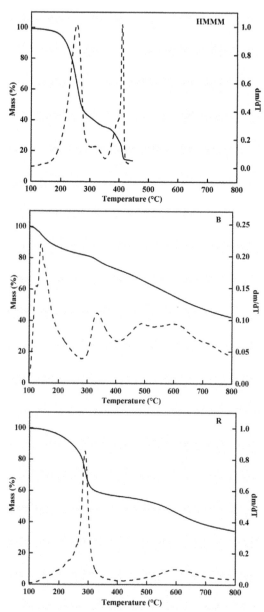

FIGURE 5.9 Thermogravimetry (TG) and differential thermogravimetry (DTG) curves for R, HMMM and blend B (β = 10°C·min⁻¹).

The DSC curve for pure 100% HMMM liquid is also presented in Figure 5.8 and shows strong multiple endotherms at around 124 and 144°C. The TG and DTG curves for pure HMMM are presented in Figure 5.9. The material starts to lose weight from 125°C and shows a major weight loss till 300°C and amounts to a weight loss of 42%. The next major weight loss starts at around 350°C and ends at 400°C and within this temperature region the amount of material lost from liquid HMMM is 22%. This particular compound shows a much complex pattern of release of volatiles after 300°C. From the DTG curve, it is possible to conclude that the liquid HMMM evaporates and starts to degrade.

The DSC curve for the blend (60% RF resin +40% HMMM) shown in Figure 5.8 provides sufficient information regarding the interaction existing between R and HMMM. The blend shows nearly strong cumulative exotherms. The exotherm starts at around 105°C and shows three clear maxima at 122, 135, and 152°C. Hence, it is reasonable to conclude that the material R and the volatiles coming out of R (most probably subliming resorcinol) are reacting exothermally with the products released from HMMM. The TG and DTG curves for the blend are shown in Figure 5.9. Compared to the pure resin and pure HMMM, the blend behaves entirely in a different manner. The material shows a combination of different weight losses in the temperature region 99–280°C with several maxima at 120, 140, and 150°C. Further weight loss is noted between 280 and 410°C, the maximum being at 328°C. From Figures 5.8 and 5.9, it is possible to conclude that blending HMMM with resin alters the nature of the TG curve indicating reasonable interactions between these two materials.

The TG and DTG curves recorded at a heating rate (β) of 10°C min^{-1} in nitrogen atmosphere for thermally treated resin R and thermally cured blend B were shown in Figure 5.10. The parameters such as onset degradation (T_s), degradation maximum (T_{max}), end set degradation (T_e), and char residues (%) at 750°C obtained from the TG and DTG are discussed further.

From Figure 5.10, it is possible to determine the reaction between the methylene acceptor R and methylene donor HMMM. Thermal curing leads to comparatively better thermal stability as evidenced by an increase in the char residue at 750°C. The material PB shows three discreet mass loss regions, the first being in between 110 and 263°C, the second being in between 263 and 403°C, and the third being in between 403 and 672°C. The three degradation maxima are found to be at 172, 308, and 569°C.

The slow and also considerably less degradation of PB clearly indicate the cross-linking reaction by the action of HMMM on resin R.

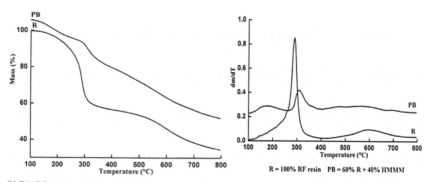

FIGURE 5.10   TG and DTG curves for R and polymer derived from R-HMMM blend ($\beta = 10°C \cdot min^{-1}$).

5.3.2   FTIR STUDIES

The FTIR spectra of the resin R, HMMM, and its cured blend PB are shown in Figure 5.11. All the important absorption bands shown by these compounds are presented in Table 5.3. The presence of absorption bands corresponding to –OH stretching of phenolic group (3000–3600 cm^{-1}), C–H stretching of methylene group (2927 cm^{-1}), C=C stretching (1606 cm^{-1}, 1508 cm^{-1}), C–O stretching (1019–1294 cm^{-1}), tri substituted benzene ring (837 cm^{-1}) in the FTIR spectrum of R confirmed its structure.

The absorption bands corresponding to C–H stretching of methylene group (2829–2993 cm^{-1}), N–C–N stretching (1555 cm^{-1}), C–N stretching in triazine ring (1329 cm^{-1}, 1094 cm^{-1}), C–O stretching (1022 cm^{-1}), triazine ring (818 cm^{-1}) in the FTIR spectrum of HMMM confirmed its structure.

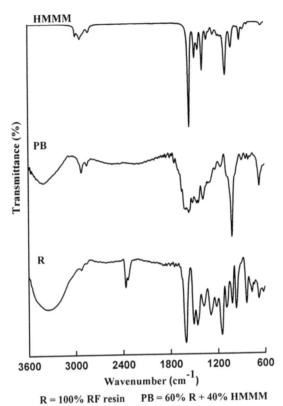

FIGURE 5.11   Fourier-transform infrared (FTIR) spectra for R, HMMM, and blend B.

TABLE 5.3   Fourier-Transform Infrared Studies: Functional Groups Identified Along with the Corresponding Wavenumber.

Wavenumber (cm⁻¹)	Functional group
Resorcinol-Formaldehyde Resin (R)	
3600–3000	–OH stretching of Phenolic methylol hydroxyl group
2927	C–H stretching of methylene group
1294, 1222, 1149, 1090, 1019	Aromatic C–O stretching of phenol and methylol groups
837	1, 2, 4-Tri-substituted benzene (CH out of plane deformation)
681, 768	1, 2, 3- Tri-substituted benzene (CH out of Plane deformation)

TABLE 5.3   *(Continued)*

Wavenumber (cm⁻¹)	Functional group
Hexamethoxymethylmelamine (HMMM)	
2993, 2933, 2829	C–H stretch of methylene group
1555	N–C–N bending and ring deformation in triazine ring
1329, 1094,	C–N stretch
1022	C–O stretch in methylol groups
818	Triazine ring
R (60%)/HMMM (40%) blend (B)	
3600–3000	–OH stretching of Phenolic methylol hydroxyl (broad brand)
2924, 2853	Symmetric stretch (–CH₃ and –CH₂– in aliphatic compounds) and CH antisymmetric stretch
1849, 1802, 1750	C=O stretch
1655	C=O stretch in quinone
1555	N–C–N bending and ring deformation in triazine ring
894, 841, 810	1,2,4-Tri substituted benzene
671	C–O–H bending vibration

The thermally cured blend PB shows decrease in the intensity of the absorption band at 3000–3600 cm⁻¹ (phenolic –OH group) indicating the utilization of the phenolic –OH groups during the curing reaction. The weak absorption at 810–894 cm⁻¹ represents the presence of tri substituted benzene rings in the resin R. The blend PB shows the absorption bands corresponding to N–C–N stretching (1555 cm⁻¹), C–N stretching in triazine ring (1382–1461 cm⁻¹), and C–O–C stretching (1165 cm⁻¹) confirming the presence of HMMM unit in the cured material. The multiple weak absorptions occurring in between 1655 and 1849 cm⁻¹ are assigned to C=O stretching. The appearance of new low intensity peak at 1665 cm⁻¹ confirms the presence of quinonemethide structures in the cured material shown in Figure 5.12. In general, the cross-linking reaction between the methylene acceptor and methylene donor leads to the formation of quinonemethide and benzoxazine structures. The band noted at 948 cm⁻¹ is useful to recognize the oxazine ring structure. This band is due to the benzene ring mode of the benzene to which oxazine is attached, but it is not the oxazine ring mode. When oxazine ring opens, this mode disappears. It is important to note that disappearance of this mode is not the evidence of oxazine polymerization, but it is a simple indication that the oxazine ring has opened. The absence of this band in the thermally cured material PB indicates the absence of benzoxazine structures in PB. Hence, from this preliminary investigation, it is possible to state that quinone-methide structures are formed in detectable quantities during the reaction between the methylene acceptor and donor. In the concentration range that has been used and from the temperature chosen for the reaction it is possible to state that benzoxazine units are not formed.

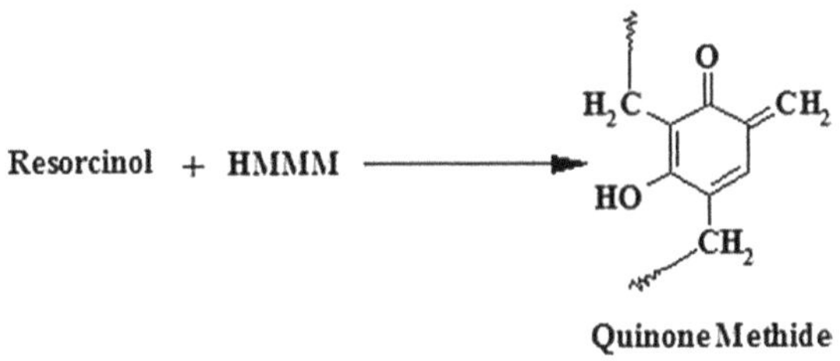

FIGURE 5.12   Quinonemethide structure.

5.4 CONCLUSION

In this study, for the dry bonding of rubber, R is used as a methylene acceptor and HMMM is used as a methylene donor. The blend B (60% R+40% HMMM) is prepared and thermally cured at 150°C for 3 h. From the DSC curves, the interaction existing between R and HMMM is explicit. The TG studies of the cured B showed that PB is thermally more stable than the resin R and the rate of degradation of the cured material is comparatively slower than the blend B. The FTIR spectral investigations indicate the presence of quinone methide structure in the thermally cured PB and there is no sufficient evidence for the formation of benzoxazine structures within the concentration region and the temperature used in the present investigation.

5.5 ACKNOWLEDGMENT

The authors would like to express their sincere thanks to Mr. Abhishek K. Agarwal, Executive Director, and Dr. Raj B. Durairaj, Technical director, TWC Group companies, Kolkata 700046, India. The authors acknowledged the Management, Principal, and the Dean of Kamaraj College of Engineering and Technology, S. P. G. C. Nagar, K. Vellakulam Post 625 701, India for providing all the facilities to carry out this work.

KEYWORDS

- **resorcinol-formaldehyde resin**
- **dry bonding system**
- **thermal studies**
- **FTIR**

REFERENCE

1. Bhakuni, S.; Grover, W. Bonding Tire Cord to Rubber. US Patent 4,031,288, June 21, 1977.
2. Carlos, B. Distrito Federal; Adhesion of metal to rubber. US Patent 3,897,583, June 30, 1970.
3. Durairaj, R. B. *Resorcinol Chemistry, Technology and Applications;* Springer: Berlin, Heidelberg, New York; 2005.
4. Durairaj, R. B.; Michael Walkup, C.; Lawrence, A. Phenolic Modified Resorcinolic Resin for Rubber Compounding. US20040116592A1, June 17, 2004.
5. Eirich, R. *Science and Technology of Rubber,* Academic Press, 111 fifth Avenue, New York, 10003, 1978.
6. Evans, M. S. *Tyre Compounding for Improved Performance,* Rapra Technology Limited, Shawbury, United Kingdom, 2002; Vol. 12; p 4.
7. Lindenmuth, B. E.; *The Pneumatic Tire,* DOT HS 810 561.
8. Pilato, L.; *Phenolic Resins: A Century of Progress*; Springer: Heidelberg Dordrecht, London, New York, 2004.
9. Rediger, A.; Edward L. Phenol Formaldehyde Novolac Resin Having Low Concentration of Free Phenol. US Patent 8,778,495, July 15, 2014.
10. Sandler S. R.; Karo W. *Polymer Syntheses – Volume -II*, ACADEMIC PRESS, INC. Harcourt Brace & Company, Publishers: Boston San Diego, New York, 1997.
11. Shay, D.; James, H. Crosslinking Agent for Powder Coating. US Patent 4,189,421, February 19, 1980.

CHAPTER 6

NOVEL BIOACTIVE STRONTIUM CONTAINING COMPOSITES FOR MEDICAL APPLICATIONS

VIBHA C. and P. P. LIZYMOL*

Biomedical Technology Wing, Sree Chitra Tirunal Institute for Medical Sciences and Technology, Poojappura, Thiruvananthapuram, Kerala 695012, India

Corresponding author. E-mail: lizymol@rediffmail.com

CONTENTS

ABSTRACT

A single-pot modified solgel preparation was used to synthesize novel bioactive strontium-incorporated inorganic–organic hybrid resins containing polymerizable tetramethacrylate groups. The purpose of the study is to investigate the viability of developing photocured composites using the synthesized inorganic–organic hybrid resin using quartz as filler. Photocured polymeric composite was prepared from novel bioactive strontium containing resins exhibited lower polymerization shrinkage with good mechanical properties. Therefore, it can be a potential aspirant in biomedical field mainly in dental, orthopedic, and coating applications.

6.1 INTRODUCTION

The apatites seen in tooth enamel, dentin, and bone have indistinctly different compositions and therefore have different physical and mechanical properties. Dental caries is a contagion, which leads to demineralization and demolition of enamel, dentin, and cementum. It is the outcome of the production of acid by bacterial fermentation of food debris accumulated on the tooth surface. As the demineralization exceeds remineralization, this natural hydroxyapatite dissipates leading to cavities, osteoporosis, and so forth. Replacement of damaged tooth is necessary for aesthetics and proper functioning of teeth. The treatment of dental caries includes dental fillings using amalgams, resin, porcelain and gold, endodontic therapy, and extraction. Resin-based composites are among the most popular restorations due to their aesthetics.[1] The essential components of a composite comprised an oligomer/monomer system, filler, coupling agent, an appropriate initiator/accelerator system, and to bond matrix and filler. Commonly, the commercially accessible composites are photocuring in which the polymerization of the oligomer/monomer system is initiated by irradiating with a visible light ensuing in a cross-linked polymer network enfolded the filler particles. Usually, a long chain dimethacrylate (Bis GMA or UDMA) is diluted by means of a low viscosity dimethacrylate (habitually TEGDMA) as the oligomer/monomer system. Modifications in the concentration of the dimethacrylate, particle size, and particle size distribution of the filler have attributed the composites with excellent aesthetics and physicomechanical properties. The mechanical properties of the existing composites can meet those of amalgams so their use in posterior restorative dentistry is growing. Composites do not adhere directly to the tooth structure and require a mediator bonding agent to obtain an interfacial bonding of composite to the tooth. The polymerization shrinkage beset with these materials limits its application, as it leads to stress at the cavity walls and finally secondary caries.

Advances in formulations, the development of new techniques, and optimization of properties have made the direct composite restoration more dependable and predictable.[2] The most significant progress such as improvement in mechanical and clinical triumph was achieved by modifying the filler composition, particle size, distribution, and quantity incorporated. The dimethacrylate resins used as the organic matrix are Bowen monomer, that is, Bis GMA, modified Bowen monomers with lower viscosity (hydroxyl-free Bis GMA) and urethane dimethacrylate (UDMA). The resin matrix has a vital influence on the properties of the composite resins besides the filler system.[3] The contemporary dental research is focusing on the development of new resin monomer. Resin composites are used for a range of applications in dentistry, counting restorative materials, cavity liners, pit and fissure sealants, cores and buildups, crowns, onlays, inlays, temporary restorations, cements for single or multiple tooth prostheses and orthodontic devices, endodontic sealers, and root canal posts.[4] The swiftness by which the materials have evolved proposed a constantly changing state of the art. Light-cured composite resins are the materials of choice for direct restorations, due to their high potential for bonding to hard tissue, improved mechanical properties, and aesthetics, even if these materials have hindrance related to marginal integrity and leakage due to the polymerization shrinkage.[5] During the polymerization, the individual monomer molecules will come closer to the polymer network results in shrinkage.[6] The composite shrinkage is well inclined by material properties,[7] cavity configuration,[8] irradiation time and intensity,[9] degree of conversion,[10] and elasticity of the dentin and enamel substrate. The occurrence of contraction stress development in dental composite restoratives is extremely complex, and of both theoretical and pragmatic importance. Milewski et al. studied stress at the boundaries and at tooth crowns in detail.[11–12]

The reduction of shrinkage has been the topic of numerous studies which can be summarized in two groups: (a) materials development—decline of the shrinkage by introducing new low-shrinking monomers,[13–19] to use different comonomers such as multimethacrylate, highly branched methacrylates,[2,20] and ormocers,[21] monomers with liquid crystalline structure which can diminish the shrinkage owing to phase transform during polymerization,[22] reducing the shrinkage-induced stress using nonbonded fillers,[23,24] and incorporation of more fillers into the matrix phase using multimodal filler particle size distributions and (b) methods development—using different light irradiation regimes to reduce the shrinkage-induced stress,[25,26] and a choice of placement techniques aspiring to relieve the stress.[27-35]

New methacryloyloxyalkylaminoalkylalkoxysilanes have been synthesized by Moszner et al.[36] It has been synthesized by Michael addition of the corresponding acryloyloxyalkyl methacrylates by means of (3-aminopropyl)triethoxysilane (APTES). These polycondensates were free-radically polymerizable and could be cured in the presence of a VL-photoinitiator system resulting in an inorganic–organic polymer network. Furthermore, these polycondensates enabled the preparation of highly filled composites. These composites were free of commercially used diluents and showed almost no extractable content of residual monomer, which is very attractive for dental filling materials. In 2008, they reported that ormocers which were synthesized from amine or amide dimethacrylate trialkoxysilanes showed improved biocompatibility in dimethacrylate-diluent-free composite restoratives.[37] They also founded that modifying ormocers among the organic moieties such as methacrylate-substituted ZrO2 or SiO2 organosol nanoparticles can improve the mechanical properties of resin-based composites.

Secondary caries occurred normally at the interface between the restoration and the cavity prepared.[38-40] The tooth structure is demineralized subsequent to assault of acid producing bacteria, such as *Streptococcus mutans*, when fermentable carbohydrates are present.[41] It has been revealed that the colonization of caries creating bacteria is enhanced with increasing surface energy and surface roughness of the composite fillings.[42-44] Therefore, an efficient antibacterial/bactericidal restorative material ought to be in the best spot to prevent secondary decay.[45-47] Numerous endeavors have been made to bring in antimicrobial properties within restorative materials,[48] but some of these attempts resulted in compromised physical properties of the novel material.[49-53] The change in composition is carried out to instigate antibacterial properties or leaching of particles, which may influence the material's mechanical properties, made it an unsuitable restorative material or restricts its use to non-load-bearing areas. Another issue maybe problems with transform in color, by the introduction of silver nanoparticles to light curable resins, which may in turn, confined the exploit of the material to posterior restorations. Photocured composites developed from zinc containing inorganic–organic hybrid resin exhibited 22.15% reduction in *Escherichia coli* ATCC 25922 compared to Bis GMA after 1 h exposure which proved that it can impart antimicrobial activity.[54]

Another property that might be malformed is the composite adhesion to the tooth. Decreasing bond strength when compared to the conservative material is not ideal, as it would raise the chances of microleakage and as a result, recurrent decay.[55]

Dental materials developed for restorative purpose ought to be capable for sealing restored interfaces preventing bacterial colonies, better mechanical properties, and bioactivity for its longevity. Recent works in dental biomaterial research targets on the development of bioactive restorations such as development of novel dental cements,[56] innovative light-curable therapeutic resin-based restorative materials to promote remineralization in mineral dwindling dental hard tissues at the bonding interface,[57] and so forth.

An autograft remains the gold standard for treating bone defects and the obligation for further surgical site persists to encourage the research for alternatives.[58] Synthetic bone graft substitutes mostly from calcium phosphate provided promising results for this scenario. Hydroxyapatite is used as coating and filling material for bone defects in biomedical purposes due to its outstanding bioactivity, biocompatibility, insolubility, and resemblance to the natural bone. In spite of this, the performance of synthetic bone grafts remains inferior to autograft. Numerous endeavors have been made to improve the properties of calcium phosphates by incorporating them with growth factors,[59,60] cells,[61,62] and inorganic additives such as magnesium, copper, strontium, and fluoride.[63,64] Strontium is used for biological benefits due to its chemical structure similar to calcium.[65,66] It is used in osteoporosis treatment for many years, and the mineral structure of the bone contains strontium, especially at the sections of the elevated metabolic turnover.[67,68] Usually, filler part is more concentrated for research due to its versatility of dopants. Improvising the resin matrix with these dopants may sometimes enhance the properties of the composite materials. Solgel method will be the appropriate technique for tailoring the properties, which are required to impart in a biomaterial. Inorganic–organic hybrid resins are synthesized through solgel method. They have enhanced thermal and physicomechanical properties owed to the assimilation of organic and inorganic compositions at molecular/nano level. The hydrolysis and condensation reactions during solgel process are inclined by different factors including pH, temperature, water to alkoxide proportion, type and quantity of catalyst, reactant's concentration, and so forth.[69,70] Either the interaction of inorganic–organic hybrid materials ensues through hydrogen bonding or weaker interaction between organic and inorganic parts.[71] The major applications of these resins were in electrochemical devices, biomaterials, coatings, and so forth.[72] Silicones are employed in various biomedical applications owing to their excellent biocompatibility, elasticity, and thermal and oxidative constancy, though these exhibits extreme hydrophobicity.[73] Inorganic–organic hybrid resins with silica base have well-controlled ionic and hydrophilic surface with outstanding resilience and adhesion onto different substrates[74] and are found to be promising materials for biomedical applications together with biomedical devices and biosensors. Inorganic–organic hybrid resins have applications in bone cement, dental composites, and denture-based materials for their low polymerization shrinkage compared to the existing resin matrix, as polymerization shrinkage is a significant concern in various dental and orthopedic applications.[75] The progress of composite resin polymerization occurs through the conversion of the monomer molecules into a polymer network, followed by the contraction of the gap between the molecules, which leads to polymerization shrinkage in the composite.[76,77] Shrinkage is the property of resin matrix and the cross-linking density of the resin directly influenced the rate of polymerization of the cured resin. Decline in cross-linking density of the resin will lead to dropping off glass transition temperature, solvent resistance, and strength of polymerized resin. If inadequate monomer conversion occurred in polymerization, quantity of leachable residual monomer amplifies. This leachable uncured monomers guide to cytotoxicity of the resin matrix.[78,79] Polymerization shrinkage, biocompatibility, and endorsement of surface hydrophilicity are the major challenges come across the biomedical fields. In dental biomaterial research, current works aimed on the development of bioactive restorations by increasing photopolymerized curative resin-based restorative materials, novel dental cements, to promote mineral precipitation at the bonding interface, and so forth, by compromising the physicomechanical properties.[53,55,80–85] Earlier studies have reported that use of inorganic–organic hybrid resins containing alkoxides of calcium in dental restoratives could diminish polymerization shrinkage and improve wear resistance.[86] This encouraged us to synthesize novel inorganic–organic hybrid resins containing alkoxide of strontium with polymerizable tetramethacrylate groups. The present chapter depicts our efforts to substitute the current organic dimethacrylate resin matrix with new inorganic–organic hybrid resins, containing alkoxide of strontium with polymerizable tetramethacrylate groups. The resin was synthesized through a modified solgel process. It was developed primarily to diminish the polymerization shrinkage and enhance bioactivity and physicomechanical properties. Physicomechanical properties of photocured composites were assessed in terms of diametral tensile strength (DTS) and flexural strength (FS) as per international standards.

6.2   MATERIALS AND METHODS

6.2.1   MATERIALS

The dimethacrylate resin 1,3-bis methacryloxy 2-(trimethoxy silyl propoxy) propane and triethylene glycol dimethacrylate (TEGDMA) used were of Aldrich Chem. Co. Milwaukee, and Laboratory Rasayan (LR) grade strontium chloride, hydrochloric acid (analytical grade), LR grade diethyl ether employed were of S. D. Fine Chemicals, Mumbai, India; diphenyl (2,4,6-trimethylbenzoyl) phosphine oxide (TPO), 2-hydroxy-4-methoxy-benzophenone, 4-(dimethylamino)phenethyl alcohol, 4-methoxy phenol, phenyl salicylate, and 2,6-di-tert-butyl-4-methyl phenol were from Sigma-Aldrich, US.

6.2.2   METHODS

6.2.2.1   SYNTHESIS OF INORGANIC–ORGANIC HYBRID RESINS

Owing to proprietary reason, formulation and detailed procedure of the synthesis of resin cannot be revealed at this stage. Inorganic–organic hybrid resins with polymerizable tetramethacrylate group were synthesized as per the patented procedure through modified solgel technique,[87] using 1,3-bis methacryloxy 2-(trimethoxy silyl propoxy) propane as the precursor. 1,3-bis methacryloxy 2-(trimethoxy silyl propoxy) propane was synthesized by reacting 1.1 mol of glycerol dimethacrylate with chloropropyl trimethoxy silane (1 mol) in presence of benzyl triethyl ammonium chloride catalyst.[86] Resins containing alkoxide of strontium were synthesized by reacting 1,3-bis methacryloxy 2-(trimethoxy silyl propoxy) propane and deionized water in the molar ratio 1:3. The silane–water mixture was subjected to vigorous stirring for 30 min. A solution of 1 ml 6N HCl/NaOH was added to the stirred mixture, followed by the addition of strontium chloride (0.08, 0.1, and 0.3% weight of silane) and kept stirred for 6 h. The hydrolyzed silane was kept at room temperature overnight for post condensation. The product obtained was then extracted with ether, washed with distilled water till alkali/acid free, and dried. To the dried sample, about 200 ppm 4-methoxy phenol was added in order to avoid self-polymerization. Various formulations with varying inorganic content were used during synthesis for the optimization of resin (Table 6.1). Varying concentrations of strontium were incorporated into the resin according to the weight percentage of the silane precursor (0.08, 0.1, and 0.3%) in presence of diethyl ether.

TABLE 6.1   Formulations of Varying Concentrations of Strontium Chloride Used for the Synthesis of Inorganic–Organic Hybrid Resins.

Serial no.	Concentration of strontium chloride (%)	Sample code
1	0.08	Sr0.08
2	0.1	Sr0.1
3	0.3	Sr0.3

6.2.2.2   PREPARATION OF PHOTOCURED COMPOSITES

The photocured composites were prepared as per the patented procedure.[88] Inorganic–organic hybrid resins [Sr0.08, Sr0.1, and Sr0.3] (60 parts) diluted with TEGDMA (40 parts) were used as the resin matrix along with diphenyl (2,4,6-trimethylbenzoyl) phosphine oxide (TPO) as the photoinitiator. The filler used for the preparation of the paste consists of 300–325 phr of silanated quartz and 12% fumed silica. Other chemicals used for the preparation of resin mixture were 4(dimethyl-amino) phenethyl alcohol, 4-methoxyphenol, phenyl salicylate, 2-hydroxy-4-methoxybenzophenone, and 2,6 di-tert-butyl-4-methylphenol (which act as inhibitors, activators, and UV stabilizers). The resin matrix, filler, diluent, initiator, accelerator, and UV absorbers were masticated in an agate mortar to get a uniform paste packed into the mold and exposed to visible light for the duration of 60 s on both sides using Prolite (Caulk/Dentsply). It is used to polymerize visible-light-activated materials with an intensity not less than 300 mw/cm^2.

6.2.3 EVALUATION OF DENTAL COMPOSITES

6.2.3.1 POLYMERIZATION SHRINKAGE

The stainless steel mold used for the sample preparation was 6 mm in diameter and 3 mm in depth. A digital vernier caliper (Mitutoyo, Japan) calculated the internal diameter of the mold accurately. Six measurements were taken and the mean value was calculated. Samples prepared were the same as that for DTS as per ADA specification 27.[89] The mold was kept on a transparent sheet on a metallic plate. The paste was packed into the mold and a second transparent sheet was kept on top followed by a second metallic plate. The mold and strip of film between the metallic plates were pressed to displace excess material. The plates were removed and the paste was exposed to photocuring for the duration of 60 s on both sides through the transparent sheet. The sample was then taken from the mold and the diameter of the cured sample was measured at six points and the mean value was calculated. The percentage shrinkage can be calculated by equation (6.1).

Polymerization shrinkage =
$$(\text{diameter of ring} - \text{diameter of composite/diameter of ring}) \times 100 \qquad (6.1)$$

6.2.3.2 DIAMETRAL TENSILE STRENGTH (DTS)

Samples for DTS were prepared as per ADA specification. The mold was kept on a transparent sheet on a metallic plate. The paste was packed into the mold and a second transparent sheet was kept on top followed by a second metallic plate. The mold and strip of film sandwiched between the metallic plates were pushed to displace excess material. The plates were removed and the paste was exposed to photocuring for the duration of 60 s on both sides through the transparent sheet. The cured samples were removed from the mold and kept at $37 \pm 1°C$ for 23 ± 1 h and $22 \pm 2°C$ for 1 h before testing. DTS was determined using the Universal Testing Machine (Instron, Model 3363, UK) with a crosshead speed of 10 mm/min. The load at which break occurs was noted, and DTS was calculated using the following equation:

$$\text{DTS}(\text{MPa}) = 2P/\pi DL \qquad (6.2)$$

where P is the load in Newtons, D is the diameter, and L is the thickness of the specimen in millimeter. Mean and standard deviation of six samples were calculated.

6.2.3.3 FLEXURAL STRENGTH (FS)

The FS test specimens were prepared as per ISO specification no. 4049–2009 (E).[90] The FS was determined using Universal Testing Machine (Instron, Model 3363, UK). Load at break was noted and FS was determined using the formula:

$$FS(MPa) = 3FL/2bd^2 \qquad\qquad (6.3)$$

where F is the load at break in Newtons, L is the length of the specimen between two metal rods at the base plate in millimeter, b is the width of the specimen in millimeter, and d is the depth of the specimen in millimeter.

6.3   RESULTS

6.3.1   DTS AND FS

We investigated the effect of inorganic content on DTS and FS of photo-cured composites based on newly synthesized inorganic–organic hybrid resins and the results were shown in Figures 6.1 and 6.2. DTS helps in investigating the strength of brittle materials with little or no plastic deformation. Hardness, elastic modulus, compressive strength, fatigue resistance, and DTS of materials are closely related to one another.[91] FS of the composite assigns the quantity of defects within the material, which are potential to cause catastrophic breakdown when subjected to loading.[92] It also specifies the constancy and endurance of a restorative material. Composite restorations were subjected to high flexural stresses in the anterior and posterior areas in clinical application.[2] Flexural properties very much depend on the anticipated clinical appliance of restorative materials, as low FS is required for Class V restorations while Class I and Class II restorations required higher FS. Inadequate FS of the restoration leads to deformation and the loss of marginal seal between the composite resin and tooth structure.[93]

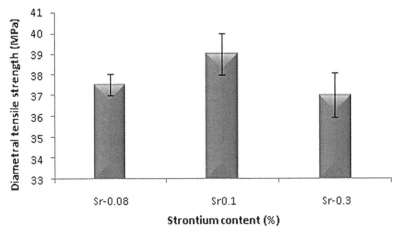

FIGURE 6.1   Effect of strontium on diametral tensile strength (DTS) of photocured composites.

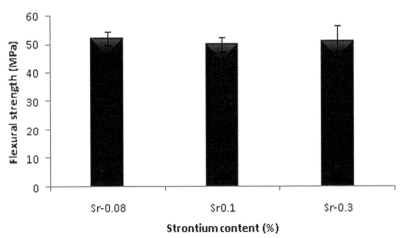

FIGURE 6.2   Effect of strontium on flexural strength (FS) of photocured composites.

The effect of strontium ion concentration on DTS and FS of photo-cured composites was investigated by changing the concentration of strontium content (0.08, 0.1, and 0.3%) in the synthesized resins. DTS values of Sr0.1 were significantly high ($p<0.05$) compared to Sr0.08 and Sr0.3 (Fig. 6.1). FS value of photocured composites did not have any significant differences among the three composites (Fig. 6.2). Both DTS and FS satisfy the stipulated value recommended as per the international standards. A composite material with high compressive strength, FS, and DTS might be clinically applied and should be resistant to masticatory forces.[94]

6.3.2 POLYMERIZATION SHRINKAGE

The results of polymerization shrinkage of photocured composites were graphically depicted in Figure 6.3. It had been reported that commercially available dimethacrylate resin-based composites exhibit percentage shrinkage of range 2.3–3.[95] Figure 6.3 implies that the shrinkage percentage of three hybrid composites were significantly lower than Bis GMA-based composite. Out of three synthesized resins, Sr0.1 resin-based photocured composites exhibited diminished polymerization shrinkage.

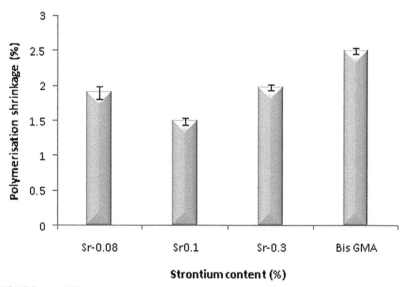

FIGURE 6.3 Effect of strontium on polymerization shrinkage of photocured composites.

6.4 DISCUSSION

6.4.1 DTS AND FS

The mechanical properties of the resin composite ought to be strong enough for its enduring clinical appliance in dental restorations. During the synthesis of inorganic–organic hybrid resins, the organic and the inorganic components are combined at a nanoscopic or molecular scale. As a result, these materials have potential to improve the properties of restorative composites.[96] Among the three formulations, Sr0.1 had the highest DTS value, which may be due to better integrated interfaces between the quartz filler and resin matrix (Fig. 6.1), compared to others. From the mechanical properties, it can be stated that the strontium incorporated within the hybrid resins has a crucial role in imparting better properties.

6.4.2 POLYMERIZATION SHRINKAGE

Figure 6.3 demonstrated that the polymerization shrinkage of all photo-cured composites was significantly lower than the reported polymerization shrinkage value of Bis GMA. In inorganic–organic hybrid resins, the inorganic Si–O–Si network was formed by the hydrolysis and polycondenzation reactions of the alkoxysilyl groups of the silane followed by organic polymerization induced by the methacrylate groups photochemically.[97] This Si–O–Si network and organic polymerization reaction have been reported to reduce the polymerization stress compared with conventional composites.[98,99] Multifunctional monomers have effective bonding with the filler which in turn reduces the polymerization shrinkage during photopolymerization. It may be the reason for low polymerization shrinkage of photo-cured composites based on synthesized resins. The inorganic and organic components are connected by covalent bonds. This approach necessitates molecular precursors that contain a hydrolytically stable chemical bond between the element that will form the inorganic network through solgel processing and the organic moieties. Inorganic–organic hybrid resins were synthesized by incorporating a polymer into the solgel process in an early (sol) stage of the process, as the silica network is being created, as a result, the inorganic and organic species interact at the nanoscale.

The synergetic effect of strontium within the resin of the formulation Sr0.1 supplements better properties for the composites. The nonexistence of reaction site and within the resin hinders the performance of other formulations. The lower polymerization shrinkage of the hybrid resin-based composites confirmed the effective bonding between the inorganic and organic parts within them. Ionic and covalent bondings present in these novel resins bind the inorganic filler and resin matrix together firmly. The inorganic content, that is, strontium in the resin has a direct influence on the structural disparity of these synthesized resins. The selection of alkoxide and specific organic monomers having functional groups responsible for effective chemical bond formation is vital because it will enhance properties of the inorganic–organic hybrid materials.

6.5   CONCLUSION

Newly developed composites from hybrid resins containing strontium exhibited good mechanical properties with low polymerization shrinkage. The synergetic effect of the inorganic content within the resin of this particular composition supplements better properties for the composites. Therefore, the application of these hybrid resins can be expanded to orthopedic, dental, and bioactive coating applications as these contains polymerizable methacrylate groups which can endure in situ polymerization to obtain bioactive polymers.

6.6   ACKNOWLEDGMENT

Financial support from Kerala State Council for Science, Technology and Environment, Government of Kerala, India is gratefully acknowledged. We thank Director, Head, and Dr. V. Kalliyana Krishnan, Head of Department of Biomaterial Science and Technology, Division of Dental Products, Biomedical Technology Wing, SCTIMST for extending the facilities in BMT Wing.

KEYWORDS

- inorganic-organic hybrid resin
- single pot
- modified sol-gel method
- photocured polymeric composites
- polymerization shrinkage

REFERENCES

1. Keith; Chan, H. S.; Mai, Y.; Kim, H.; Tong, K. C. T.; Desmond, NG; Hsiao, J. C. M. Review: Resin Composite Filling. *J. Mater.* **2010**, *3*, 1228–1243.

2. Bayne, S. C.; Heymann, H. O.; Swift, E Jr. Update on Dental Composite Restoration. *J. Am. Dent. Assoc.* **1994**, *125*, 687–670.
3. Kawaguchi, M.; Fukushima, T.; Horiba, T. Effect of Monomer Structure on the Mechanical Properties of Light Cured Composite Resin. *Dent. Mater. J.* **1989**, *81*, 40–45.
4. Ferracane, J. L. Resin Composite—State of the Art. *Dent. Mater.* **2011**, *27*, 29–38.
5. Davidson, C. L., Feilzer, A. J. Polymerization Shrinkage and Polymerization Shrinkage Stress in Polymer-Based Restoratives. *J. Dent.* **1997**, *25*(6), 435–440.
6. Ferracane, J. L. Developing a More Complete Understanding of Stresses Produced in Dental Composites During Polymerization. *Dent. Mater.* **2005**, *21*(1), 36–42.
7. Charton, C.; Colon, P.; Pla, F. Shrinkage Stress in Lightcured Composite Resins: Influence of Material and Photoactivation Mode. *Dent. Mater.* **2007**, *23*(8), 911–920.
8. Choi, K. K.; Ryu, G. J.; Choi, S. M.; Lee, M. J.; Park, S. J.; Ferracane, J. L. Effects of Cavity Configuration on Composite Restoration. *Oper. Dent.* **2004**, *29*(4), 462–469.
9. Anders, A.; Peutzfeldt, A.; van Dijken, J. W. V. Effect of Power Density of Curing Unit, Exposure Duration, and Light Guide Distance on Composite Depth of Cure. *Clin. Oral. Invest.* **2005**, *9*(2), 71–76.
10. Braga, R. R.; Ferracane, J. L. Contraction Stress Related to Degree of Conversion and Reaction Kinetics. *J. Dent. Res.* **2002**, *81*(2), 114–118.
11. Milewski, G.; Hille, A. Experimental Strength Analysis of Orthodontic Extrusion of Numan Anterior Teeth. *Acta Bioeng. Biomech.* **2012**, *14*(1), 15–21.
12. Milewski, G. Numerical and Experimental Analysis of Effort of Human Tooth Hard Tissues in Terms of Proper Occlusal Loadings. *Acta Bioeng. Biomech.* **2005**, *7*(1), 47–59.
13. Ahn, K.-D.; Chung, C.-M.; Kim, Y.-H. Synthesis and Photopolymerization of Multifunctional Methacrylates Derived from Bis-GMA for Dental Applications. *J. Appl. Polym. Sci.* **1999**, *71*, 2033–2037.
14. Klee, J. E; Neidhart, F.; Flammersheim, H.-J.; Mulhaupt, R. Monomers for Low Shrinking Composites, 2a: Synthesis of Branched Methacrylates and Their Application in Dental Composites. *Macromol. Chem. Phys.* **1999**, *200*, 517–523.
15. Chung, C. M.; Kim, J. G.; Kim, M. S.; Kim, K. M.; Kim, K. N. Development of a New Photocurable Composite Resin with Reduced Curing Shrinkage. *Dent. Mater.* **2002**, *18*, 174–178.
16. Chung, C. M.; Kim, M. S.; Kim, J. G.; Jang, D. O. Synthesis and Photopolymerization of Trifunctional Methacrylates and Their Application as Dental Monomers. *J. Biomed. Mater. Res.* **2002**, *62*, 622–627.
17. Khatri, C. A.; Stansbury, J. W.; Schultheisz, C. R.; Antonucci, J. M. Synthesis, Characterization and Evaluation of Urethane Derivatives of Bis-GMA. *Dent. Mater.* **2003**, *19*, 584–588.
18. Kim, Y.; Kim, C. K.; Cho, B. H.; Son, H. H.; Um, C. M.; Kim, O. Y. A New Resin Matrix for Dental Composite Having Low Volumetric Shrinkage. *J. Biomed. Mater. Res. Part B Appl. Biomater.* **2004**, *70B*, 82–90.
19. Chen, M. H.; Chen, C. R.; Hsu, S. H.; Sun, S. P.; Su, W. F. Low Shrinkage Light Curable Nanocomposite for Dental Restorative Material. *Dent. Mater.* **2006**, *22*(2), 138–145.

20. Culbertson, B. M.; Wan, Q.; Tong, Y. Preparation and Evaluation of Visible Light-Cured Multi-Methacrylates for Dental Composites. *J. Macromol. Sci. Pure Appl. Chem.* **1997**, *34*, 2405–2421.
21. Wolter, H.; Storch, W. A New Silane Precursor with Reduced Polymerization Shrinkage. *J. Sol-Gel Sci. Tech.* **1994**, *2*, 93–96.
22. Satsangi, N.; Rawls, H. R.; Norling, B. K. Synthesis of Low-Shrinkage Polymerizable Liquid-Crystal Monomers. *J. Biomed. Mater. Res. Part B Appl. Biomater.* **2004**, *71B*, 153–158.
23. Watts, D. C.; Al Hindi, A. Intrinsic 'soft-start' Polymerization Shrinkage-Kinetics in an Acrylate-Based Resin-Composite. *Dent. Mater.* **1999**, *15*, 39–45.
24. Lim, B. S.; Ferracane, J. L.; Sakaguchi, R. L.; Condon, J. R. Reduction of Polymerization Contraction Stress for Dental Composites by Two-Step Light-Activation. *Dent. Mater.* **2002**, *18*, 436–444.
25. Condon, J. R.; Ferracane, J. L. Reduction of Composite Contraction Stress Through Non-bonded Microfiller Particles. *Dent. Mater.* **1998**, *14*, 256–260.
26. Condon, J. R.; Ferracane, J. L. Reduced Polymerization Stress Through Non-bonded Nanofiller Particles. *Biomaterials* **2002**, *23*, 3807–3815.
27. Torstenson, B.; Oden, A. Effects of Bonding Agent Types and Incremental Techniques on Minimizing Contraction Gaps Around Resin Composites. *Dent. Mater.* **1989**, *5*(4), 218–223
28. Prati, C.; Montanari, G. Comparative Microleakage Study Between the Sandwich and Conventional Three-Increment Techniques. *Quintessence Int.* **1989**, *20*(8), 587–594.
29. Saunders, W. P.; Strang, R.; Ahmad, I. Effect of Composite Resin Placement and Use of an Unfilled Resin on the Microleakage of Two Dentin Bonding Agents. *Am. J. Dent.* **1990**, *3*(4), 153–156.
30. Krejci, I.; Lutz, F. Marginal Adaptation of Class V Restorations Using Different Restorative Techniques. *J. Dent.* **1991**, *19*(1), 24–32.
31. Crim, G. A. Microleakage of Three Resin Placement Techniques. *Am. J. Dent.* **1991**, *4*(2), 69–72.
32. Santini. A.; Plasschaert, A. J.; Mitchell, S. Effect of Composite Resin Placement Techniques on the Microleakage of Two Self-Etching Dentin-Bonding Agents. *Am. J. Dent.* **2001**, *14*(3), 132–136.
33. St Georges, A. J.; Wilder, A. D. Jr; Perdigao, J.; Swift, E. J. Jr. Microleakage of Class V Composites Using Different Placement and Curing Techniques: An in Vitro Study. *Am. J. Dent.* **2002**, *15*(4), 244–247.
34. Mullejans, R.; Lang, H.; Schuler, N.; Baldawi, M. O.; Raab, W. H. Increment Technique for Extended Class V Restorations: An Experimental Study. *Oper. Dent.* 2003, *28*(4), 352–356.
35. Loguercio, A. D.; Reis, A.; Ballester, R. Y. Polymerization Shrinkage: Effect of Constraint and Filling Technique in Composite Restorations. *Dent. Mater.* **2004**, *20*(3), 236–243.
36. Moszner, N.; Völkel, T.; von Clausbruch, S. C.; Elisabeth, G.; Nadine, B.; Volker, R. Sol-Gel Materials, 1. Synthesis and Hydrolytic Condensation of New Cross-Linking Alkoxysilane Methacrylates and Light-Curing Composites Based upon the Condensates. *Macromol. Mater. Eng.* **2002**, *5*, 287.

37. Moszner, N.; Gianasmidis, A.; Klapdohr, S.; Urs, K. F.; Volker, R. Sol–gel Materials: 2. Light-Curing Dental Composites Based on Ormocers of Cross-Linking Alkoxysilane Methacrylates and Further Nano-components. *Dent. Mater.* **2008**, *24*(6), 851–856.

38. Xie, D.; Weng, Y.; Gua, X.; Zhao, J.; Gregory, R. L.; Zheng, C. Preparation and Evaluation of a Novel Glass-Ionomer Cement with Antibacterial Functions. *Dent. Mater.* **2011**, *27*, 487–496.

39. Morrier, J. J.; Barsotti, O.; Blanc-Benon, J.; Rocca, J. P.; Dumont, J. Antibacterial Properties of Five Dental Amalgams: An *in Vitro* Study. *Dent. Mater.* **1989**, *5*, 310–313.

40. Cher, F.; Josette, C. Antimicrobial Properties of Conventional Restorative Filling Materials and Advances in Antimicrobial Properties of Composite Resins and Glass Ionomer Cements—A Literature Review. *Dent. Mater.* **2015**, *31*, e89–e99.

41. Zalkind, M. M.; Keisar, O.; Ever-Hadani, P.; Grinberg, R.; Sela, M. N. Accumulation of Streptococcus Mutans on Light-Cured Composites and Amalgam: An In Vitro Study. *J. Esthet. Dent.* **1998**, *10*, 187.

42. Auschill, T. M.; Arweiler, N. B.; Brecx, M.; Reich, E.; Sculean, A.; Netuschil, L. The Effect of Dental Restorative Materials on Dental Biofilm. *Eur. J. Oral. Sci.* **2002**, *110*, 48–53.

43. Beyth, N.; Domb, A. J.; Weiss, E. I. An *In Vitro* Quantitative Antibacterial Analysis of Amalgam and Composite Resins. *J. Dent.* **2007**, *35*, 201–206.

44. Yoshida, K.; Tanagawa, M.; Atsuta, M. Characterization and Inhibitory Effect of Antibacterial Dental Resin Composites Incorporating Silver-Supported Materials. *J. Biomed. Mater. Res.* **1999**, *47*, 516–522.

45. Yoshida, K.; Tanagawa, M.; Matsumoto, S.; Yamada, T.; Atsuta, M. Antibacterial Activity of Resin Composites with Silver-Containing Materials. *Eur. J. Oral. Sci.* **1999**, *107*, 290–296.

46. Willershausen, B.; Callaway, A.; Ernst, C. P.; Stender E. The Influence of Oral Bacteria on the Surfaces of Resin-Based Dental Restorative Materials—an *In Vitro* Study. *Int. Dent. J.* **1999**, *49*, 231–239.

47. Chen, L.; Shen, H.; Suh, B. I. Antibacterial Dental Restorative Materials: A State-of-the-Art Review. *Am. J. Dent.* **2012**, *25*, 337–346.

48. Yesilyurt, C.; Er, K.; Tasdemir, T.; Buruk, K.; Celik, D. Antibacterial Activity and Physical Properties of Glass-Ionomer Cements Containing Antibiotics. *Oper. Dent.* **2009**, *34*, 18–23.

49. Takahashi, Y.; Imazato, S.; Kaneshiro, A. V.; Ebisu, S.; Frencken, J. E.; Tay, F. R. Antibacterial Effects and Physical Properties of Glass-Ionomer Cements Containing Chlorhexidine for the ART Approach. *Dent. Mater.* **2006**, *22*, 647–652.

50. Botelho, M. G. Compressive Strength of Glass Ionomer Cements with Dental Antibacterial Agents. *SADJ* **2004**, *59*, 51–53.

51. Deepalakshmi, M.; Poorni, S.; Miglani, R.; Rajamani, I.; Ramachandran, S. Evaluation of the Antibacterial and Physical Properties of Glass Ionomer Cements Containing Chlorhexidine and Cetrimide: An *In Vitro* Study. *Indian J. Dent. Res.* **2010**, *21*, 552–556.

52. Sebastian, W.; Iris, H.; Galina, H.; Rainer, A.; Matthias, K. Biaxial Flexural Strength of New Bis-GMA/TEGDMA-Based Composites with Different Fillers for Dental Applications. *Dent. Mater.* **2016**, *32*, 1073–1078.

53. Sauro, S.; Osorio, R.; Watson, T. F.; Osorio, E.; Toledano, M. Novel Light-Curable Materials Containing Experimental Bioactive Micro-fillers Remineralise Mineral-Depleted Bonded-Dentine Interfaces. *J. Biomater. Sci.* **2013**, *24*(8), 940–956.

54. Vibha, C.; Lizymol, P. P. Effect of Inorganic Content on Thermal Stability and Anti-microbial Properties of Inorganic–Organic Hybrid Dimethacrylate Resins. *Int. J. Sci. Res.* **2015**, *4*(8), 59–60.

55. Sauro, S.; Osorio, E.; Watson, T. F.; Toledano, M. Therapeutic Effects of Novel Resin Bonding Systems Containing Bioactive Glasses on Mineral-Depleted Areas Within the Bonded-Dentine Interface. *J. Mater. Sci. Mater. Med.* **2012**, *23*, 1521–1532.

56. Birgani, Z. T.; Malhotra, A.; van Blitterswijk, C. A.; Habibovic, P. Human Mesenchymal Stromal Cells Response to Biomimetic Octacalcium Phosphate Containing Strontium. *J. Biomed. Mater. Res. A* **2016**, *104A*, 1946–1960.

57. Vo, T. N.; Kurtis, K. F.; Mikos, A. G. Strategies for Controlled Delivery of Growth Factors and Cells for Bone Regeneration. *Adv. Drug. Deliv. Rev.* **2012**, *64*, 1292–1309.

58. Liu, Y.; de Groot, K.; Hunziker, E. B. BMP-2 Liberated from Biomimetic Implant Coatings Induces and Sustains Direct Ossification in an Ectopic Rat Model. *Bone* **2005**, *36*, 745–757.

59. Malhotra, A.; Pelletier, M.; Oliver, R.; Christou, C.; Walsh, W. R. Platelet-Rich Plasma and Bone Defect Healing. *Tissue Eng. Part A* **2014**, *20*, 2614–2633.

60. Jager, M.; Hernigou, P.; Zilkens, C.; Herten, M.; Li, X.; Fischer, J.; Krauspe, R. Cell Therapy in Bone Healing Disorders. *Orthop. Rev.* **2010**, *2*, 79–87.

61. Mooney, D. J.; Vandenburgh, H. Cell Delivery Mechanisms for Tissue Repair. *Cell Stem Cell* **2008**, *2*, 205–213.

62. Boanini, E.; Gazzano, M.; Bigi, A. Ionic Substitutions in Calcium Phosphates Synthesized at Low Temperature. *Acta Biomater.* **2010**, *6*, 1882–1894.

63. Yang, L.; Perez-Amodio, S.; Barrère-de Groot, F. Y.; Everts, V.; van Blitterswijk, C. A.; Habibovic, P. The Effects of Inorganic Additives to Calcium Phosphate on In Vitro Behavior of Osteoblasts and Osteoclasts. *Biomaterials.* **2010**, *31*,2976–2989.

64. Yıldız, Y. O.; Fatih, E. B.; Fatih, U. Synthesis and Characterization of Strontium-Doped Hydroxyapatite for Biomedical Applications. *J. Therm. Anal. Calorim.* **2016**, *125*, 745–750.

65. Guo, D.; Xu, K.; Zhao, X.; Han Y. Development of Strontium Containing Hydroxyapatite Bone Cement. *Biomaterials.* **2005**, *19*, 4073–4083.

66. Lala, S.; Brahmachari, S.; Das, P. K.; Das, D.; Kar, T.; Pradhan, S. K. Biocompatible Nanocrystalline Natural Bonelike Carbonated Hydroxyapatite Synthesized by Mechanical Alloying in a Record Minimum Time. *Mater. Sci. Eng. C* **2014**, *42*, 647–656.

67. Tredwin, C. J.; Young, A. M. E.; Neel, A. A.; Georgiou, G.; Knowles, J. C. Hydroxyapatite, Fluor-Hydroxyapatite and Fluorapatite Produced via the Sol–Gel Method: Dissolution Behaviour and Biological Properties After Crystallisation. *J. Mater. Sci. Mater. Med.* **2014**, *25*, 47–53.

68. Bigi, A.; Boanini, E.; Capuccini, C.; Gazzano, M. Strontium-Substituted Hydroxyapatite Nanocrystals. *Inorg. Chim. Acta* **2007**, *360*, 1009–1016.

69. Pereiro, I.; Rodríguez-Valencia, C.; Serra, C.; Solla, E. L.; Serra, J.; González, P. Pulsed Laser Deposition of Strontium-Substituted Hydroxyapatite Coatings. *Appl. Surf. Sci.* **2012**, *258*(23), 9192–9197.

70. Brinker, C. J.; Scherer, G. W. Eds. Sol–Gel Science, the Physics and Chemistry of Sol–Gel Processing; Academic Press: San Diego, 1990.

71. Schmidt, H.; Sieferling, B.; Philip, G.; Deichman, K. Development of Organic–Inorganic Hard Coatings by Sol–Gel Method. In *Ultrastructure Processing of Advanced Ceramics*; Mackenzie, J. D., Ulrich, D. R. Eds.; Wiley: USA, 1988; p 651.

72. Novak, B. M. Hybrid Nanocomposite Materials—Between Inorganic Glasses and Organic Polymers. *Adv. Mater.* **1993**, *5*, 422–433.

73. Chiang, C. L.; Ma, C.-C. M. Synthesis, Characterization and Thermal Properties of Novel Epoxy Containing Silicon and Phosphorus Nanocomposites by Sol–Gel Method. *Eur. Polym. J.* **2002**, *38*, 2219–2224.

74. Joseph, R.; Zhang, S.; Ford, W. Structure and Dynamics of a Colloidal Silica–Poly(methyl methacrylate) Composite by^{13}C and^{29}Si MAS NMR Spectroscopy. *Macromolecules* **1996**, *29*, 1305–1312.

75. Canto, C. F.; de A Prado, L. A. S.; Radovanovic, E.; Yoshida, I. V. P. Organic–Inorganic Hybrid Materials Derived from Epoxy Resin and Polysiloxanes: Synthesis and Characterization. *Polym. Eng. Sci.* **2008**, *48*, 141–148.

76. Murthy, R.; Cox, C. D.; Hahn, M. S.; Grunlan, M. A. Protein-Resistant Silicones: Incorporation of Poly(ethylene oxide) via Siloxane Tethers. *Biomacromolecules* **2007**, *8*, 3244–3252.

77. Jana, S.; Lim, M. A.; Baek, I. C.; Kim, C. H.; Seok, S. I. Non-hydrolytic Sol–Gel Synthesis of Epoxysilane-Based Inorganic–Organic Hybrid Resins. *Mater. Chem. Phys.* **2008**, *112*, 1008–1014.

78. Gilbert, J. L.; Hasenwinkel, J. M.; Wixson, R. L.; Eugene, P. A Theoretical and Experimental Analysis of Polymerization Shrinkage of Bone Cement: A Potential Major Source of Porosity. *J. Biomed. Mater. Res. A* **2000**, *52*, 210–218.

79. Sakar-Deliormanli, A.; Guden, M. Microhardness and Fracture Toughness of Dental Materials by Indentation Method. *J. Biomed. Mater. Res. Part B Appl. Biomater.* **2006**, *76*, 257–264.

80. Chung, S. M.; Yap, A. U. J.; Chandra, S. P.; Lim, C. T. Flexural Strength of Dental Composite Restoratives: Comparison of Biaxial and Three-Point Bending Test. *J. Biomed. Mater. Res. Part B Appl. Biomater.* **2004**, *71*, 278–283.

81. Keulemans, F.; Palav, P.; Moustafa, M. N. A.; van Dalen, A.; Kleverlaan, C. J.; Feilzer, A. Fracture Strength and Fatigue Resistance of Dental Resin-Based Composites. *J. Dent. Mater.* **2009**, *25*, 1433–1441.

82. Kim, Y. K.; Mai, S.; Mazzoni, A.; Liu, Y.; Tezvergil-Mutluay, A.; Takahashi, K.; Zhang, K.; Pashley, D. H.; Tay, F. R. Biomimetic Remineralization as a Progressive Dehydration Mechanism of Collagen Matrices—Implications in the Aging of Resin-Dentin Bonds. *Acta Biomater.* **2010**, *6*, 3729–3739.

83. Hench, L. L. Bioceramics and the Origin of Life. *J. Biomed. Mater. Res.* 1989, 23, 685–703.

84. Dickens, S. H.; Flaim, G. M.; Takagi, S. Mechanical Properties and Biochemical Activity of Remineralizing Resin-Based Ca–PO4 Cements. *Dent. Mater.* **2003**, *19*, 558–566.

85. Sauro, S.; Osorio, R.; Fulgencio, R.; Watson, T. F.; Cama, G.; Thompsona, I.; Toledanob, M. Remineralisation Properties of Innovative Light-Curable Resin-Based Dental Materials Containing Bioactive Micro-fillers. *J. Mater. Chem.* **2013**, *1*, 2624–2638.

86. Lizymol, P. P. Studies on Shrinkage, Depth of Cure, and Cytotoxic Behavior of Novel Organically Modified Ceramic Based Dental Restorative Resins. *J. App. Pol. Sci.* **2010**, *116*, 2645–2650.

87. Lizymol P. P.; Vibha C. A Process for the Synthesis of Inorganic–Organic Hybrid Resins Comprising of Alkoxides or Mixture of Alkoxides of Calcium, Magnesium, Zinc, Strontium, Barium and Manganese with Polymerizable (Di/Tetra) Methacrylate Groups, 4027/CHE/2014, India, 2014.

88. Lizymol P. P., Vibha C. A Visible Light Cure Dental Restorative Composite with Excellent Remineralization Ability with Good Physico-mechanical Properties and Low Shrinkage Based on a Novel Calcium Containing Inorganic Organic Hybrid Resin with Polymerizable Methacrylate Groups, 4996/CHE/2014, India, 2014.

89. American National Standard. American Dental Association Specification n. 27 for Resin-Based Filling Materials, 1993.

90. International Organization for Standardization: ISO 4049. *Dentistry —Polymer-Based Filling, Restorative and Luting Materials,* 4th ed; 2009.

91. Della, B. A.; Benetti, P.; Borba, M.; Cecchetti, D. Flexural and Diametral Tensile Strength of Composite Resins. *Braz. Oral. Res.* **2008**, *22*(1), 84–89.

92. Marchan, S.; White, D.; Smith, W.; Coldero, L.; Dhuru, V. Comparison of the Mechanical Properties of Two Nano-filled Composite Materials. *Rev. Clín. Pesq. Odontol.* **2009**, *5*, 241–246.

93. Shamszadeh, S.; Vaghareddin A. Z.; Maryam, M., Maryam, A. T.; Sanambar Y. Comparison of Flexural Strength of Several Composite Resins Available in Iran. *J. Dent. Sch.* **2013** *31*(3),170–176.

94. Palin, W. M.; Fleming, G. J.; Burke, J. F.; Marquis, P. M.; Randall, R. C. The Reliability in Flexural Strength Testing of a Novel Dental Composite. *J. Dent.* **2003**, *31*(8), 549–57.

95. Ruili, W.; Mo, Z.; Shuang, B.; Fengwei, L.; Xiaoze, J.; Meifang, Z. Synthesis of Two Bis-GMA Derivates with Different Size Substituents as Potential Monomer to Reduce the Polymerization Shrinkage of Dental Restorative Composites. *J. Mater. Sci. Res.* **2013**, *2*(4), 12–22.

96. Rathbun, M. A.; Craig, R. G.; Hanks, C.T.; Filisko, F. E. Cytotoxicity of a BIS-GMA Dental Composite Before and After Leaching in Organic Solvents. *J. Biomed. Mater. Res.* **1991**, *25*(4), 443–457.

97. Wolter, H.; Storch, W.; Ott, H. In *New Inorganic/Organic Copolymers (ORMOCERs) for Dental Applications*, Materials Research Society Symposium Proceedings, **1994**, *346*, 143–149.

98. Fleming, G. J. P.; Hall, D.; Shortall, A. C. C.; Burke, F. J. T. Cuspal Movement and Microleakage in Premolar Teeth Restored with Posterior Filling Materials of Varying Reported Volumetric Shrinkage Values. *J. Dent.* **2005**, 33, 139–146.

99. Hahne, S.; Dowling, A. H.; El-Safty, S.; Fleming, G. J. P. The Influence of Monomeric Resin and Filler Characteristics on the Performance of Experimental Resin-Based Composites (RBCs) Derived from a Commercial Formulation. *Dent. Mater.* **2012**, *28*, 416–423.

CHAPTER 7

HYBRID NATURAL FIBERS REINFORCED POLYMER COMPOSITE: THERMAL ANALYSIS

M. K. GUPTA* and R. K. SRIVASTAVA

Department of Mechanical Engineering, Motilal Nehru National Institute of Technology Allahabad 211004, India

Corresponding author. E-mail: mkgupta@mnnit.ac.in, mnnit.manoj@gmail.com

CONTENTS

ABSTRACT

Recently, hybridization has become a well-known technique to overcome the limitations such as low impact strength, high moisture absorption, and poor durability of single natural fiber reinforced polymer composites. In the present work, hybridization technique is used by incorporating the sisal fibers into jute fiber reinforced epoxy composite in order to study thermal properties. The hybrid composites are prepared by hand lay-up technique with varying weight ratios of fibers keeping constant 30 wt.% as total fibers content. The viscoelastic properties such as storage modulus E', loss modulus (E''), and damping (Tan δ) of prepared hybrid composites are carried out using dynamic mechanical analysis (DMA) at 1 Hz frequency. Thermal stability of prepared composites is studied by thermogravimetric analysis (TGA) whereas differential scanning calorimetry (DSC) is used to measure glass transition temperature (T_g), crystallization temperature (T_c), and decomposition temperature (T_d). The results show that hybrid composite with 50% jute fibers has the higher value of thermal stability, E', E'', and T_g.

7.1 INTRODUCTION

Natural fibers are being used in place of glass and other synthetic fibers for medium strength applications due to advantages such as low-cost, low density, abundant availability, environmental-friendly, non-toxicity, high flexibility, renewability, biodegradability, relative non-abrasiveness, high specific strength and modulus, and easy processing.[1–5] In view of the above advantages, natural fiber reinforced polymer composites have been used in many industries such as automotive, packaging, construction, household, toys, and furniture. In spite of the above advantages, natural fibers have some limitations also such as low impact strength and thermal stability, poor fiber-matrix adhesion, and high moisture absorption properties.[6–10] These limitations can be overcome to some extent by using hybridization technique. The incorporation of additional fibers with single fiber into polymer matrix has led to the development of hybrid composite.[11] The advantages of one type of fiber could compliment what is lacking in the other fiber using hybrid composites. The properties of hybrid composites are decided by many factors such as fiber content, fiber length and orientation, fiber to matrix bonding, and so forth.[12] Strength of the hybrid composite is also dependent on the failure strain and modulus of individual fiber.[13]

Thermal analysis is an essential technique to find out the thermal stability, glass transition temperature, crystallization temperature, melting temperature, enthalpy of the polymer composites, and so forth.[14] Many researchers have studied the thermal properties of hybrid fiber reinforced polymer composite and suggested that hybridization can improve the thermal properties of single fiber reinforced polymer composite. Thermal properties in terms of storage modulus, loss modulus, and damping using dynamic mechanical analysis were proposed by many researchers.[15-19] Idicula et al.[17] reported studies on the DMA of hybrid banana/sisal fiber reinforced polyester composites. DMA of the hybrid composite was evaluated in terms of storage modulus, loss modulus, and damping parameter under the three-point bending mode within a temperature range of 30–150°C at different frequencies. The composite with 40 vol.% fiber content has the maximum storage modulus and peak width of (Tan δ) but the minimum height of (Tan δ). Shanmugam and Thiruchitrambalam[18] proposed studies on dynamic mechanical properties of alkali-treated hybrid Palmyra palm leak stalk/jute fiber reinforced polyester composite. The results indicated that addition of jute fibers to Palmyra palm leak stalk fiber and alkali treatment of fibers has enhanced storage and loss modulus. The maximum damping behavior was observed for composite with higher jute fiber content. Jawaid et al.[19] studied the dynamic mechanical analysis of hybrid jute/oil palm fiber reinforced epoxy composite. It was reported that the storage modulus of palm fiber reinforced epoxy composite was increased due to incorporation of jute fibers.

Thermal properties in terms of thermal stability, thermal conductivity, glass transition temperature, and so forth were reported by other researchers.[11,16,20-22] Boopalan et al.[9] studied the thermal properties of hybrid jute/banana fiber reinforced epoxy composite. They suggested that the hybrid composite J50B50 (50% jute and 50% banana fiber) of 30 wt.% shows higher thermal stability as compared to other composites. Naidu et al.[20] studied the thermal properties of sisal/glass fiber reinforced polyester composite. The thermal conductivity of hybrid sisal/glass fiber was found higher than sisal fiber reinforced composite but lower than glass fiber reinforced polyester composite. Dhakal et al.[21] worked on thermal properties of hybrid hemp/glass fiber reinforced polyester composite. They reported that the sodium hydroxide (NaOH) treated fiber composite shows the better thermal stability as compared to an untreated composite which shows the positive effect of alkali treatment. Saw and Datta[22] studied the thermal properties of hybrid jute/bagasse fiber reinforced epoxy composite. They suggested that the chemically treated fiber composites show better thermal stability than untreated composites as degradation temperature was shifted from 438 to 475°C.

The aim of the present study is to overcome the disadvantages of single natural fiber reinforced polymer composites. Therefore, in the present study, hybrid jute/sisal fiber reinforced epoxy composites are prepared using hand lay-up technique with varying weight percentage of fibers and subjected to thermal properties.

7.2   MATERIALS AND METHODS

7.2.1   MATERIALS

Jute and sisal fibers are used as reinforcements and epoxy AY 105 is the matrix. Jute and sisal fibers were purchased from Women's Development Organization, Dehradun, India, whereas epoxy resin with hardener was purchased from Bakshi Brothers/Universal Enterprises, Kanpur, Uttar Pradesh, India. The properties of jute and sisal fibers are given in Table 7.1.

TABLE 7.1   Properties of Jute and Sisal Fiber.[1]

Properties	Jute fiber	Sisal fiber
Density (g/m³)	1.3	1.5
Diameter (µm)	25–200	50–200
Elongation at break (%)	1.5–1.8	2–2.5
Tensile strength (MPa)	393–773	511–700
Young's modulus (GPa)	26.5	9.4–22
Cellulose (%)	61–71	65
Lignin (%)	12–13	9.9
Microfibrillar angle	8°	22°
Wax (%)	0.5	2
Hemi-cellulose (%)	14–20	22

7.2.2   FABRICATION METHOD OF COMPOSITES

Unidirectionally continuous aligned bilayer hybrid composites are fabricated by reinforcing the sisal and jute fibers in the epoxy matrix by hand lay-up technique.[1] The epoxy resin AY 105 and corresponding hardener HY

951 are mixed in a ratio of 10:1 by weight as recommended by the supplier (Bakshi Brothers/Universal Enterprises, Kanpur, Uttar-Pradesh, India). The mixture is stirred manually to disperse the resin and the hardener in the matrix. A stainless steel mold having dimensions of $300 \times 200 \times 3$ mm is used for casting of the composites. Silicon spray is used to facilitate easy removal of the composite from the mold after curing. The cast of each composite is cured under a load of 50 kg for 24 h before it is removed from the mold. Dimensions of the specimens are cut as per ASTM D5023 by using a diamond cutter for analysis of thermal properties of the hybrid composite. The composites manufactured with varying weight percentage of fibers have been given notations as shown in Table 7.2.

TABLE 7.2   Notation for Composites Keeping 30 wt.% of Total Fibers Content.

Composites	Jute fibers content (%)	Sisal fibers content (%)
J100S0	100	0
J75S25	75	25
J50S50	50	50
J25S75	25	75
J0S100	0	100

7.3   CHARACTERIZATIONS OF COMPOSITES

7.3.1   DYNAMIC MECHANICAL ANALYSIS

The viscoelastic properties of epoxy and hybrid composites are studied using the dynamic mechanical analyzer (Seiko instruments DMA 6100). The viscoelastic properties are determined with a three-point bending test at 1 Hz frequency as a function of temperature. The composites are cut into samples having dimensions of $50 \times 13 \times 3$ mm according to ASTM D 5023. Experiments are carried out in the temperature range of 30–200°C at the heating rate of 10°C/min.

7.3.2   THERMOGRAVIMETRIC ANALYSIS

Thermal stability of the prepared composites is assessed by performing thermogravimetric analysis (Perkin Elmer TGA 4000). The TGA tests are carried out on 15–25 mg of a sample placed in a platinum pan, heated from 30–800°C at a heating rate of 10°C/min with a flow rate of 20 ml/min. The nitrogen atmosphere is used to avoid unwanted oxidation.

7.3.3   DIFFERENTIAL SCANNING CALORIMETRY

The thermal properties such as crystallization temperature, glass transition temperature, and decomposition temperature are obtained using differential scanning calorimetry (Perkin Elmer model DSC 4000). The experiments are carried out in the temperature range of 30–400°C at the heat flow rate of 10°C/min under nitrogen atmosphere purged at 20 ml/min and using an aluminum pan.

7.4   RESULTS AND DISCUSSIONS

7.4.1   DYNAMIC MECHANICAL ANALYSIS

7.4.1.1   STORAGE MODULUS

Storage modulus is defined as the amount of the maximum energy stored by the material during one cycle of oscillation.[8] It also gives an approximation of temperature-dependent stiffness behavior and load-bearing capability of the composite material.[23] The variation of the storage modulus of epoxy and hybrid jute/sisal composites as a function of temperature at 1 Hz frequency is shown in Figure 7.1. Mainly three significant regions: glassy, transition, and rubbery are seen in the variation of storage modulus versus temperature curve. In the glassy region, it can be observed that storage modulus of composites is close to each other because at a lower temperature, stiffness is not much affected by fiber contents. In the glassy region, a higher value of E' is found to be 3.87 GPa for the hybrid composite J50S50. In this region, increase in storage modulus of composites follows the order: J50S50 > J75S25 > J25S75 > J100S0 > J0S100.   The storage modulus of epoxy and composites is found to decrease as temperature increases due to a loss in stiffness of fibers.[18] In the transition region, it can be observed that the composites have a gradual fall in the value of E' except epoxy which has a sudden fall in the value of E' with increase in temperature. This trend is due to the incorporation of high modulus fibers in the epoxy matrix.[18] In the rubbery region, the hybrid composite J50S50 shows the higher value (0.672 GPa) of storage modulus whereas epoxy has the lower value (0.05 GPa) of storage modulus. In the rubbery region, increase in storage modulus follows the order: J50S50>J25S75>J100S0 >J75S25>J0S100>Epoxy. Epoxy has the lowest value of E' which shows an increase in molecular mobility at higher temperature.[23] A higher value of storage modulus for hybrid composite J50S50 is due to proper stress transfer from matrix to fibers. It can also be observed that values of storage moduli of composites are not close to each other in the rubbery region. This is because at high temperature the fibers control the stiffness of materials.[24] Equation 7.1 is used to calculate the effectiveness constant of reinforcement ($\in$) of composites.

$$\in = \frac{\left(\dfrac{E'_g}{E'_r}\right)_{Composite}}{\left(\dfrac{E'_g}{E'_r}\right)_{Epoxy}} \qquad (7.1)$$

Where, E'_g and E'_r are the storage moduli in the glassy and rubbery region, respectively. A lower efficiency of reinforcement shows the higher value of ϵ and vice versa. The calculated values of effectiveness constant of reinforcement are given in Table 7.3. The lowest value of effectiveness constant of reinforcement is found for the composite J25S75 followed by J50S50 composite. It may be due to the addition of high strength sisal fibers which cause uniform stress transfer.[25]

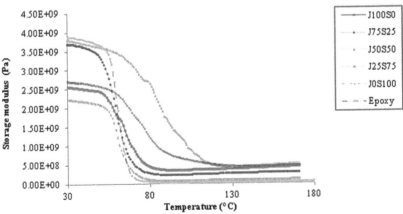

FIGURE 7.1 Variation of storage modulus with temperature of epoxy and hybrid composites.

TABLE 7.3   The Values of ∈ and T_g(°C) from E'' and Tan δ curve.

Composites	∈	T_g from E'' curve	T_g from (Tan δ) curve
J100S0	0.284	64.79	71.82
J75S25	0.319	60.12	67.84
J50S50	0.251	85.32	95.01
J25S75	0.226	79.51	89.87
J0S100	0.452	61.56	72.58
Epoxy	–	60.95	71.77

7.4.1.2   LOSS MODULUS

Loss modulus is the amount of energy dissipated in the form of heat by a material during one cycle of sinusoidal load.[26] It represents the viscous response of the polymer-based composite materials.[18] The variation of the loss modulus as a function of temperature at 1 Hz frequency is shown in Figure 7.2. On increasing the temperature, the value of E'' is found to be increased up to glass transition temperature and then decreased. The maximum value of E'' is found 0.525 GPa for the composite J50S50. The glass transition temperature of the hybrid composite J50S50 is shifted to higher temperature due to the incorporation of sisal fibers into jute composite which causes a decrease in the mobility of polymer matrix. The value of T_g for epoxy and hybrid composites which are obtained from loss modulus curve is given in Table 7.3.

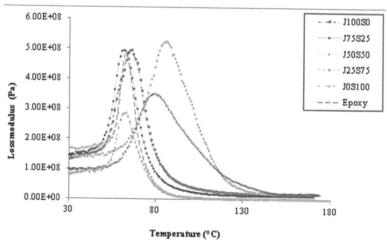

FIGURE 7.2   Variation of loss modulus with temperature of epoxy and hybrid composites.

7.4.1.3   DAMPING

The damping or Tan δ of the polymer materials is the ratio of loss modulus and storage modulus. It is related to the degree of molecular mobility in polymer material.[26] Temperature corresponding to the Tan δ peak represents the glass transition temperature of the composites. The glass transition temperature is the temperature range where the state of polymer materials changes from glassy (hard, rigid) to rubbery (flexible, yielding). The effect of damping parameter on epoxy and hybrid composites as a function of temperature is shown in Figure 7.3. The maximum value of Tan δ (0.617) for epoxy shows better damping properties as compared to all the other composites. The hybrid composite J50S50 shows lower value (0.278) of Tan δ which shows good load bearing capacity due to strong adhesion between fibers and matrix. The value of T_g obtained from the curve for epoxy and composites are given in Table 7.3.

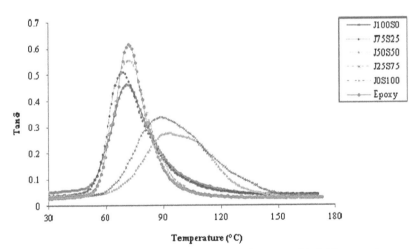

FIGURE 7.3   Variation of Tan δ with temperature of epoxy and hybrid composites.

7.4.2   THERMOGRAVIMETRIC ANALYSIS

Thermal stability of prepared composites is assessed by performing TGA. The TGA results of hybrid jute/sisal fiber reinforced epoxy composites are plotted in Figure 7.4. Three significant regions of weight loss, that is, initial, major, and final can be observed due to rise in temperature. The initial weight loss of composites is due to remove the solvent from the composites. Initial weight loss of composites takes place at the temperature between 280–300°C. Major weight loss is due to degradation and volatization of epoxy along with fibers present in the composites. Major weight loss of hybrid composites J50S50 takes place at higher temperature of 515°C. Here, degradation is shifted to higher temperature for hybrid composite J50S50 which shows the better thermal stability as compared to all other composites. It can be due to strong interface between fibers and polymer matrix. Moreover, the final weight loss in which char is left as residue requires higher temperature for subsequent degradation. The temperature corresponding to final weight loss of composites lies between 650–700°C.

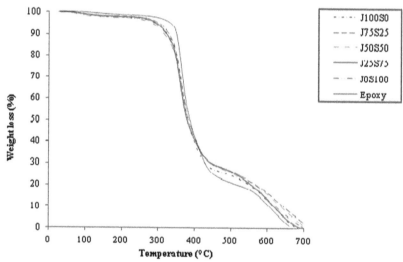

FIGURE 7.4   Variation of weight (%) of epoxy and hybrid composites with temperature.

7.4.3 DIFFERENTIAL SCANNING CALORIMETRY

Thermal properties of epoxy and hybrid composites such as glass transition temperature, crystallization temperature, and decomposition temperature are obtained from differential scanning calorimetry (DSC) studies as shown in Figure 7.5. It is observed that glass transition temperature of composites lies between 68–73°C, crystallization temperature lies between 114–119°C, and decomposition temperature lies between 348–357°C. It is observed that hybrid composite J50S50 has the higher values of glass transition, crystallization, and decomposition temperature as compared to all other composites. These results show that hybridization plays an important role in increase in thermal properties but the effect is not significant as seen from the small changes of the crystallization and glass transition temperature when compared to epoxy.

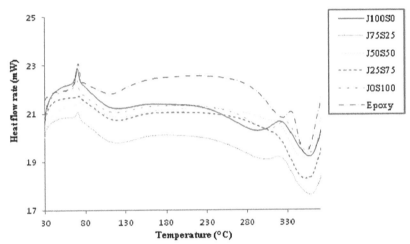

FIGURE 7.5   DSC analysis of epoxy and hybrid composites.

7.5   CONCLUSIONS

Unidirectional bilayer hybrid jute/sisal fiber reinforced epoxy composites are successfully developed by hand lay-up technique. Their mechanical and thermal properties are studied and following conclusions are drawn:

- The positive effect of hybridization is observed as an increase in dynamic mechanical properties. The hybrid composite J50S50 shows the higher value for storage modulus, loss modulus, and glass transition temperature than all other composites.
- Epoxy has the higher value of damping whereas lower value is shown by hybrid composite J50S50.
- The lower value of effectiveness constant of reinforcement is given by hybrid composite J25S75 followed by J50S50.
- The hybrid composite J50S50 exhibits higher thermal stability than all other composites.

- Glass transition, crystallization, and decomposition temperature are found to increase due to hybridization as compared to pure jute and sisal composites.
- The potential applications of jute-based hybrid composites in packaging, automobiles, and constructions are going to increase in the near future.

7.6   ACKNOWLEDGMENT

The authors would like to thank the head of Mechanical Engineering Department of Motilal Nehru National Institute of Technology, Allahabad, India, for his support in allowing us to perform the tests.

The study is partially supported by Cumulative Professional Development Allowances (CPDA) and R and C fund for teachers of Motilal Nehru National Institute of Technology Allahabad.

KEYWORDS

- differential scanning calorimetry
- dynamic mechanical analysis
- hybrid composite
- polymer matrix
- thermal properties

REFERENCES

1. Gupta, M. K.; Srivastava, R. K. Mechanical Properties of Hybrid Fibers Reinforced Polymer Composite: A Review. *Polym. Plast. Technol. Eng.* **2016,** *55,* 626–642.
2. Gupta, M. K.; Srivastava, R. K. Effect of Sisal Fiber Loading on Dynamic Mechanical Analysis and Water Absorption Behaviour of Jute Fiber Epoxy Composite. *Mater. Today: Proc.* **2015,** *2,* 2909–2917.
3. Nair, K. C. M.; Thomas, S.; Groeninckx, G. Thermal and Dynamic Mechanical Analysis of Polysterine Composites Reinforced with Short Sisal Fiber. *Compos. Sci. Technol.* **2001,** *61,* 2519–2529.

4. Gupta, M. K.; Srivastava, R. K. Mechanical and Water Absorption Properties of Hybrid Sisal/Glass Fiber Reinforced Epoxy Composite. *Am. J. Polym. Sci. Eng.* **2015,** *3,* 208–219.

5. Gupta, M. K.; Srivastava, R. K. Effect of Sisal Fiber Loading on Wear and Friction Properties of Jute Fiber Reinforced Epoxy Composite. *Am. J. Polym. Sci. Eng.* **2015,** *3,* 198–207.

6. Gupta, M. K.; Srivastava, R. K.; Bisaria, H. Potential of Jute Fiber Reinforced Polymer Composites: A Review. *Int. J. Fiber. Text. Res.* **2015,** *5,* 30–38.

7. Gupta, M. K.; Srivastava, R. K. A Review on Characterization of Hybrid Fiber Reinforced Polymer Composite. *Am. J. Polym. Sci. Eng.* **2016,** *4,* 1–7.

8. Gupta, M. K. Dynamic Mechanical and Thermal Analysis Hybrid Jute/Sisal Fiber Reinforced Epoxy Composite. *Proc. Inst. Mech. Eng., Part L* DOI: 10.1177/1464420716646398.

9. Gupta, M. K.; Srivastava, R. K. Properties of Sisal Fiber Reinforced Epoxy Composite. *Indian. J. Fib. Text. Res.* **2016,** *41,* 235–241.

10. Mourya, H.; Gupta, M. K.; Srivastava, R. K.; Singh, H. Study on The Mechanical Properties of Epoxy Composite Using Short Sisal Fiber. *Mater. Today Proc.* **2015,** *2,* 1347–1355.

11. Boopalan, M.; Niranjana, M.; Umapathy, M. J. Study on The Mechanical Properties and Thermal Properties of Jute and Banana Fiber Reinforced Epoxy Hybrid Composites. *Compos: Part B.* **2013,** *51,* 54–57.

12. Kumaresan, M.; Sathish, S.; Karthi, N. Effect of Fiber Orientation on Mechanical Properties of Sisal Fiber Reinforced Epoxy Composites. *J. Appl. Sci. Eng.* **2015,** *18,* 289–294.

13. Sreekala, M. S.; George, G.; Kumaran, M. G.; Thomas, S. The Mechanical Performance of Hybrid Phenol-Formaldehyde-Based Composite Reinforced with Glass and Oil Palm Fibers. *Compos. Sci. Technol.* **2002,** *62,* 339–353.

14. Cheng, S.; Lau, K.; Liu, T.; Zhao, Y.; Lam, P.; Yin, Y. Mechanical and Thermal Properties of Chicken Feather Fiber/PLA Green Composites. *Composites Part B.* **2009,** *40,* 650–654.

15. Nair, K. C. M.; Thomas, S.; Groeninckx, G. Thermal and Dynamic Mechanical Analysis of Polysterine Composites Reinforced with Short Sisal Fiber. *Compos. Sci. Technol.* **2001,** *61,* 2519–2529.

16. Muralidhar, B. A. Study of Flax Hybrid Preforms Reinforced Epoxy Composites. *Mater. Des.* **2013,** *52,* 835–840.

17. Idicula, M.; Malhotra, S. K.; Josheph, K.; Thomas, S. Dynamic Mechanical Analysis of Randomly Oriented Intimately Mixed Short Banana/Sisal Fiber Reinforced Polyester Composites. *Compos. Sci. Technol.* **2005,** *65,* 1077–1087.

18. Shanmugam, D.; Thiruchitrambalam, M. Static and Dynamic Mechanical Properties of Alkali Treated Unidirectional Continuous Palmyra Palm Leaf Stalk Fiber/Jute Fiber Reinforced Hybrid Polyester Composites. *Mater. Des.* **2013,** *50,* 533–542.

19. Jawaid, M.; Khalil, H. P. S. A.; Hassan, A.; Dungani, R.; Hadiyani, A. Effect of Jute Fiber Loading on Tensile and Dynamic Mechanical Properties of Oil Palm Epoxy Composites. *Composites Part B.* **2013,** *45,* 619–624.

20. Naidu, V. N. P.; Kumar, A. M.; Ramachandra, R. G. Thermal Conductivity of Sisal/ Glass Fiber Reinforced Hybrid Composites. *Int. J. Fib. Text.* **2011,** *1,* 28–30.
21. Dhakal, H. N.; Zhang, Z. Y.; Bennett, N. Influence of Fiber Treatment and Glass Fiber Hybridization on Thermal Degradation and Surface Energy Characteristics of Hemp/Unsaturated Polyester Composites. *Composites Part B.* **2012,** *43,* 2757–2761.
22. Saw, S. K.; Datta, C. Thermo Mechanical Properties of Jute/Bagasse Hybrid Fiber Reinforced Epoxy Thermoset Composites. *BioResources.* **2009,** *4,* 1455–1476.
23. Mohanty, S.; Verma, S. K.; Nayak, S. K. Dynamic Mechanical and Thermal Properties of MAPE Treated Jute/HDPE Composites. *Compos. Sci. Technol.* **2006,** *66,* 538–547.
24. Pothan, L. A.; Thomas, S. Polarity Parameters and Dynamic Mechanical Behaviour of Chemically Modified Banana Fiber Reinforced Polyester Composite. *Compos. Sci. Technol.* **2003,** *63,* 1231–1240.
25. Shinoj, S.; Visvanathan, R. S.; Panigrahi, S.; Varadharaju, N. Dynamic Mechanical Properties of Oil Palm Fiber (OPF)-Linear Low Density Polyethylene (LLDPE) Biocomposites and Study of Fiber-Matrix Interactions. *Biosys. Eng.* **2011,** *109,* 99–107.
26. Gupta, M. K.; Srivastava, R. K. Tribological and Dynamic Mechanical Analysis of Epoxy Based Hybrid Sisal/Jute Composite. *Indian. J. Eng. Mater. Sci.* **2016,** *23,* 37–44.

CHAPTER 8

MECHANICAL AND THERMAL ANALYSIS OF UNIDIRECTIONAL JUTE-FIBER-REINFORCED POLYMER COMPOSITE

M. K. GUPTA* and R. K. SRIVASTAVA

Department of Mechanical Engineering, Motilal Nehru National Institute of Technology Allahabad 211004, India

*Corresponding author. E-mail: mkgupta@mnnit.ac.in,
mnnit.manoj@gmail.com*

CONTENTS

ABSTRACT

In recent past decades, natural fibers have been considered as suitable alternative reinforcement materials opposite to glass and other synthetic fibers for polymer matrix due to their benefits such as biodegradability, eco-friendliness, recyclability, and acceptable mechanical properties. The aim of the present chapter is to investigate the mechanical and thermal properties of unidirectional jute-fiber-reinforced epoxy composite. The composites were prepared by hand lay-up method using varying weight percentage (10, 20, 30, and 40) of jute fibers into the epoxy matrix. The mechanical properties were studied in terms of tensile, flexural, and impact tests whereas thermal analysis was carried out in terms of storage modulus (E'), loss modulus (E''), and damping parameter (Tan δ) using dynamic mechanical analysis (DMA). In addition, other thermal properties such as crystallization temperature (T_c), glass transition temperature (T_g), and decomposition temperature (T_m) were also studied using differential scanning calorimetry (DSC). The results showed that mechanical and thermal properties were found to increase due to increase in fiber content up to 30 wt.% then decrease due to poor interfacial adhesion between fibers and matrix. Moreover, morphological analysis of prepared jute composites was studied using scanning electron microscope (SEM). The present prepared composite can be used in packaging, lightweight automotive parts and construction applications.

8.1   INTRODUCTION

Natural fiber-reinforced polymer composite (NFRPC) is a class of fiber-reinforced polymer composite (FRPC) including natural fibers such as jute, banana, sisal, hemp, kenaf, bamboo, cotton, flax, abaca, palm, and so forth as a reinforcement, and thermosets, thermoplastics and biopolymers as polymer matrix. Thermosets include epoxy, polyester, vinyl ester and phenolic resin, whereas thermoplastics include polypropylene, polyethylene, polyetheretherketone, and so forth. In addition, biopolymers include poly(lactic acid) (PLA), polyhydroxybutyrate, and so forth. The enormous use of synthetic fibers in thousands of applications has been a burden for researchers and scientists because these fibers are neither biodegradable nor eco-friendly. Therefore, researchers and scientists have been motivated toward natural fibers which have advantages over synthetic fibers such as biodegradability and eco-friendliness. In addition, these fibers have some other advantages such as low density, low cost, recyclability, easy processing, and acceptable mechanical properties.[1-5] On the other hand, these fibers exhibit some limitations also, such as higher moisture absorption, low impact strength, poor thermal stability and poor adhesion with polymers.[6-8] These NFRPCs are being used in many applications such as packaging, building and constructions, electrical, electronics, sports, and so forth.[9-12]

Study about mechanical properties of NFRPCs is a basic investigation which is being carried out by many researchers. Hossain et al.[13] evaluated the effect of fiber orientation on tensile properties of jute fiber-reinforced epoxy laminated composite. The composite was prepared by vacuum assisted resin infiltration technique with varying stacking sequences, that is, (0/0/0/0), (0/+45°/−45°/0) and (0/90°/90°/0) keeping constant 25 vol.% of fibers. They concluded that composite with stacking sequences 0/0/0/0 and 0/+45°/−45°/0 showed higher values of tensile strength in longitudinal direction as compared to transverse direction, whereas values of tensile strength were very close to each other in both directions for 0/90°/90°/0 composites. Ray et al.[14] investigated the mechanical properties of jute-fiber-reinforced vinyl ester resin composites. The composites were prepared by pultrusion method with varying vol.% of fibers such as 8, 15, 23, 30 and 35. They reported that the mechanical properties improved with increasing the jute fiber content. It was also reported that the composite with 35 vol.% fiber content had the optimum mechanical properties as compared to all other composites. Kaewkuk et al.[15] studied the mechanical properties of sisal fiber-reinforced polypropylene composite. It was reported that the tensile strength and modulus were found to increase with increase in fiber content. However, impact strength and elongation at break decreased with increase in fiber content.

Thermal analysis is one important study about FRPCs which is carried out by many researchers. Pothan et al.[16] investigated the dynamic mechanical analysis of banana fiber-reinforced polyester composites and highlighted that the composite with 40 vol.% fiber loading showed the maximum value of storage modulus whereas lower value of loss modulus and damping parameters. Nair et al.[17] presented studies on thermal and dynamic mechanical analysis of short sisal fiber-reinforced polystyrene composites. They observed that the sisal fiber-reinforced composite showed higher thermal stability as compared to neat polystyrene and sisal fibers. The storage modulus decreased upon increasing the temperature and the glass transition temperature of composite shifted toward lower temperature compared to neat polystyrene. Mohanthy et al.[18] investigated an experimental study on the viscoelastic behavior of jute fiber-reinforced high-density polyethylene composites and observed that the storage modulus was found to increase with increase in fiber loading up to 30% whereas damping parameters was decreased as compared to epoxy.

In this chapter, unidirectional jute-fiber-reinforced epoxy composites are prepared using hand lay-up technique subjected to mechanical tests (tensile, flexural, and impact) and thermal tests using dynamic mechanical analysis (DMA) and differential scanning calorimetry (DSC). The mechanical properties of present jute composite were found better in performance when compared with earlier published work.

8.2　MATERIALS

Jute fibers were used as reinforcement and epoxy AY 105 with hardener HY 951 is the matrix. Jute fiber was purchased from Women's Development Organization, Dehradun, India whereas epoxy resin, with hardener, was purchased from Bakshi Brothers/Universal Enterprises, Kanpur, Uttar Pradesh, India.

8.2.1　MATRIX

Epoxy resin is prepared by reacting epichlorohydrin with bisphenol A. The standard name of diglycidyl ether of bisphenol A is epoxy resin. Chemical formula of epoxy resin is given below.

To convert epoxy resin to epoxy plastic, a hardener is required. The most widely used curing agents are primary and secondary amines. Chemical formula of hardener (diethylene triamine) is given below.

$$H_2N\!\!-\!\!\!-\!\!H_2C\!\!-\!\!\!-\!\!H_2C\!\!-\!\!\!-\!\!NH\!\!-\!\!\!-\!\!H_2C\!\!-\!\!\!-\!\!H_2C\!\!-\!\!\!-\!\!NH_2$$

In present study, epoxy resin of grade AY105 having density 1.11 g/cm³ and dynamic viscosity (at 25°C) 11.79 Pa.s was used. Epoxy was cured with hardener HY951 in a ratio of 10:1 as recommended by the suppliers.

8.2.2 REINFORCEMENT

Jute is a bast fiber whose scientific name is *Corchorus capsularis* of Tiliaceae family. Plant of jute takes about 3 months to grow up to a height of 12–15 ft. Jute plant is cut and kept immersed in the water for retting process, during season. The inner and outer stem gets separated and the outer plant gets individualized to form fibers. India and Bangladesh are the top producers of jute fiber. According to facts collected from Food and Agriculture Organization of the United Nations (FAOSTAT)-2012, the total production of jute (tonnes) in the world was 3,461,964. The physical and mechanical properties and chemical composition of jute fiber are given in Table 8.1.

TABLE 8.1   Properties of Jute Fiber.

Density (g/m³)	Diameter (µm)	Elongation at break (%)	Tensile strength (MPa)	Young's modulus (GPa)	Cellulose (%)	Lignin (%)	Microfibrillar angle	Wax (%)	Hemi-cellulose (%)
1.3	25–200	1.5–1.8	393–773	26.5	61–71	12–13	8	0.5	14–20

8.3 FABRICATION OF COMPOSITES

The composites were fabricated by reinforcing unidirectional jute fibers into epoxy matrix using hand lay-up technique followed by static compression. A stainless steel mold having dimensions of $300 \times 200 \times 3$ mm³ was used to maintain the 3 mm thickness of laminates. A releasing agent was used to facilitate easy removal of the composite's laminates from the mold after curing. The cast of each composite was cured under a load of 50 kg for 24 h before it was removed from the mold. Dimension of specimens were cut as per ASTM standard using a diamond cutter. The composites manufactured with varying wt.% of fibers are designated as J10 (10 wt.% of jute fiber), J20 (20 wt.% of jute fiber), J30 (30 wt.% of jute fiber) and J40 (40 wt.% of jute fiber).

8.4 EXPERIMENTAL PROCEDURE

The fabricated composites were tested for mechanical and thermal properties. The experimental procedure of each test is given in subsequent text.

8.4.1 MECHANICAL CHARACTERIZATION

8.4.1.1 TENSILE TEST

Tinius Olsen H10 K-L (Biaxial testing machine) was used to obtain tensile properties of the composite samples with a crosshead speed of 2 mm/min operated at room temperature. Tests were conducted as per ASTM D638 with dimension of $165 \times 20 \times 3$ mm. Five specimens of each composite were tested and their average values and standard deviations are reported.

8.4.1.2 FLEXURAL TEST

Flexural properties of the composite were measured using a three point bending test on Tinius Olsen H10K-L (Bi-axial testing machine). The dimension of sample for the flexural test was $80 \times 12.7 \times 3$ mm as per ASTM D790. The flexural test was carried out at room temperature with the crosshead speed of 2 mm/min. Flexural strength and flexural modulus were both calculated using following equation:

$$\text{Flexural strength} = \frac{3FL}{2bd^2} \text{ and flexural modulus} = \frac{mL^3}{4bd^3} \qquad (8.1)$$

Where F is ultimate failure load (N), L is span length (mm), b and d are width and thickness of the specimen in (mm) respectively, and m is slope of tangent to the initial line portion of the load–displacement curve. Five specimens of each composite were tested and their average values and standard deviations are reported.

8.4.1.3 IMPACT TEST

Tinius Olsen Impact 104 machine was used to obtain impact properties of jute fiber composite samples. The sample was prepared for the impact test, with dimension of $65 \times 12.7 \times 3$ mm and 2.5 mm notch thickness as per ASTM D256. Five specimens of each composite were tested and their average values and standard deviations are reported.

8.4.2   STATISTICAL ANALYSIS

Statistical analysis is required to know if the test is significant or not. T-test and analysis of variance (ANOVA) are used to find out the statistical analysis of tensile, flexural, and impact test. Probability value $p = 0.05$ is considered as an analytical of significance compared to the control composite (J10).

8.4.3   DYNAMIC MECHANICAL ANALYSIS (DMA)

The dynamic mechanical properties in terms of storage modulus, loss modulus, and damping of jute-fiber-reinforced epoxy composite were studied using the dynamic mechanical analyzer (Seiko instruments DMA 6100). The dynamic mechanical properties were determined with a 3-point bending test as a function of temperature. The composites were cut into samples having dimensions of $50 \times 13 \times 3$ mm as per ASTM D 5023. Experiments were carried out in the temperature range of 25–200°C at 1 Hz frequency.

8.4.4   DIFFERENTIAL SCANNING CALORIMETRY (DSC)

The thermal properties, such as glass transition temperature, crystallization temperature, and decomposition temperature of jute composites, were studied by using DSC (Perkin Elmer model DSC 4000). The experiments were carried out in the temperature range of 30–400°C at the heat flow rate of 10°C/min under nitrogen atmosphere purged at 20 ml/min, and by using aluminum pan. Nitrogen was used for efficient heat transfer and removal of volatiles from the samples.

8.4.5   SCANNING ELECTRON MICROSCOPY (SEM)

The fracture surface of the composite samples was studied with a scanning electron microscope (Carl Zeiss EVO MA 15). The samples were mounted on an aluminum stub using double-sided tape and all specimens were coated with a very thin layer of gold to prevent electric charging during examinations. The SEM micrographs were obtained under conventional SEM conditions.

8.5 RESULTS AND DISCUSSIONS

8.5.1 MECHANICAL TESTS

8.5.1.1 TENSILE TEST

Tensile strength and tensile modulus of epoxy and jute composites are given in Table 8.2. Tensile properties of jute composites are found to increase on increasing jute fiber content up to 30 wt.% in epoxy matrix, then decrease due to improper stress transfer between fibers and matrix. The composite J30 has the maximum value for tensile strength and tensile modulus due to strong interface between fibers and epoxy matrix which is the cause of effective stress transfer. Tensile strength and tensile modulus are found to be maximum for composite J30, such as 87.73 MPa and 1.74 GPa respectively, which are found to be more than 112 and 110% respectively, as compared to epoxy. It is observed that tensile strength of composite J30 is 38, 20, and 13% more than composites J10, J20, and J40 respectively, whereas tensile modulus is 33, 14, and 12% more than composites J10, J20 and J40 respectively. According to statistical analysis, the value of tensile strength and tensile modulus is found to be significant as compared to the J10 composite. The results of ANOVA also show the significant difference between composites (Table 8.2). The tensile properties of present jute composite are found better in performance as compared to the earlier published work as shown in Table 8.3.

TABLE 8.2   Tensile, Flexural, and Impact Properties of Epoxy and Jute Composites.

Composite	Tensile strength (MPa)	Tensile modulus (GPa)	Flexural strength (MPa)	Flexural modulus (GPa)	Impact strength (kJ/m²)
Epoxy	41.32±2.95	0.83±0.04	105.47±6.39	4.88±0.12	6.83±0.25
J10	63.43±2.98	1.31±0.04	115.60±5.20	5.87±0.22	8.45±0.60
J20	73.32±2.69	1.53±0.25	132.35±7.34	7.17±0.43	9.32±0.90
J30	87.73±3.89	1.74±0.11	158.52±8.16	10.03±0.45	10.77±0.47
J40	77.45±5.79	1.55±0.15	146.84±9.52	8.96±0.36	8.89±0.58

The data represents the mean ± standard deviation.

TABLE 8.3   Comparison of Mechanical Properties of Present Composites with Published Work.

Reinforcement	Matrix	Tensile strength (MPa)	Tensile modulus (GPa)	Flexural strength (MPa)	Flexural modulus (GPa)	Impact strength (J/m)	Ref.
Pineapple leaf fiber	Polyester	52.90	2.29	80.20	2.76	80.30	[20]
Kenaf	Polypropylene	26.90	2.70	43.10	2.30	43.80	[21]
Borassus fruit fiber	Polypropylene	29.29	2.58	45.34	1.46	28.61	[22]
Coir	Polypropylene	26.14	2.21	42.67	1.83	21.99	[22]
Jute	Polypropylene	29.40	2.49	48.76	1.70	22.53	[22]
Sisal	Polypropylene	28.05	2.64	48.77	1.43	25.98	[22]
Banana	Epoxy	16.12	0.64	57.33	8.92	39.75	[23]
Sisal	Epoxy	21.20	0.72	62.04	9.34	67.62	[23]
Jute	Epoxy	87.73	1.74	158.52	10.03	32.31	Present work

8.5.1.2   FLEXURAL TEST

Flexural strength and flexural modulus of epoxy and jute composites are given in Table 8.2. Flexural properties of jute composite are also found to be increased on increasing jute fiber content in epoxy matrix. The composite J30 offers the maximum value of flexural strength and flexural modulus due to strong fiber–matrix adhesion and minimum possibility of voids in the composite. Flexural strength and flexural modulus are found as maximum for jute composite J30 such as 158.52 MPa and 10.03 GPa respectively which is 50 and 106% higher than that of epoxy. It is seen that flexural strength of composite J30 is more than 37, 20 and 8% as compared to composites J10, J20 and J40 respectively whereas flexural modulus is 71, 40 and 12% higher than composites J10, J20 and J40 respectively. According to statistical analysis, the value of flexural strength and flexural modulus is found to be significant as compared to the J10 composite. The results of ANOVA also show the significant difference between composites (Table 8.2). The flexural properties of present jute composite are found better in performance as compared to earlier published work as shown in Table 8.3.

8.5.1.3   IMPACT TEST

Variations of impact strength on increasing jute fiber content are given in Table 8.2. Similar trend, as seen in tensile and flexural tests, is observed for impact test of jute-fiber-reinforced epoxy composite. Impact strength is found to be increased with increase in jute fiber content up to 30 wt.% in epoxy matrix then, it decreases. The maximum value of impact strength is found to be 10.77 kJ/m^2 for jute composite J30, which is 58% more than that of epoxy. The composite J30 shows the maximum impact strength due to strong adhesion between jute fibers and epoxy matrix. The impact strength of J30 is found 27, 16 and 21% higher than those of composites J10, J20, and J40 respectively. According to statistical analysis, the value of impact strength is found to be significant as compared to jute composite J10. The results of ANOVA also show the significant difference between jute composites (Table 8.2). The impact strength of present jute composite is found to be comparable with earlier published work as shown in Table 8.3.

8.5.2   MORPHOLOGICAL ANALYSIS

Morphological analysis of prepared jute composites was carried out using SEM. Figure 8.1 shows the SEM image of unfractured jute composite J30 whereas fracture behavior after tensile test can be seen in Figure 8.2. SEM images clearly indicate the fiber–matrix adhesion, nonuniform distribution of fibers in matrix, and fracture of fibers after tensile test. In addition, fracture of fiber after flexural test can be observed in Figures 8.3 and 8.4, which shows the SEM image of composite J30 subjected to impact test. This image shows the fracture of fibers and cracking of matrix after impact test along with uniform distribution of fibers in matrix.

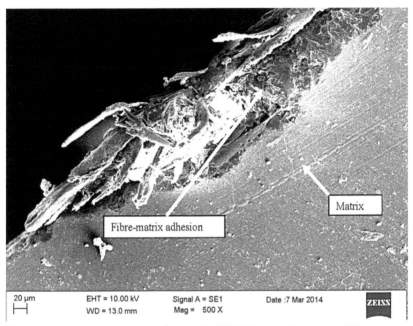

FIGURE 8.1   Scanning electron microscopic (SEM) image of composite J30.

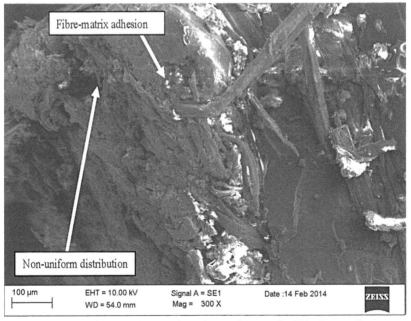

FIGURE 8.2   SEM image of fractured specimen of composite J30 after tensile test.

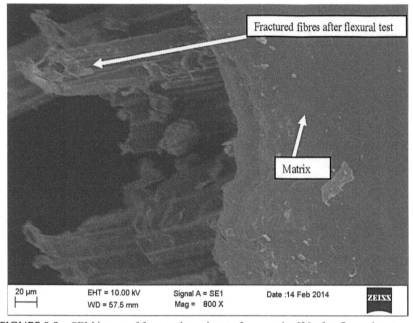

FIGURE 8.3   SEM image of fractured specimen of composite J30 after flexural test.

FIGURE 8.4   SEM image of fractured specimen of composite J30 after impact test.

8.5.3   DYNAMIC MECHANICAL ANALYSIS

8.5.3.1   STORAGE MODULUS

Storage modulus is a real part of complex modulus which is defined as amount of the maximum energy stored by material during one cycle of oscillation. It also gives an estimate of temperature-dependent stiffness behavior and load-bearing capability of the composite material.[19] Figure 8.5 shows the variation of the storage modulus of epoxy and jute composites as a function of temperature at 1 Hz frequency. On comparing the different composites it is found that the value of E' increases with increase in the weight fraction of jute fibers in the composites up to 30 wt.%. The value of E' is found to be 3.86 GPa for epoxy in the glassy region, but this value increases up to 5.01 GPa for the jute composite J30. It is due to reinforcement of high strength jute fibers in epoxy matrix. In all cases, storage modulus of epoxy and jute composites is found to decrease as temperature increases due to loss in stiffness of fibers.[24] In transition region, it can be seen that the jute composites have a gradual fall in the value of E' when temperature is increased as compared to the neat epoxy which has a very steep fall as shown in Figure 8.5. In the rubbery region, storage modulus of epoxy is much lower than jute composites. In this region, the value of E' for epoxy is 0.05 GPa whereas with incorporation of jute fiber this value increased to 0.87 GPa for composite J30. This is attributed to the reinforcement of jute fibers which allowed uniform stress transfer from matrix to fibers.[25]

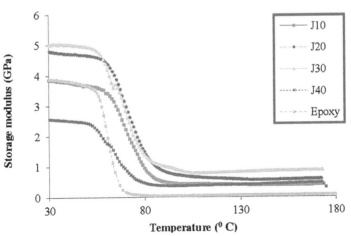

FIGURE 8.5   Variation of storage modulus with temperature of epoxy and jute composites.

8.5.3.2   LOSS MODULUS

Loss modulus is an imaginary part of complex modulus which is defined as amount of energy dissipated in the form of heat by material during one cycle of oscillation. It represents the viscous response of the materials. The peak of loss modulus curve of polymer material is known as dynamic glass transition temperature.[26] The variation of the loss modulus as a function of temperature is shown in the Figure 8.6. It can be observed that the value of E'' increased and then decreased with increase in temperature. The value of E'' is found 0.46 GPa for epoxy but this value is found to increase up to 0.83 GPa for the composite J30, due to incorporation of high strength jute fibers. The T_g of composite J10 has shifted to higher temperature which shows its better thermal stability. This observation may be due to decrease in mobility of matrix due to incorporation of jute fibers. In addition, jute composite J40 shows lower thermal stability due to poor interface between fibers and matrix. The value of glass transition temperature for epoxy and jute composites obtained from loss modulus curve is given in Table 8.4.

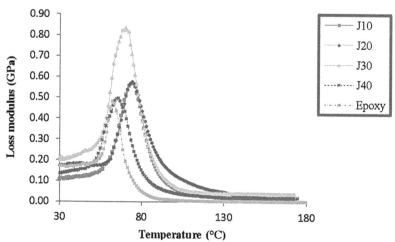

FIGURE 8.6   Variation of loss modulus with temperature of epoxy and jute composites.

8.5.3.3   DAMPING (Tan δ)

Damping property of the polymeric materials is the ratio of loss modulus and storage modulus. It is related to degree of molecular mobility in polymer material.[25] The effect of damping parameter on epoxy and jute composites as a function of temperature is shown in the Figure 8.7. The peak of Tan δ curve occurs in glass transition region, where composite changes from rigid to more elastic state due to movement of molecules in polymer structure. The maximum value of Tan δ is found (0.617) for epoxy as expected. In addition, lower value of Tan δ is found for jute composite J20 followed by J30. The lower value of Tan δ shows good load-bearing capacity due to strong adhesion between the fibers and matrix. Furthermore, the values of T_g for epoxy and jute composites obtained from peak of Tan δ curve are given in Table 8.4. The value of T_g obtained from Tan δ curve is found higher than that obtained from loss modulus curve.

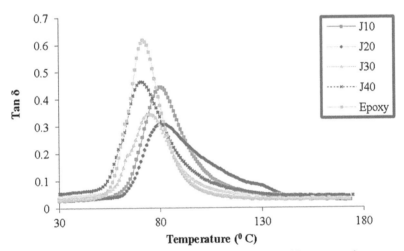

FIGURE 8.7   Variation of Tan δ with temperature of epoxy and jute composites.

TABLE 8.4   Glass Transition Temperature (°C) from Loss Modulus and Tan δ Curve.

Composite	Loss modulus curve	Tan δ curve
J10	74.12	80.75
J20	73.49	81.16
J30	70.15	75.24
J40	64.79	71.85
Epoxy	60.95	71.77

8.5.4 DIFFERENTIAL SCANNING CALORIMETRY (DSC)

Thermal properties of epoxy and jute composites such as glass transition temperature, crystallization temperature, and decomposition temperature are obtained from DSC studies as shown in Figure 8.8. The results suggested that glass transition temperature lies between 64–69°C, crystallization temperature lies between 111–116°C, and decomposition temperature lies between 335–345°C. There is no significant change observed in values of crystallization and glass transition temperatures of jute composites when compared to epoxy. This observation can be explained, as at lower temperature fibers do not much affect the thermal properties.

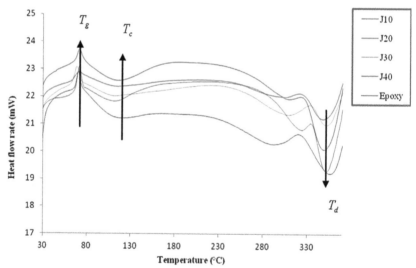

FIGURE 8.8   DSC analysis of epoxy and jute composites.

8.6 CONCLUSION

Unidirectional jute-fiber-reinforced epoxy composites are successfully developed. Their mechanical and thermal properties are studied and following conclusions are drawn:

- Mechanical properties (tensile, flexural, and impact) are observed to increase on increasing the jute fiber content up to 30 wt.% then the properties decrease.

- Statistical analysis is carried out using T-test and ANNOVA, and the significant difference among composites is observed.
- Jute composite J30 has higher values of storage modulus and loss modulus, whereas higher value of glass transition temperature is shown by jute composite J10.
- A significant change is not observed in values of crystallization and glass transition temperatures of jute composites when it was compared to epoxy resin.

8.7 ACKNOWLEDGMENT

The authors would like to thank the Head of Mechanical Engineering Department of Motilal Nehru National Institute of Technology, Allahabad, India, for his support in allowing us to perform the tests in Materials Characterization Lab. The study is partially supported by Cumulative Professional Development Allowances (CPDA) and R and C fund for teachers of Motilal Nehru National Institute of Technology, Allahabad.

KEYWORDS

- dynamic mechanical analysis
- jute fiber
- mechanical properties
- polymer matrix
- thermal properties

REFERENCES

1. Gupta, M. K.; Srivastava, R. K. Tensile and Flexural Properties of Sisal Fiber Reinforced Epoxy Composite: A Comparison between Unidirectional and Mat Form of Fibers. *Procedia Mater. Sci.* **2014,** *5*, 2434–2439.
2. Mourya, H.; Gupta, M. K.; Srivastava, R. K.; Singh, H. Study on The Mechanical Properties of Epoxy Composite Using Short Sisal Fiber. *Mater. Today Proc.* **2015,** *2*, 1347–1355.

3. Bisaria, H.; Gupta, M. K.; Sandilya, P.; Srivastava, R. K. Effect of Fiber Length on Mechanical Properties of Randomly Oriented Short Jute Fiber Reinforced Epoxy Composite. *Mater. Today Proc.* **2015**, *2*, 1193–1199.

4. Gupta, M. K.; Srivastava, R. K. Investigations of Structural Materials Properties Using Uni-Axial Tensile Test. *Int. J. Appl. Eng. Res.* **2011**, *6*, 1295–1302.

5. Idicula, M.; Boudenne, A.; Umadevi, L.; Ibos, L.; Candau, Y.; Thomas, S. Thermo Physical Properties of Natural Fiber Reinforced Polyester Composites. *Compos. Sci. Technol.* **2006**, *66*, 2719–2725.

6. Gupta, M. K.; Srivastava, R. K. Mechanical and Water Absorption Properties of Hybrid Sisal/Glass Fiber Reinforced Epoxy Composite. *Am. J. Polym. Sci. Eng.* **2015**, *3*, 208–219.

7. Gupta, M. K.; Srivastava, R. K. Potential of Jute Fiber Reinforced Polymer Composites: A Review. *Int. J. Fiber Text. Res.* **2015**, *5*, 30–38.

8. Gupta, M. K.; Srivastava, R. K. Mechanical Properties of Hybrid Fibers Reinforced Polymer Composite: A Review. *Polym. Plast. Technol. Eng.* **2016**, *55*, 626–642.

9. Venkateshwaran, N.; ElayaPerumal, A. Mechanical and Water Absorption Properties of Woven Jute/Banana Hybrid Composites. *Fibers Polym.* **2010**, *13*, 907–914.

10. Bisaria, H.; Gupta, M. K.; Sandilya, P.; Srivastava, R. K. Effect of Fiber Length on Mechanical Properties of Randomly Oriented Short Jute Fiber Reinforced Epoxy Composite. *Mater. Today: Proc.* **2015**, *2*, 1193–1199.

11. Gupta, M. K.; Srivastava, R. K. A Review on Characterization of Hybrid Fiber Reinforced Polymer Composite. *Am. J. Polym. Sci. Eng.* **2016**, *4*, 1–7.

12. Gowda, T. M.; Naidu, A. C. B.; Rajput, C. Some Mechanical Properties of Untreated Jute Fabric-Reinforced Polyester Composites. *Compos. Part A.* **1999**, *30*, 277–284.

13. Hossain, M. R.; Islam, M. A.; Vuurea, A. V.; Verpoest, I. Effect of Fiber Orientation on The Tensile Properties of Jute Epoxy Laminated Composite. *J. Sci. Res.* **2013**, *5*(1), 43.

14. Ray, D.; Sarkar, B. K.; Rana, A. K.; Bose, N. R. The Mechanical Properties of Vinylester Resin Matrix Composites Reinforced with Alkali-Treated Jute Fibers. *Compos. Part A.* **2001**, *32*, 119–127.

15. Kaewkuk, S.; Sutapun, W.; Jarukumjorn, K. Effects of Interfacial Modification and Fiber Content on Physical Properties of Sisal Fiber/Polypropylene Composites. *Compos. Part B.* **2013**, *45*, 544–549.

16. Pothan, L. A. Oommen, Z.; Thomas, S. Dynamic Mechanical Analysis of Banana Fiber Reinforced Polyester Composites. *Compos. Sci. Technol.* **2003**, *63*, 283–293.

17. Nair, K. C. M.; Thomas, S.; Groeninckx, G. Thermal and Dynamic Mechanical Analysis of Polystyrene Composites Reinforced with Short Sisal Fibers. *Compos. Sci. Technol.* **2001**, *61*, 2519–2529.

18. Mohanty, S.; Verma, S. K.; Nayak, S. K. Dynamic Mechanical and Thermal Properties of MAPE Treated Jute/HDPE Composites. *Compos. Sci. Technol.* **2006**, *66*, 538–547.

19. Gupta, M. K. Dynamic Mechanical and Thermal Analysis Hybrid Jute/Sisal Fiber Reinforced Epoxy Composite. *J. Mater: Des. Appl. Part-L.* DOI: 10.1177/1464420716646398.

20. Devi, L. U.; Bhagnani, S. S.; Thomas, S. Mechanical Properties of Pineapple Leaf Fiber-Reinforced Polyester Composites. *J. Appl. Polym. Sci.* **1998,** *64*, 1739–1748.
21. Karnani, R.; Krishnan, M.; Narayan, R. Biofiber-Reinforced Polypropylene Composite. *Polym. Eng. Sci.* **1997,** *37*, 476–483.
22. Sudhakaraa, P.; Jagadeesh, D.; Wang, Y.; Prasad, C. V.; Devi, A. P. K.; Balakrishnana, G.; Kim, B. S.; Song, J. I. Fabrication of Borassus Fruit Lignocellulose Fiber/PP Composites and Compression with Jute, Sisal and Coir Fibers. *Carbohydr. Polym.* **2013,** *98*, 1002–1010.
23. Venkateshwaran, N.; ElayaPerumal, A.; Alavudeen, A.; Thiruchitrambalam, M. Mechanical and Water Absorption Behaviour of Banana/Sisal Reinforced Hybrid Composites. *Mater. Des.* **2011,** *32*, 4017–4021.
24. Jawaid, M.; Khalil, H. P. S. A.; Hassan, A.; Dungani, R.; Hadiyani, A. Effect of Jute Fiber Loading on Tensile and Dynamic Mechanical Properties of Oil Palm Epoxy Composites. *Composites Part B.* **2013,** *45*, 619–624.
25. Gupta, M. K.; Srivastava, R. K. Tribological and Dynamic Mechanical Analysis of Epoxy Based Hybrid Sisal/Jute Composite. *Indian J. Eng. Mater. Sci.* **2016,** *23*, 37–44.
26. Gupta, M. K.; Srivastava, R. K. Properties of Sisal Fiber Reinforced Epoxy Composite. *Indian J. Fiber Text. Res.* **2016,** *41*, 235–241.

CHAPTER 9

A COMPARATIVE STUDY OF METACHROMASY INDUCED IN ACRIDINE ORANGE BY ANIONIC POLYELECTROLYTES

NANDINI R.[1*] and VISHALAKSHI B.[2]

[1]*Department of Chemistry, MITE, Moodabidri (DK), Karnataka 574226, India*

[2]*Department of Chemistry, Mangalore University, Mangalagangotri, Mangalore, Karnataka 574199, India*

Corresponding author. E-mail: hodche@mite.ac.in

CONTENTS

ABSTRACT

The interaction of cationic dye, acridine orange (AO) with two anionic polyelectrolytes, namely, sodium heparinate (NaHep) and sodium alginate (NaAlg) has been investigated by spectrophotometric method. The polymer induced metachromasy in the dyes resulting in the shift of the absorption maxima of the dyes towards shorter wavelengths. The stability of the complexes formed between AO and NaAlg was found to be greater than that formed between AO and sodium heparinate. This fact was further confirmed by reversal studies using alcohols, urea surfactants, and electrolytes. The interaction parameters revealed that binding between AO and anionic polyelectrolytes was mainly due to electrostatic interaction. A comparative account of stoichiometry, stability of complexes formed, and thermodynamic parameters of interaction have been reported. The nature of the interaction between the polymer and the dyes has been discussed in detail. Moreover, the fluorescence spectra of the dye with the polymers have also been reported and the values of Stern–Volmer constant have been evaluated.

9.1   INTRODUCTION

Metachromasy is a well-known phenomenon in the case of dye-polymer interactions and is generally found in the case of aggregation of cationic dyes on anionic polymers.[1,5] Conformation of the polyanions controls the induction of metachromasy of aqueous dye solution, although there are several reports on metachromasy of various classes of acidic polysaccharides and different synthetic polyanions.[6,7] Several physicochemical properties of macromolecule-surfactants are quite relevant in this context. Formulation procedures, based on suitable mixture have appealing applications.[8–10] It has been noted that oppositely charged surfactant binds to polymer surfaces through both electrostatic and hydrophobic interaction.[11] Different techniques for the isolation and stability determination of metachromatic complexes have been reported in the literature.[12] The phenomena of reversal of metachromasy by addition of urea, alcohol, neutral electrolytes,[13] and also by increasing the temperature of the system may be used to determine the stability of the metachromatic compounds. The interaction between polyelectrolytes and oppositely charged surfactants[14] has been investigated due to its importance in both fundamental and applied fields.[15] Comprehensive studies of a variety of cationic surfactants with anionic polyelectrolytes have been reported.[16–20] The interaction of various bacterial polysaccharides such as *Klebsiella* K-17, K-10, K-20, K-14, and K-16 with cationic dyes such as acridine orange (AO), pinacyanol chloride, phenosafranin, etc. has been studied by Chakraborty et al. and the thermodynamic parameters of interaction have been reported.[21,22]

The studies on synthetic polyanions inducing metachromasy in different cationic dyes are available in the literature.[23] Measurement of fluorescence intensity of the dye molecules in presence of polymers has also been used as a tool to understand the polymer-dye interactions.[24]

9.2 MATERIALS METHODS

9.2.1 APPARATUS

The spectral measurements were carried out using Shimadzu UV-2500 Spectrophotometer using a 1 cm quartz cuvette.

9.2.2 REAGENTS

Acridine orange (AO) was obtained from HiMedia, Germany, and used as received. NaAlg and NaHep (Loba Chemie, India) were used without further purification; Methanol, (MeOH) ethanol (EtOH) and 2-Propanol (PrOH) (Merck, India) were distilled before use. Urea, sodium chloride, and potassium chloride were obtained from (Merck, India). Sodium lauryl sulfate and sodium dodecylbenzene sulfonate (Lobachemie, India) were obtained from (Loba Chemie, India).

9.2.3 METHODS

To a fixed volume of polymer solution, varying amounts of polymer solution were added and absorbances were measured using Shimadzu UV-2500 spectrophotometer.

9.3 RESULTS AND DISCUSSION

The orange indicating the presence of a monomeric dye species in the concentration range was studied. On adding increasing amounts of polymer solution the absorption maxima shifts to 457 nm in case of AO-NaHep and at 451 nm in case of AO-NaAlg complex. The blue-shifted band is attributed to the stacking of the dye molecules on the polymer backbone and this reflects high degree of cooperation in binding.[25] The appearance of multiple banded spectra proposed that the polymer might have a random coil structure in solution. Whereas at higher concentration of the polymer, almost a single banded spectrum was observed due to a possible change from random coil to helical form.[26] The absorption spectra of polymer-dye complexes at various P/D ratios are shown in Figures 9.1 and 9.2, respectively.

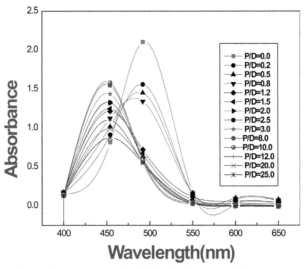

FIGURE 9.1   Absorption spectrum of AO-NaAlg at various P/D ratios.

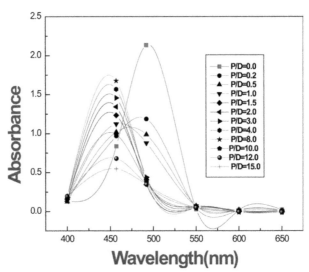

FIGURE 9.2   Absorption spectrum of AO-NaHep at various P/D ratios.

9.3.1 DETERMINATION OF STOICHIOMETRY

To determine the stoichiometry of the polymer-dye complex, a plot of A_{Mono}/A_{Meta} versus the polymer/dye ratio was made for AO-NaHep system. A similar procedure was repeated with AO-NaHep complex, too. The stoichiometry of AO-NaAlg complex was found 2:1 which indicates that the binding is at alternate anionic sites. This indicates that every potential anionic site of the polyanion was associated with the dye cation, and aggregation of such dye molecules was expected to lead to the formation of a card pack stacking of the individual monomers on the surface of the polyanion so that the allowed transition produces a blue-shifted metachromasy.[27] The results were in good agreement with the reported values for the interaction of similar dyes with polyanions.[28] While in case of AO-NaHep complex the stoichiometry is 1:1, and the binding is at adjacent anionic sites. This indicates that there is lesser overcrowding and more aggregation of the bound dyes on the polymer chain in the latter case than in the former case. Similar results were reported in case of binding of pinacyanol chloride on poly (methacrylic acid) and poly (styrene sulfonate) systems.[29] The stoichiometry results are obtained by plotting A_{meta}/A_{Mono} versus P/D ratio in each case. The results are shown in Figure 9.3.

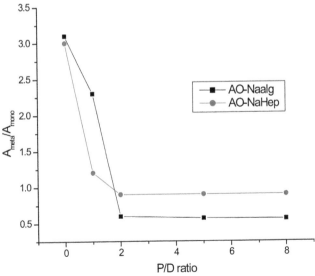

FIGURE 9.3   Stoichiometry of AO-NaHep and AO-NaAlg complex.

9.3.2   REVERSAL OF METACHROMASY USING ALCOHOLS AND UREA

The metachromatic effect is presumably due to the association of the dye molecules with the polyanion which may involve both electrostatic and hydrophobic interactions. The destruction of metachromatic effect may occur on addition of low molecular weight electrolytes, alcohols, or urea. The destruction of metachromasy by alcohol and urea attributed to the involvement of hydrophobic bonding has already been established.[30,31] Efficiency of alcohols in disrupting metachromasy was found to be in the order of methanol<ethanol<2-propane, indicating that reversal becomes quicker with increasing hydrophobic character of the alcohols. The above facts are further confirmed in the present system. On addition of increasing amount of alcohol to the polymer/dye system at P/D=2.0, the original monomeric band of dye species is gradually restored. The efficiency of the alcohols namely, methanol, ethanol, and 2-propanol on the destruction of metachromasy was studied. In case of AO-NaHepsystem, 40% methanol, 30% ethanol, 20% 2-propanol were sufficient to reverse metachromasy. 60% methanol, 50% ethanol, 40% 2-propanol were required to reverse metachromasy in AO-NaAlg system. From the plot of A_{451}/A_{492} or A_{457}/A_{492} (Figs. 9.4 and 9.5) against the percentage of alcohols or molar concentration of urea, the percentage of alcohols or molar concentration of urea needed for complete reversal has been determined. The concentration of urea to reverse metachromasy is found to be as high as 4 M in AB-NaHep system and 5 M in case of AB-NaAlg system (Fig. 9.6). This indicates that the metachromatic complex formed between AB-NaAlg is more stable than that between AB-NaHep complexes. Similar reports are available in literature for reversal of metachromasy in anionic polyelectrolyte/cationic systems by addition of alcohols or urea.[32,33]

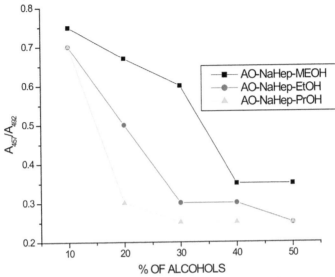

FIGURE 9.4 Reversal of metachromasy on the addition of alcohols.

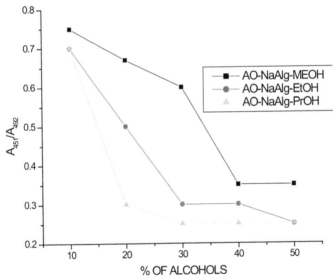

FIGURE 9.5 Reversal of metachromasy on the addition of alcohols.

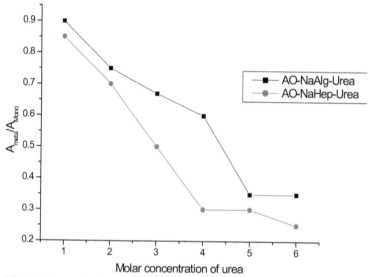

FIGURE 9.6   Reversal of metachromasy on the addition of urea.

9.3.3   REVERSAL OF METACHROMASY USING SURFACTANTS

The strength and nature of the interaction between water-soluble polyelectrolytes and oppositely charged surfactants depend on the characteristic features of both the polyelectrolytes and the surfactants. The charge density, flexibility of the polyelectrolyte, the hydrophobicity of the nonpolar part, and the bulkiness of the polar part also plays a vital role in the case of polysaccharide-surfactant interaction.[34–36] On adding increasing amounts of sodium lauryl sulfate and sodium dodecylbenzene sulfonate to AB-NaHep complex, the molar concentrations of sodium lauryl sulfate and sodium dodecylbenzene sulfonate needed to cause reversal was found to be 1×10^{-4} and 1×10^{-5} M in case of AO-NaHep, and 1×10^{-5} and 1×10^{-6} M in case of AO-NaAlg system. These results agree with those reported earlier in literature.[37–38] Thus, the addition of surfactants causes the production of micelles. Thus, the surfactant molecules interacted with the polymer by replacing the cationic dye. The release of dye molecules from the dye-polymer complex in the presence of cationic surfactants revealed that surfactants interacted electrostatically[39] with the anionic site of the polymer, and thus the dye became free. The ease of reversal of metachromasy can be correlated with its chain length. Thus, the binding between oppositely charged polymer surfactants is primarily by electrostatic forces which is reinforced by hydrophobic forces. From the plot of absorbance at A_{meta}/A_{mono} in case of AO against molar concentration of surfactants, the molar concentration of surfactants needed for a complete reversal of metachromasy has been determined. The results are shown in Figures 9.7 and 9.8.

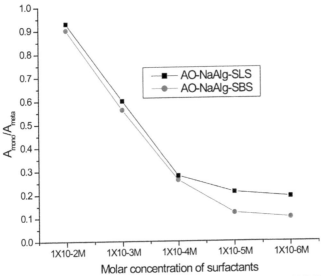

FIGURE 9.7   Reversal of metachromasy on the addition of surfactants (AO-NaAlg).

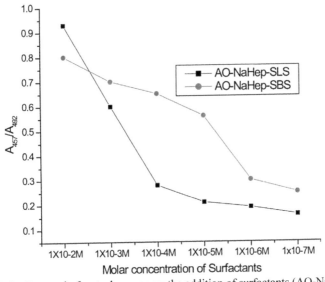

FIGURE 9.8   Reversal of metachromasy on the addition of surfactants (AO-NaHep).

The interaction constant K_C for the complex formation between AO and NaHep, and AO-NaAlg was determined by absorbance measurements at the metachromatic band at four different temperatures, taking different sets of solutions, containing varying amounts of polymers (C_S) in a fixed volume of the dye solution (C_D). The value of K_C was obtained from the slope and intercept of the plot of $C_D C_S/(A-A_0)$ against Cs as shown in Figure 9.9. In each case, the thermodynamic parameters of interaction, namely ΔH, ΔG and ΔS were also calculated. The results are given in Table 9.1, polymers (C_S) in a fixed volume of the dye solution (C_D). The value of K_C was obtained from the slope.[40]

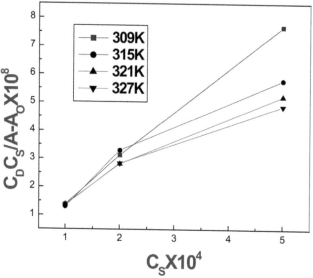

FIGURE 9.9 Effect of temperature on AO-NaAlg system.

TABLE 9.1 Thermodynamic Parameters of Interaction of AO-NaAlg and AO-NaHep systems.

System	Temp (K)	K_C ($dm^3 \cdot mol^{-1}$)	ΔG ($kJ \cdot mol^{-1}$)	ΔH ($kJ \cdot mol^{-1}$)	ΔS ($J \cdot mol^{-1} \cdot K^{-1}$)
AO-NaHep	309	5400	−22.1	−32.6	−18.5
	315	3200	−21.4		
	321	1600	−19.7		
	327	1062	−19.0		

TABLE 9.1 *(Continued)*

System	Temp (K)	K_C ($dm^3 \cdot mol^{-1}$)	ΔG ($kJ \cdot mol^{-1}$)	ΔH ($kJ \cdot mol^{-1}$)	ΔS ($J \cdot mol^{-1} \cdot K^{-1}$)
AO-NaAlg	309	3122	−21.6	−38.7	−25.0
	315	2144	−21.4		
	321	1977	−20.1		
	327	1322	−19.5		

9.3.3.1 FLUORESCENCE MEASUREMENTS

Fluorescent studies were performed with AO-polyanion systems and it was found that the fluorescent intensity of AO decreased on the addition of increasing amounts of the polymer solution as evidenced from AO-NaHep (Fig. 9.10) and AO-NaAlg (Fig. 9.11) systems, respectively. Finally, to study the interaction between the polymer and dye, the fluorescence data was fitted to Stern–Volmer equation 68, $F0/F = 1 + Ksv[Q]$, where Fo is the fluorescence intensity of the dye solution and F is that of the dye-polymer mixture, and [Q] is the molar concentration of the polymer, Ksv is the Stern–Volmer constant.[41-48] The Stern–Volmer constant for AO-NaAlg and AO-NaHep systems was reported to be 3.02×10^4 and 3.27×10^4 $dm^3 \cdot mol^{-1}$, respectively.

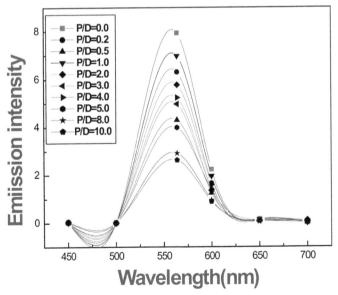

FIGURE 9.10 Fluorescence spectra of AO-NaHep system.

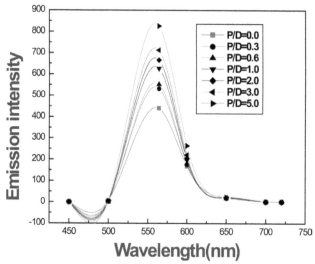

FIGURE 9.11   Fluorescence spectra of AO-NaAlg system.

9.4   CONCLUSION

The polymers sodium alginate and sodium carboxymethyl cellulose induced metachromasy in the dye AO. The monomeric band occurs at 492 nm while the metachromatic band occurs at 457 nm in the case of AO-NaHep, and at 451 nm in the case of AO-NaAlg. These results are further confirmed by the reversal studies using alcohols, urea, electrolytes, and surfactants. The thermodynamic parameters further confirm the above fact. It is, thus, evident from the above studies that both electrostatic and nonionic forces contribute towards the binding process. Based on the results, it can be concluded that AO is more effective in inducing metachromasy in sodium alginate than in sodium heparinate. Cooperativeness in binding is observed to occur due to neighbor interactions among the bound dye molecules at lower P/D ratios leading to stacking. The stacking tendency is enhanced by the easy availability and close proximity of the charged sites.

KEYWORDS

- cationic dye
- metachromasy
- anionicpolyelectrolytes
- electrostatic interaction
- aggregation

REFERENCES

1. Bakshi, M. S.; Sachar, S. *Colloid Polym. Sci.* **2005**, *283,* 671.
2. Bakshi, M. S.; Varga, I.; Gilanyi, T. *J. Phys. Chem. B.* **2005**, *109,* 13538.
3. Balomenou, L.; Bokias, G. *Langmuir* **2005**, *21,* 9038.
4. Basu, S.; Gupta, A. K.; Rohatgi-Mukherjee, K. K. *J. Indian Chem. Soc.* **1982,** *59,* 578.
5. Bergeron, J. A.; Singer, M. *J. Biophys. Biochem. Cytol.* **1958**, *4,* 433.
6. Berret, J. F.; Cristobal, G.; Hervel, P.; Oberdisse, J.; Grillo, I. *Eur.Phys. J.* **2002**, *E9,* 301.
7. Bruning, W.; Holtzer, A. *J. Am. Chem. Soc.* **1961**, *83,* 4865.
8. Chakraborty, A. K.; Nath, R. K. *Spectrochim. Acta.* **1989**, *45A,* 981.
9. Chatterjee, A.; Moulik, S. P.; Majhi, P. R.; Sanyal, S. K. *Biophys Chem.* **2002**, *98,* 313.
10. Dasgupta, S.; Nath, R. K.; Biswas, S.; Hossain, J.; Mitra, A.; Panda, A. K. *Colloids Surf. A.* **2007**, *302,* 17.
11. Frank, H. S.; Evans, M. W. *J. Chem. Phys.* **1945**, *17,* 507.
12. Frank, H. S.; Quist, A. S. *J. Chem. Phys.* **1961**, *34,* 604.
13. Fundin, J.; Hansson, P.; Brown, W.; Lidegran, I. *Macromolecules* **1997**, *30,* 1118.
14. Honda, C.; Kamizono, H.; Matsumoto, K.; Endo, K. *J. Colloid Interface Sci.* **2004**, *278,* 310.
15. Horbin, R. W. *Biochemie. Biochem.* **2002**, *77,* 3.
16. Jain, N.; Trabelsi, S.; Guillot, S.; Meloughlin, D.; Langevin, D.; Leteiller, P.; Turmine, M. *Langmuir* **2004**, *20,* 8496.
17. Kauzmann, W. *Adv. Protein.Chem.* **1959**, *14,* 1.
18. Konradi, R.; Ruhe, J. *Macromolecules* **2005**, *38,* 6140.
19. Lee, J.; Moroi, Y. *J. Colloid Interface Sci.* **2004**, *273*(2), 645.
20. Levine, A.; Schubert, M. *J. Am. Chem. Soc.* **1958**, *74,* 5702.
21. Mata, J.; Patel, J.; Jain, N.; Ghosh, G.; Bahadu, P. *J.Colloid Interface Sci.* **2006**, *297,* 797.
22. Mesa, C. L. *J. Colloid Interface Sci.* **2005**, *286,* 148.
23 Meszaos, R.; Varga, I.; Gilanyi, T. *J. Phys. Chem. B.* **2005**, *117,* 13538.
24. Mitra, A.; Chakraborty, A. K. *J. Photochem. Photobiol. A.* **2006**, *178,* 198.
25. Mitra, A.; Nath, A. R.; Chakraborty, A. K. *Colloid Polym Sci.* **1993**, *271,* 1042.
26. Mitra, A.; Nath, R. K.; Biswas, S.; Chakraborty, A. K.; Panda, A. K.; Mitra, A. *J. Photochem. Photobiol., A.* **1997**, *111,* 157.

27. Monteux, C; Williams, C. E.; Meunier, J.; Anthony, O.; Bergeron, V. *Langmuir* **2004,** *20,* 57.

28. Moulik, S. P.; Gupta, S.; Das, A. R. *Macromol. Chem.* **1980,** *181,* 1459.

29. Mukerjee, P.; Ray, A. *J. Phys. Chem.* **1963,** *67,* 190.

30. Nandini, R.; Vishalakshi, B. *Sci. J. Chem.* **2014,** *2,* 1.

31. Nandini, R.; Vishalakshi, B. *Sci. J. Chem.* **2014,** *2,* 8.

32. Norden, B.; Kubista, M. In *Polarized Spectroscopy of Ordered Systems;* Samori, B., Thulstrup, E. W. Eds.; Kulwer Academic Publisher: Dordrecht, Holland, 1988; Vol. 242., p 243

33. Pal, M. K.; Ghosh, B. K. *Macromol. Chem.* **1979,** *180,* 959.

34. Pal, M. K.; Ghosh, B. K. *Macromol. Chem.* **1980,** *181,* 1459.

35. Pal, M. K.; Schubert, M. *J. Histochem. Cytochem.* **1961,** *9,* 673.

36. Pal, M. K.; Schubert, M. *J. Phys. Chem.* **1963,** *17,* 182.

37. Panda, A. K.; Chakraborty, A. K. *J. Colloid Interface Sci.* **1998,** *203,* 260.

38. Rabinowitch, E.; Epstein, L. F. *J. Am. Chem. Soc.* **1941,** *63,* 69.

39. Romani, A. P.; Gehlen, M. H.; Itri, R. *Langmuir* **2005,** *21,* 127.

40. Rose, N.J.; Drago, R. S. *J. Am. Chem. Soc.* **1959,** *81,* 6138.

41. Sabate, R.; Esterlich, J. *J. Phys. Chem B* **2003,** *107,* 4137.

42. Sabate, R.; Gallardo, M.; De la Maza, A.; Esterlich, J. *Langmuir* **2001,** *17,* 6433.

43. Sjogren, H.; Ericsson, C. A.; Evenas, J.; Ulvenlund, S. *Biophys. J.* **2005,** *89,* 4219.

44. Villeti, M.; Borsali, A.; Crespo, R.; Soldi, V.; Fukada, K. *Macromol. Chem. Phys.* **2004,** *205,* 907.

45. von Berlepsch, H.; Kirstein, S.; Bottcher, C. *Langmuir* **2002,** *18,* 769.

46. Wang, C.; Tam, K. C. *Langmuir* **2002,** *18,* 6484.

47. Whitney, P. L.; Tanford, C. *J. Biol. Chem.* **1962,** *237,* 1735.

48. Zhu, D. M.; Evans, R. K. *Langmuir* **2006,** *22,* 3735.

CHAPTER 10

RECLAMATION OF NATURAL RUBBER WASTE

S. A. YAWALKAR[1], P. S. JOSHI[2], P. V. PATIL[2], J. M. GADGIL[2], and H. V. JOSHI[1,*]

[1]M. E. Polymer Engineering Department, MIT, Pune, India

[2]Polymer and Rubber Consultants, Pune, India

*Corresponding author. E-mail: sharayu1@gmail.com

CONTENTS

ABSTRACT

Reclaimed rubber (RR) is produced by breaking down the vulcanized structure of rubber using heat, chemicals, and mechanical techniques. Devulcanization and reclaiming of rubber waste is carried out by physical and chemical processes. Physical process involves application of mechanical,

thermomechanical, microwave, or ultrasound energy to partially devulcanized rubber, whereas chemical reclaiming processes use various chemicals and devulcanizing agents.

The chemical system under study was based on natural rubber (NR)
waste. Diphenyl disulfide (DPDS) and garlic oil were used as devulcanizing
agents. Devulcanization of NR waste was carried out by three modes—(a)
treatment with microwave radiation (thermal), (b) shearing rubber and
vulcanizing agent system in roll mill (mechanochemical), and (c) microwave treatment followed by shearing in roll mill (thermal and mechanochemical). The RR samples are tested for mechanical properties (tensile,
percent elongation at break), curing time (Mooney viscometry), processing
time, cross-linking density (sol-gel testing), and Fourier-transform infrared
spectroscopy (FTIR). The promising results are obtained for combination
of thermal and mechanochemical modes of devulcanization process.

10.1   INTRODUCTION

10.1.1   GENERAL OVERVIEW

Every day the consumption of tire increases very rapidly and due to
which pollution also increases tremendously. Achieving aims to protect
the environment and to save natural resources of rubber is very important while considering future needs and troubles. The global consumption of natural and synthetic rubber was around 12.14 and 16.76 million
metric tons (MMT), respectively in 2014.[1] Because of high pollution and
less resources, waste rubber is retreated or reused, and hence by now the
consumption is 6.110 MMT of NR and 8.45 MMT of synthetic rubbers till
midyear (refer Fig. 10.1).

To overcome this situation, it is important and necessary to find latest
solution to recycle and new ways to utilize recovered rubber. Recently
available main routes to achieve rubber recycling are:

1. Retreading or reapply: The use of whole tires as fenders or in
   agriculture
2. Rubber recycling: It is based again on the use of powder or reclaimed
   in rubber items such as in mats, road construction, or floorings

3. Pyrolysis process: Generates back feedstock, that is, recovering or regeneration of carbon black, gas, and oil.
4. Incineration: It is another way to recover the valuable energy from tires and other rubber waste.[2]

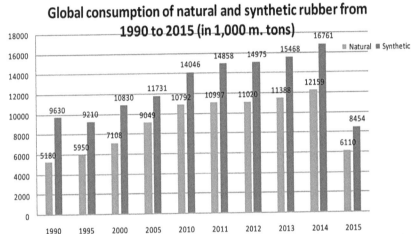

FIGURE 10.1   Global consumption of natural and synthetic rubber (overall consumption) from 1990 to 2015.[1]

In overall rubber waste, tires have 75% of the total rubber consumption. Hence, it is important to concentrate on tire waste. Figures 10.2 and 10.3 shows the scrap tire disposal routes in the Europe and US.

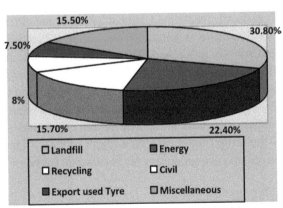

FIGURE 10.2   Scrap tire disposal routes in the US, 2000–2001.[2]

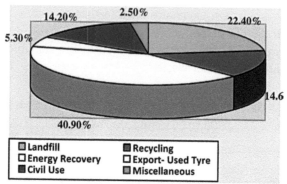

FIGURE 10.3   Scrap tire disposal routes in the Europe 2000–2001.[2]

Nowadays, the primary focus of ongoing research is on devulcanization of waste rubber by means of various methods such as mechanical, thermal, chemicals, or their combinations to achieve good quality material having better properties comparable with the virgin (raw) rubber.[3]

In our work, we have used the mechanochemical and thermal methods of devulcanization by varying devulcanization aids, conditions, and techniques. This devulcanized rubber is then again vulcanized by making blends of NR, styrene-butadiene rubber (SBR), and other conventional additives.[4]

10.2   LITERATURE REVIEW

Vast quantities of rubbers are used as tires for aeroplanes, vans, motors, wheelers, and so forth. Reclaimed rubber (RR) is the product ensuing, and the waste vulcanizate is handled to produce a plastic/rubber material which is effortlessly processed. Then it is formulated and vulcanized without or with the incorporation of virgin rubber either with natural or synthetic rubbers. Reproduction arises because of breaching the accessible cross-links inside the vulcanizate or through selling scission of the main chain of the rubber or by combining one or more techniques. Reclaiming or recovery of the scrap rubber may be a tough process.[3] There are lots of numerous motives, to recover and/or reclaim waste rubber:

- It is a high-quality manner to put off undesirable rubber products that are often tough.

- The petroleum products which are nonrenewable and reconserved are used to manufacture synthetic rubbers. The cost of recovered rubber is half of NR or synthetic rubber.
- This recycled elastomer may have unique properties, which are superior than that of vulcanizate of virgin rubber.
- It requires less energy to produce rubber from reclaim for the total production process than that of virgin material.
- Such recycling activities generates work and jobs in many countries.
- Lots of constructive products and articles can be obtained from reused tires and other recovered rubber wastes.
- The incineration of tires can reproduce substantial quantities of valuable energy and power.[5]

10.2.1 WHAT IS DEVULCANIZATION?

Devulcanization is breaking of the polysulfide (C–Sx–C), disulfide (C–S–S–C), and monosulfide (C–S–C) linking network of the cured elastomer. It was found that in the sulfur system, arrangement of every carbon–sulfur (C–S) bond and sulfur–sulfur (S–S) bond acquires gap, and in devulcanization those bonds are broken. Devulcanization is nothing but the opposite process of vulcanization. With respect to the present analysis, reclamation is not similar to devulcanization owing to the cleavage of the C–C bonds of the chains of polymer. From this point of view, the presently used physical or chemical processes, referred to as reclamation in preference to devulcanization, convert waste rubber product into processable and reusable form.[6,7] In this, the damaged polymer chains produced within the reclaiming techniques have an effect on the properties which decreases the quality of the recovered rubber. When the rubber is devulcanized, preferably simplest S cross-links are damaged, whereas the polymer chains stay integral. Therefore, the devulcanized material is similar in structure and qualities to the original material (Fig. 10.4).[8]

The popular methods of different types of devulcanization of rubber are thermomechanical, mechanochemical, physical such as microwaves, ultrasound, thermochemical, and biochemical. All the methods involve complex transformations which result in depolymerization, mostly, the degradation/decomposition of polymeric chains, and oxidation of rubber, along with the reduction in viscosity of it.[9]

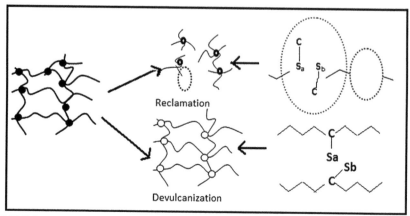

FIGURE 10.4   Reclamation process.[9]

10.2.2   DIFFERENT TYPES OF RECLAIMING PROCESSES

There are a number of processes to convert rubber into reclaim. In these processes, the first step is that scrap rubber is shredded or crumbed, that is, grounded into powder to allow chemicals or reclaiming agents to act on the vulcanizate structure which provide high-quality heat transfer, and damage the linking because of mechanical or chemical actions. All methods are mentioned below.[8,9] Figure 10.5 indicates the basic classification of the devulcanization methods.

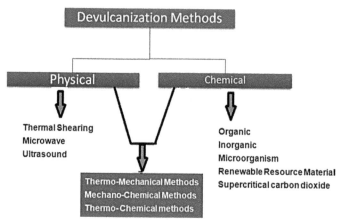

FIGURE 10.5   Classification of devulcanization methods.

10.2.2.1   MECHANICAL RECLAIMING PROCESS

Many different ways are used in this mechanical reclaiming process and most of them are continuous processes. Cross-link less crumb elastomer is mixed with reclaiming agents and poured incessantly into the extruder for 5–10 min at nearer to 200°C. Due to heating element of extruder, elements and friction of the powder heat is generated. The extruded devulcanized product from the machine is in a dry state. Then it is sent for refining.

In this reclaiming process, scrap is placed in an open two roll mill, and processed out at high temperatures. Molecular weight breakdown takes place drastically at high temperatures because of mechanical shearing, that is, above 200°C. This physical process of reclaiming of vulcanizate and the devulcanized rubber are mentioned in a US patent by Maxwell.[10] Crumb of the vulcanized rubber is processed by addition of reclaiming agents by passing the samples between the smooth stator. To provide an axial shear zone, an essentially cylindrical rotor is arranged. Because of the rotor action, the rubber is frictionally propelled in cylinder. De et al.[11] also described the mechanical reclaiming method of curing NR. The NR reclaimed was prepared from vulcanized rubber on a two roll mill at about 80°C. After processing, it produced a sheet band on the roll. Afterward, various additives were mixed with rubber.

In other examples, blend of virgin rubber with RR with varied proportions and study of its mechanical properties, curing characteristics, and other related properties were studied. In this the Mooney viscosity of the RR was found to be very high >200, that is, elasticity of rubber must be very low due to high cross-linked rubber.[9]

10.2.2.2   THERMAL RECLAIM PROCESSES

In this process, reclaiming of the rubber is done by using chemicals and heat to plasticize the scrap rubber waste. It is given below.[12]

10.2.2.2.1   Alkaline Process

In this type, the links in the waste are grasped by means of sodium hydroxide with high concentration approximately 7%, the cellulose is hydrolyzed from the strand containing scrap. Then it must be washed to remove the excess sodium hydroxide or other defiberizing agent. Then this crumb is dried and refined. In alkaline reclaiming processes for vulcanized SBR, some *N, N*-dialkyl aryl amine sulfides[13] were found to be highly effective reclaiming agents.

10.2.2.2.2 Digester Process

In the digester process, where the waste rubber is assorted with plasticizing oils, water, fiber dissolving agents, and if required, reclaiming agents. This is a batch process. They are heated in an autoclave for 8–12°h with a steam pressure of 15 bar (180°C). The material dissolves and the rubber softens. Drying off the regenerate and refining the aqueous phase is the final stage.

Reclaiming possible chemicals used to dissolve the fibers in this process include zinc and calcium chlorides jointly with a very complicated combination of liquids, hydrocarbon resins, softening oils, reclaim catalysts, and pine tar. If the digester process is being used, then more quantity of outstanding network remains with the reclaim.

10.2.2.2.3 Heater or Pan Process

This is the simplest and oldest method used for the rubber recovery process. Lastly, ground scrap rubber which is free from fibers by means of mechanical cyclones, is added with reclaiming chemicals. Afterward, it is poured in the pans of an autoclave. Then it is heated with live steam at the pressure of 15 bars at 180°C for 4–12 h. The remaining residual H_2O is removed. Reclaiming chemicals that are added include aromatic oils, disulfides, and aromatic thiols. This gives shorter reclaiming times, lower temperatures, and production of a product having greater mechanical properties. In this process, the chemicals used are almost alike to those used in the pan method, only difference is that zinc or calcium chlorides are excluded.

10.2.2.2.4 Neutral Process

The advancement to beat the hardening drawback is the main development of the neutral process. During this method, zinc chloride and pine oil are used.[14] The heater, alkaline digester, and neutral methods need long reaction cycles, and characterization time is also longer. Calcium chloride or zinc chloride is mixed in the process for the hydrolysis of the fiber. Most reactive reclaiming agents for vulcanized in neutral reclaiming processes shown to the SBR are N, N-dialkyl aryl amine sulfides.

10.2.2.2.5   High-pressure Steam Process

These are recent developments expected to shorten time reactions. Within the elevated pressure steam procedure[14] with reclaiming agents, linking-free, crumb is added. Reclaiming is done under high pressure and heated for 1–10 min at an approximate temperature of 280°C.

10.2.2.2.6   Engelke Process

The Engelke method[15] consists of an autoclave, in which rubber crumb, together with peptizers and plasticizing oil is heated to elevated temperatures of more than 250°C for around 15 min. Refining and straining is done afterward.

10.2.2.2.7   Continuous Steam Process

This steam devulcanization method in a hydraulic column requires temperatures nearer 260°C and also high pressures. To avoid combustion because of heat and pressure, the rubber is ground and for carrying medium, water is used to save the material from unrelated oxygen. The reason for extensive breakdown in the suspension of rubber is the reaction between injected chemical agents with processing aids such as heat and pressure.

10.2.2.3   THERMOMECHANICAL RECLAMATION

In many mechanical techniques, an increase in heat takes place that results in demeaning of the cross-linking structure of rubber. This recycling of rubber is because of combination of breaking C–C bonds and S-cross-linking which produced development of sol fraction by branched structures and gel fraction.[14] Most of the new rubber recycling techniques are based on thermomechanical regeneration processes.

For thermal and mechanical breakdown, recycling chemicals and oils are usually used. Generally used recycling chemicals, such as thiols, disulfides, amines, and other unsaturated compounds, are poured in quantities of around one weight percentage.[14]

Softeners are used to minimize the heat degradation resistance of a rubber to decline the interaction between rubber chains and filler.

10.2.2.3.1 *Milling Process*

In this thermomechanical method, vulcanizate network degradation of the rubber takes place. In an appropriate solvent, the vulcanizate is swollen to prepare a fine powder (~20 μm·diameter), and then milling is done. By means of curing ingredients, this powdered rubber is revulcanized. The products produced are evidence of somewhat poorer properties to those of the original rubber.[16]

10.2.2.3.2 *Reclamation by High Speed Mixing*

Simple and small duration, that is, 15–20 min is required for this process. It is one of the fast thermomechanical recycling processes. As the name suggests, the rubber is stirred at a high speed of 500 rpm and the temperature rises to 200°C.[5]

10.2.2.4   *MECHANOCHEMICAL METHODS*

By applying a reclaiming accelerator, plasticization can be enhanced. In general, a reclaiming oil, process oil, and reclaiming catalyst are used together with the devulcanizing agent at low temperatures. Methyl halide and peroxide in combination, act as a powerful radical initiator for a redox system. In this process, tributylamine and cuprous chloride as well as phenylhydrazine and ferrous chloride create a complex with each other which is easily degraded because of oxidation initiators. In the existence of oxygen, these oxidation systems degrade diene-based rubber at room temperature. Peptizing agents used are dioxylyldisulfide, Renacit 7, and 2, 20-dibenzamidodiphenyldisulfide.

Vulcanization accelerator generally used such as N-cyclohexylbenzo-thiazole-2-sulfenamide (CBS) and N-isopropyl-N_0-phenyl-p-phenylene diamine worked as antioxidants. The rubber accelerators triphenylphosphine and tetraethylthiuram disulfide were used to increase the elasticity of reclaim.

In the mechanochemical reaction, the function of these reagents is to act as radical acceptors for the polymer radicals that are produced.

10.2.2.4.1   Trelleborg Cold Reclaiming (TCR) Method

Trelleborg cold reclaiming (TCR) is the process in which there is incorporation of devulcanizing agents into cryogenically waste rubber grind. At room temperature or a little higher temperature, in a powder mixer, a small treatment is carried out, 1, 3-diphenyl guanidine or phenylhydrazine-methyl halide reacts with the vulcanizate.[16]

10.2.2.4.2   De-link Process

A new technology suggested by Kohler[17] of sulfur cure elastomers for the devulcanization using a chemical called De-Vulc agent was innovated by Sekhar[18] and is called as De-Link process. The employment of ground rubber powder along with processing accelerators, sulfur, and other aids on a mill is employed within the proprietary of de-link method. The method is not only appropriate solely for NR, but also additionally for ethylene propylene diene monomer (M-class) rubber (EPDM). In this process, in an open two roll mill two to six parts of De-Vulc reactant is mixed with 100 phr of 40 mesh size or finer crumb at temperature below 50°C for around 10 min.

In this method, Tetramethyl thiuram disulfide can also be added as agent. It is presumed that the method relies on reaction of a nucleon transfer. Here, the quantity of network linking is reduced by a factor of two. Because the nip gap of the mill was initiated to own a major effect on properties, the matrix breakdown is maybe because of mechanical shearing.

The wide variety of network links is minimized by a factor of two. The nip at starting of the mill is observed to have a high impact on its properties. The vulcanizate linking breakage is mostly resulting from mechanical breakdown.

10.2.2.5   MICROWAVE RECYCLING

In this recycling methodology, a fixed quantity of microwave electromagnetic energy with specific frequency is applied to segregate the S–S or C–C bonds in the vulcanized rubber crumb.[5] Thus with no depolymerization to an elastomer, material waste can be reclaimed. This can be recompounded and revulcanized and has physical characteristics basically equivalent to the former vulcanizate. With usage of microwaves, the material temperature enhances very quickly to attain 260–350°C finally.[19] A former condition for this devulcanization is that it should consist of carbon black to this temperature level to the vulcanizate which makes them relevant for this process. It should have little polarity so that microwave energy can create heat required to devulcanize, which usually takes place in the normal economical procedures that are currently in practice. Here, they mentioned that for microwave devulcanization, the elastomer which is in sulfur vulcanization containing polar groups, is applicable. Tyler et al.[20] reported that the microwave devulcanization procedures are methodologies of pollution controlled recovery of sulfur vulcanizate rubber consisting polar groups. Rubber consisting of carbon black has been vulnerable to ultrahigh frequency in the chamber of microwave, free electrons, generated at an interface of various phases with a diverse dielectric constant, and conductivity are accumulated because of interface or ion polarization.

The modified microwave consists of a domestic-type microwave oven and stirring apparatus. The power of the magnetron is put to 800 W, and 100 g of the ground rubber waste is put in a beaker and stirred at a speed of 6 rpm. The exposure time in microwave is varied from 1 to 5 min. The temperature after treatment of each sample was measured (Fig. 10.6).

FIGURE 10.6   Microwave devulcanization system.

This method is helpful because it gives an ecologically, economical best method of reusing waste rubber to get it back to originally generated items and it provides a product like the original and with similar physical properties. The characteristics of this RR are claimed to be superior to the rubber generated by other recovering techniques. It is more suitable for EPDM and isobutylene isoprene rubber (IIR) process.

10.2.2.6   ULTRASOUND RECYCLING

In ultrasound recycling, ultrasonic energy is used for vulcanized rubbers. Devulcanization, which was first reported by Pelofsky[21] in 1977 for bulk rubber efficiently dissolves and disintegrates into the liquid. The ultrasonic radiation is approximately about 20 kHz and higher than intensity 100 W power. In 1987, the ultrasonic recycling of NR with vulcanization was documented by Okuda and Hatano.[22] They kept the NR vulcanizate for 20 min to 50 kHz ultrasonic energy to gain recovery. The recycling process needs a very high energy frequency to break S–S and C–S bonds. In different media, an ultrasonic field generates extension–contraction stresses in high frequency.

A continuous ultrasonic devulcanization of unfilled EPDM rubber and SBR was also reported by Yun and Isayev.[23] Gel fraction, mechanical and dynamic characteristics, and cross-link density were calculated for the virgin rubber, the devulcanized rubber, and for the revulcanized rubber by ultrasonic recycling.

This type of devulcanization practice is done by a reactor built for the reason of ultrasound devulcanization. This ultrasonic reactor is nothing but an extruder of 1.5 in. with a coaxial cone such as ultrasonic die attachment which has L/D ratio of 1:1. This apparatus consists of three temperature-controlled zones. For this process, ultrasonic power supply of a 3 kW with a 1:1 booster, a 3 in cone-tipped horn and an acoustic converter are utilized. The horn vibrates vertically at 5–10 mm amplitude and at a 20 kHz frequency. To fix the complete unit, the extruder flange is mounted on four rigid bars.

This continuous ultrasonic reclaiming method was introduced 15 years before by Isayev and coworkers. The theoretical and experimental study reports have been given in articles.[21,22] The application of high-power electromagnetic radiation from ultrasonic die in a rubber extruder is shown in Figure 10.7. This method was demonstrated as very efficient, simple, and fast. It was unaffected in solvents and many other chemicals, whereas in the other ultrasonic processes, rubber networks go down by cavitations, which are formed in the existence of pressure and heat by high-intensity ultrasonic waves. The ultrasonic energy can specifically break the C–S and S–S bonds as they are weaker than C–C bonds.

FIGURE 10.7 Ultrasonic devulcanization.

10.2.2.7 RECLAMATION BY ORGANIC REAGENTS

Efforts have been made at the start of 20th century for reclaiming of waste rubber products. To find out the relative quantities of mono-, di-, and polysulfidic cross-links, certain chemicals that cause scission of sulfur bonds are used as reagents. These reagents are also referred to as chemical probes.

Hexanethiol is located to break di- and polysulfidic, and 2-propane thiol selectively breaks polysulfidic cross-links with piperidine as a base in a nucleophilic displacement reaction.[5] Thiols together with natural bases are able to break sulfur cross-links. For this reason, many chemical reagents for natural and synthetic rubbers such as dibenzyl disulfide, diphenyl disulfide (DPDS), butyl mercaptan, diamyl disulfide, thiophenols xylene thiols, bis (alkoxy aryl) disulfides and phenol sulfides, other disulfides, and mercaptans have evolved.

The thiol–amine mixture provides a composite, probably an ion pair of piperidinium propane-2-thiolate, where the sulfur atoms increased nucleophilic characteristics, which are able to break higher organic polysulfides and trisulfides at 20°C within 30 min.[24] Disulfides react at a slower rate of a factor of 1000. The polysulfide breakage is quicker because Pπ–dπ delocalization of the moved s-electron pair of RSS—as outlined.

For vulcanized SBR in neutral as well as alkaline reclaiming processes, N, N-dialkyl aryl amine sulfides[5] are found to be more effective reclaiming agents, where the phase of reclaiming was physically reviewed throughout tack evaluation, body, and thickness. In 1967, a technology and science evaluation of RR was reported by Le Beau.[25]

The movement of diaryl disulfide in the synthetic and NR wastes of technical items was reported by Knorr.[26] In this process with the accumulation of reclaim oil and Aktiplast 6 (includes disulfides), the finely ground fabric and linking free scrap is warmed up in a saturated steam at an excessive temperature, that is, 150–180°C.

First, finely ground changed into a very well combination with reclaim oil and diallyl disulfide (DADS), and then allowed it to swell for a minimum of 12 h. The mixture was then positioned on trays of talcum powder. Proper distribution of steam and air is required for the process. Then by sealing the lid, an autoclave was kept under pressure, with 4 bar of air–steam, and for desirable air circulation the fan was turned on. The compressed air valve became closed and to achieve the temperature of 190°C, steam of 8–9 bar was pressured in it. The reclaiming time depends on the type of scrap ranging from 3 to 5 h per 200 kg charge.

Rubber network without any difficulty can be swollen by methyl iodide which may be removed from it by heating under vacuum. In the NR vulcanizate, the exploit of methyl iodide to calculate monosulfide linkages approximately is added by Meyer and Hohenemser.[24] The concentration of monosulfide groups would reflect the height of network bound iodine postreaction for 2–3 days as simple saturated monosulfide group reaction is given below.

$$R_2S + CH_3I \rightarrow R + 2SCH_3I^-$$

Anderson[27] patented reclaim of rubber which is sulfur vulcanized in presence of water vapor, oil, and aryl disulfide peptizer at increased temperature range of approximately 175–195°C and at pressure range of approximately 230–260 psi for 1–4 h. The aryl disulfide is a mixture of dicresyl disulfide DPDS and dixylyl disulfide.

In a latest study, Bilgili et al.[28] suggested a two-step process of recycling to reuse the waste of rubber. First, the smaller particles of waste were pulverized into little particles with the use of solid state shear extrusion (SSSE) technique in single screw extruder. Then the created powder was compression molded without raw rubber. With different molding conditions slabs were tested for microscopic, mechanical, and chemical tests. The test result shows that slabs have more elongation and less tensile strength. Self-adhesion of rubber molecules, compressive creep of the powder, and interchangeable actions of polysulfidic cross-links were suggested as basis of particle bonding. These have exhibited that rubber slabs are often made with pulverizing waste of rubber using SSSE method and afterwards molding that powder. The conditions of molding and characteristics of powder considerably impacted failure properties as well as the slabs' cross-link density. Increased pressures and temperatures, high cross-link density, and also the existence of enormous quantity of coarse particles usually increase the particle bonding. In this rubber molecules with self-adhesion, compressive creep of the powder, and interchangeable reactions of polysulfidic cross-links seem to control particle bonding.

10.2.2.8 RECLAIMING BY INORGANIC COMPOUNDS

Unwanted aged tire factory waste and tires were recovered by the method of desulfurization, with sodium, crumb of suspended vulcanized rubber (10–30 mesh) within a solvent namely, naphtha, toluene, cyclohexane, benzene, and so forth.[9] The alkaline metal breaks the polysulfidic, di and mono cross-linkages of rubber crumb which swells and is suspended at approximately 300°C without any oxygen. Various authors proposed that this treatment produced a polymer which has same molecular weight as the rubber weight before vulcanization. Devulcanized rubber can be used for revulcanization with no extraction of polymer from solvent by including the correct curing composition. And carbon black might also be recovered for further use. The procedure might not be suitable economically, as the process includes swelling of crumb of vulcanized rubber within the organic solvent.

Yamashita et al.[29] have successfully RRs with the usage of the catalyst iron oxide with a phenylhydrazine base. In the procedure, powdered rubber from waste tires is processed along with $FeCl_2$ and phenylhydrazine, ozonized and dealt with H_2O_2 to present higher viscosity and higher give up of liquid rubber. It is then compared with same products without the phenylhydrazine–$FeCl_2$ treatment. For this 100 gm powdered rubber from waste tires, 0.25 g $FeCl_2$ within 5 ml MeOH, and 0.5 g phenylhydrazine within 10 ml benzene were mixed neatly and kept for a whole day at normal room temperature and rolled for 10 min. This plasticized rubber was ozonized and dealt with H_2O_2 to provide fluid rubber with intrinsic viscosity at near 30°C in benzene with 13–25% yield of 0.05–0.11 dl/g, compared with 15–20% yield and 0.03–0.05 dl/g for liquid rubber acquired with no phenylhydrazine–$FeCl_2$ remedy. The cross-link density and sol content were important for the sulfur vulcanized and ZnO– tetramethylthiuram disulfide (TMTD) vulcanized samples along with cleavage at the cross-linking sites. The scission in main chain was clearly visible in the scenario of peroxide cured vulcanizate. It proves that in the reclaiming process not only the breakage of C–S or S–S bonds occur but the scission of C–C bonds also takes place.

10.2.2.9 RECLAIMING BY THE MISCELLANEOUS CHEMICALS

The scraps of vehicle tire consisting polyisoprene rubber, PBR, SBR were devulcanized via phase transfer catalyst at lower temperature. Each of the devulcanizing agent constitution and the method was patented. The novelty of this approach lies in the usage of less temperature phase transfer catalyst, and the method processed at lesser than 150°C. This invention of devulcanizing rubber is differentiable from traditional RR in the way that rubber in devulcanization is considerably loose from polysulfide cross-link which selectively cleaves through the method with negligible predominant chain scission (Fig. 10.8).[5,9]

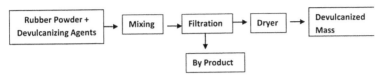

FIGURE 10.8   Layout for chemical devulcanization system.

Kasai et al.[30] stated the usage of reclaiming agent thiocarboxylic acid for cross-linked rubber. In this method, powder of rubber waste of 30 mesh element size is blended with 0.5–10.0 wt.% base on the rubber of 10% thioacetic acid solution within benzene which is left at some point at room temperature, stripped of benzene which is then rolled at <120°C in the air to provide high mechanical force resulting in RR. The rubber compound 200 phr containing five parts of ZnO, three parts of sulfur, one part of stearic acid, and 0.7 parts of vulcanizing accelerators was warmed at 145°C for approximately 30 min to give a rubber which has tensile property of 9.0 MPa and an elongation at spoil 410%. The inflamed rubber in combined solvent is carried out through the 2 mm extruder's orifice at 2–6 kN/cm² pressure to provide a paste which is used in tire treads manufacturing. A rubber compound consists of 40–60 parts of created paste which is mentioned above, and 100–150 parts butadiene rubber and isoprene rubber.

SBR was warmed for about 15 min at 160°C to provide vulcanizate which consists of elongation at spoil 500%, 300% modulus of 3 MPa, and Shore hardness of 62. As a remedy for disposal of the waste tire problem, used tire waste was recovered by drenching it in the organic solvent namely, 1,3,5-trimethyl benzene for a needed time when it decreased the tensile strength by approximately 50%. Then the drenched rubber was separated by applying a shear pressure to provide the recovered rubber.[31]

Digester process can reclaim scrap rubber consisting synthetic and NR by using reclaimed oil with the molecular weight from 200 to 1000, comprising of alkyl benzene, alkylated indanes, and benzene. The Bryson has patented the compound of the reclaimed oil and the better digester procedure using this reclaimed oil.[32] Used rubber was also reclaimed with the method of transformation derivatives and metal alloys.[9] In the current approach, organic solvent was used for inflaming vulcanized rubber and then its size was decreased by processing it using agents such as transition metal alloys with their derivatives.

10.2.2.10   DEVULCANIZATION BY PYROLYSIS

Pyrolysis is a type of thermal method for regenerating basic substances and energy from waste rubber. It has been a debatable, capital-intensive, large-scale, complex, high technology, and experimental approach for recycling of tires. In principle, the concept 'pyrolysis' indicates all types of heat decomposition, containing combustion even though practically, it is considered to signify the thermal decomposition in a nonreactive (without oxygen or anaerobic) atmosphere. Pyrolysis has been used for the revival of such materials such as oils, carbon black, glasses, and metal, for their value of energy.

10.2.2.11   EXPERIMENTS USING RENEWABLE RESOURCE MATERIAL (RRM)

Rubber vulcanizate and ground rubber crumb are milled in a two roll mill with continuous inclusion of DADS with spindle oil or renewable resource material (RRM) with spindle oil, independently. The reclamation is executed with varying reclaiming agent concentrations at two separate temperatures for various milling times. With progressive milling, on the roll surface band creation takes place, and the whole sticky mass is obtained.

The outcomes demonstrate that milling of vulcanizate by using RRM or DADS for 15 min gives the most reduced sol content, highest Mooney viscosity, and the lowest molecular weight of sol of the recovered rubber. Milling for 35 min gives the highest sol content, lowest Mooney viscosity, and the highest molecular weight. About 10 phr of RRM or 2 phr DADS was enough to acquire a sensible quantity of sol with highest molecular weight and in addition, least Mooney viscosity of the reclaim was recovered after milling for 35 min at 60°C. The sol content is enhanced by increasing milling duration, where the highest sol content is received after processing for 35 min; thus, demonstrating the noteworthy reliance of sol fraction material on the duration of processing, because at some point of milling, the rubber samples undergo significant mechanical shearing. Therefore, right precaution is important to set the reclaiming conditions so that there is less effect on breakage of the sol due to thermal or mechanical shearing. The reason behind the growth in sol fraction with revolutionary milling depends on the interaction of DADS either included externally or which are already present in RRM. With increase in temperature, the DADS breaks the radicals by mechanical shearing[29] in addition with the broken polymer chain radical, and is the reason to increase in sol content with increase in duration of milling.

10.2.2.12 BIOTECHNOLOGICAL METHODS OF RUBBER RECLAMATION

Another way to clear off waste rubber can be by degrading it using microorganisms. On NR latex, the biological attack is entirely effortless. Man has created several rubber products as a cause for the microbial attack.[29] Apparently, nature is ready to attend to its own waste difficulties, but at certain times, man grows concerned, and a biological attack is minimized when the NR matrix is converted into a technical substance by sulfur vulcanization with the addition of other ingredients.[32]

Many research experiments have been conducted in the handling of microorganisms of the class *Actinomycetes*. The most interesting method expressed recently in a German patent[33] was to use a chemolithotroph in a liquid suspension for assaulting elastomeric powder at the surface most effectively. Mixing this with the vulcanized rubber facilitated diffusion of soluble chains of polymer. Through vulcanization the bonding becomes possible.

Biosurfactants have been believed to facilitate degradation of the rubber at some point of the adhesive boom. Superior growth of microorganisms is found with microbial enrichment cultures for many months for NR with acceptance of tire crumb material. Latex type vulcanized rubber is an easily degrading alternative rather than rubbers from used truck tires which are more challenging.[34] Growth with microbial acceptance on polymer material and linkage of cosubstrate leads to more degradation. This chain scission and oxidation of the polymer backbone are the mechanisms behind microbial rubber degradation,[35] which does not bring about materials suitable for reuse. The substitution of those compounds causes decreased entrée of the microorganisms into the polymer matrix.

10.2.2.13 RECLAMATION BY SUPERCRITICAL CARBON DIOXIDE (scCO$_2$)

The latest devulcanization method was established by using supercritical carbon dioxide (scCO$_2$) in addition to devulcanizing reagents.[36] One-of-a-kind cross-link networks were prepared from unfilled polyisoprene rubber vulcanizate by controlling time of the treatment and the quantity of curative.

All of the batches of vulcanized matrix were sent for Soxhlet extraction with the usage of azeotropic acetone or chloroform to do away residual curative.[37]

The reclamation finishes in the existence of scCO$_2$ for 60 min at varying temperatures (140–200°C). The product differentiates into a sol and gel fraction, and the molecular weight of the sol aspect and the cross-link density of the gel fraction was measured.[38] Among several devulcanizing reagents considered, the most effective was the thiol–amine reagent. With an incremental pressure of CO$_2$, the yield of the sol fraction was increased. The vulcanizate was extra efficiently devulcanized in the scCO$_2$ in liquid state, than in general gaseous state of CO$_2$.[36]

10.3 DEVULCANIZATION AGENTS

10.3.1 GARLIC OIL

Garlic is a renewable source as it is a natural herb. Garlic oils from natural herbs consist of components such as esters of glycerol and fatty acids. Garlic oil contains the amino acid alliin (C$_6$H$_{12}$NO$_3$S). The odor of freshly cut garlic comes from (C$_6$H$_{10}$OS$_2$), also called as diallyl disulfide sulfoxide, which is generated because of enzymatic action and further reactions result in powerful odorant such as diallyl disulfide and vinyl-[4H]-dithiines as given in Figures 10.9 and 10.10. Table 10.1 below provides details about the chemical composition of essential oil of *Allium sativum*, that is, garlic oil.[39,40]

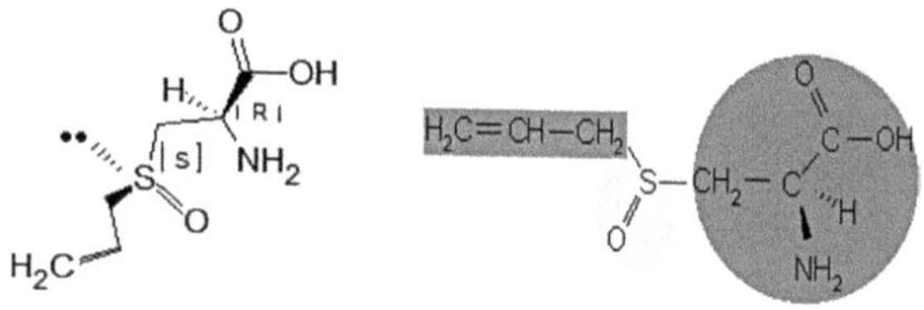

FIGURE 10.9   Structure of alliin.[39,40]

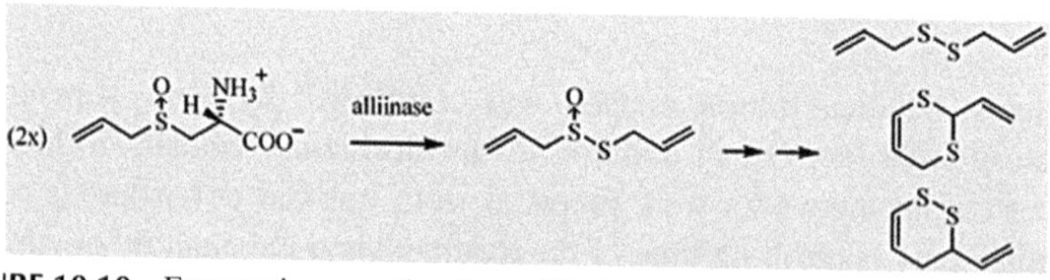

FIGURE 10.10   Enzymatic generation from alliin into and diallydisulfide vinyl-dithiines.[39]

TABLE 10.1 Chemical Composition of Essential oil of *Allium sativum*.

Component of garlic	Abbreviation	Percentage in garlic	Structure
Diallyl disulfide	DADS	30–32	
Allyl methyl trisulfide	AMTS	32–35	
Allyl methyl disulfide	AMDS	9–10	
Diallyl sulfide	DAS	6–7	
Diallyl tetrasulfide	DATTS	4–5	
Diallyl trisulfide	DATS	1–2	

General properties are:
   Color of extracted oil: Light yellow
   Oil yield: 21.6%
   Specific gravity: 0.90 g±0.01
   Odor: Pronounced odor
   Boiling point: 115°C

10.3.2 DIPHENYL DISULFIDE (DPDS)

DPDS is the chemical reagent having formula $C_6H_5S_2$. This is a crystalline and colorless material. This is one of the best reclaiming agents for many rubbers as well as unknown for the ground tire rubber (GTR) crumb. Figure 10.11 presents the structure of DPDS and Table 10.2 presents the properties of DPDS.

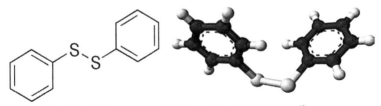

FIGURE 10.11 Chemical structure of diphenyl disulfide (DPDS).[42]

TABLE 10.2   Diphenyl Disulfide Property.[41]

Property	Value
Mp	58–60°C
Bp	310°C
Density	1.353
Refractive index	1441–1444

10.4   DEVULCANIZATION BY GARLIC OIL AND DIPHENYL DISULFIDE (DPDS)

Devulcanization is a process of breaking C–S–S–C, C–S–C, and C–Sx–C cross-links in the cured rubber. It is the exchange of waste rubber into reproducible and reusable material after using physical/chemical processes because of the main chain and cross chain scission. The following possibilities cause breakage of the cross-links which are responsible for devulcanization process.[2–8]

1. Opening by oxidation
2. Opening by heat or shear
3. Opening with nucleophilic reagents
4. Opening by rearrangement
5. Opening by substitution

Opening by oxidation: During the degradation of the cross-linking network, thiols and disulfides react with radicals framed in the response and restrain gel formation because radicals combine together with each other. It is stated that both of these kick off an oxidative breakdown of C–S–S–C, C–S–C and C–Sx–C cross-links, and hence vulcanized rubber is degraded.[6] Rubber which is recycled with thiols and disulfides gives a high degree of cross-linking network breakup. Both thiols and disulfides are evenly reactive and responsible for oxidations in thermomechanical method (Fig. 10.12).[9]

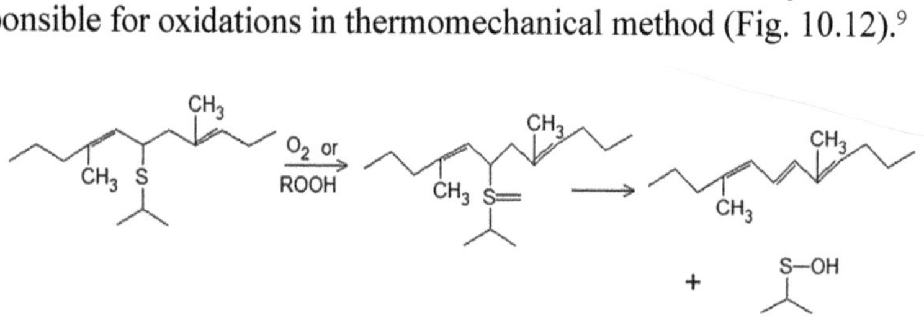

FIGURE 10.12   Opening of sulfur cross-links (ROOH = organic hydroperoxide) by oxidation.

Opening of links by heat or shear: It is found that in between reclaiming process, the thermally created polymer radicals are rummaged by the sulfur (-S-) radicals which had prevented reunion of polymer radicals as shown in Figure 10.13.

This preventive action is because of the attachment of the rubber molecules with DADS wreckage molecules. The sequences of above mentioned reactions may occur in particular of any reclaiming agent or any reclaiming temperature.

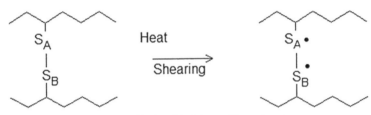

FIGURE 10.13   Reclamation with the aid of thermal heat shearing.

Opening with nucleophilic reagents: It has been found that amines are the strongest nucleophiles that are known as best reclaiming agents because primary and secondary amines have to break octasulfur (S_8) in cyclic structure. Suitable reclaiming agents such as thiols and amines often have a lone pair of electrons. Nucleophilic compounds such as thiols can react according to a radical mechanism and can also act as hydrogen transfer agents (Fig.10.14).[9]

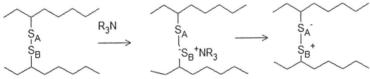

FIGURE 10.14   Opening of sulfur cross-links by nucleophilic reagents.

Opening by rearrangement: This may be supported by mastication of virgin rubbers; radicals of polymer are prohibited to combine together to generate radicals because of shearing action and the peptizing reaction of thiol organic compounds (Fig.10.15).[5]

FIGURE 10.15   Opening of sulfur cross-links by rearrangement.

Opening by substitution: The substitution of reclaiming agents is respon-sible for chain opening by substituting reactive group with active radicals. In sulfur vulcanization analysis study of rubber before and after the trial with diallyl disulfide,[6] it is found that there is increase of bound sulfur of the treated rubber (Fig. 10.16).

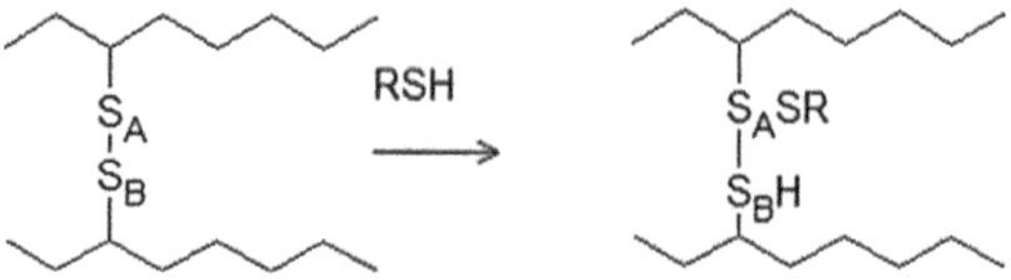

FIGURE 10.16   Opening of sulfur cross-links by substitution.

10.4.1   DEVULCANIZATION BY GARLIC OIL

Stage 1: In the initial step, unstable radicals are formed because of opening of link network or the scission of the polymer chains by means of thermal and shearing forces. Usually, this scission occurs approximately at low temperature but is of greater importance with increasing temperature. Afterward, these unstable radicals further react in secondary reaction steps ((2), (3), and (4)) in the presence of oxygen as shown in Figure 10.17. Depending on the conditions of devulcanization, these three reactions may occur equally.

Stage 2: In succeeding oxidative aging of the vulcanized rubber, a hydroperoxide is formed. The reaction of this hydroperoxide with the various mono-, di-, polysulfide groups results in a formation of different oxidized organic sulfur groups. The sulfenic acids (RSOH) and thiosulf-oxylic acids (RSSOH) generated from the oxidized sulfur vulcanized network, would be expected to react with hydroperoxides to form sulfinic (RSO_2H), sulfonic (RSO_3H), and thiosulfuric (RS_2O_3H) acids. It works as antioxidants because all of these acids can demolish hydroperoxides to inactive nonradical products.[9]

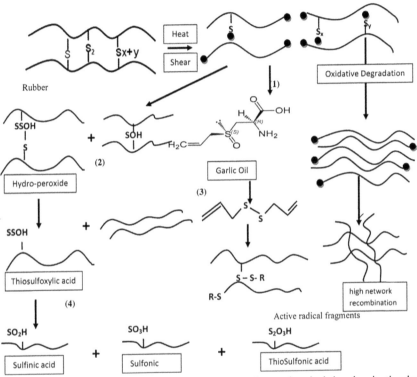

FIGURE 10.17 Simplified reaction for thermo–mechano–chemical devulcanization by garlic oil.

Stage 3: In this step, the active radical bits react with radicals of disulfide base, which prohibit reunion of rubber molecules by scavenging sulfur radical. And that is because it enhances the probability of combination of one radical with another elastomer radical, as the diffusion speed of this disulfide into the rubber matrix increases at higher temperatures.

Stage 4: Degradation causes main chain scission. At higher temperature treatment, the possibility of the generated reactive groups to go through oxidative degradation increases by activating the developing reactions in the polymer.

10.4.2 DEVULCANIZATION BY DPDS

A mechanism, which is frequently projected toward the response of sulfur vulcanizate with disulfides is the scission of chains or the opening of cross-links by thermal action and shearing forces, and its response with disulfides which counteracts recombination. Other remaining compounds with a stabilizing effect are called as antioxidants. Air oxygen is, additionally, capable of avoiding recombination by balancing out or stabilizing the radical sites. The outcome is a reduction in molecular weight of the rubber. In thermal degradation, there is formation of hydrogen sulfide and thiols.

10.4.2.1 REACTIONS FOR NATURAL RUBBER

Step 1: This represents the formation of sulfide radicals because of the hemolytic cleavage of the DPDS at around 100–120°C due to mechanical shearing.

    Step 2: As shown in the step 1 of Figure 10.18, produced sulfide radicals react with sulfur cross-linking sites which are present in polymeric main chain of cured rubber to break the cross-linking.

    Step 3: Here, another sulfide radical reacts with hydrogen atom from the main chain of polymer as shown in Figure 10.18 and there is formation of a new active cross-linking site in the main polymeric chain of revulcanized rubber.

    Step 4: Revulcanization step is where the revulcanization with sulfur and curing accelerators at 145°C forms new active cross-linking sites.[43]

1. **Process of Thermal degradation of DPDS**

2. **Breaking of sulfur bond**

3. New active site formation

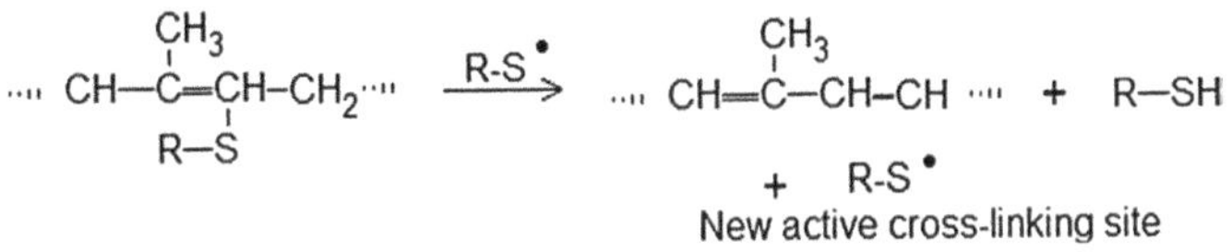

4. Revulcanization Process

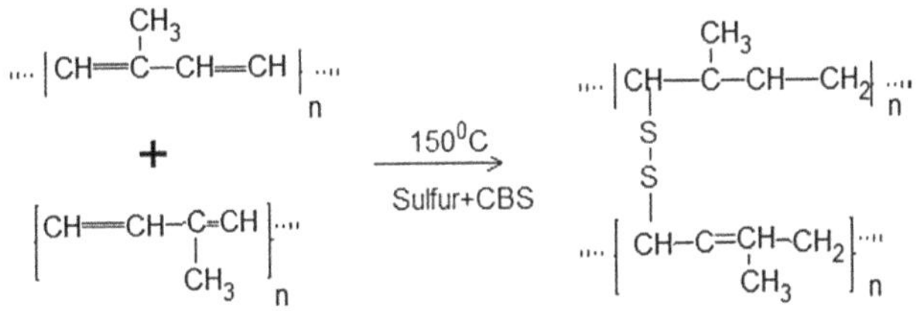

FIGURE 10.18   Reactions for devulcanization by DPDS.

Production of sulfur radicals from DPDS and breaking of sulfur bond of rubber because of thermal shearing and heat occurs in case of SBR devulcanization. Then there is formation of new active sites, which can react with additives used for revulcanization process.[44]

10.5   EXPERIMENT WORK

10.5.1   INTRODUCTION

The consumption of rubber in tire application is increasing rapidly day by day due to which there has been rise in pollution as well as environment-related issues. To protect the environment, it is very important to consider recycling and reuse of rubber waste. Throughout the globe many researchers are working on developing various methods/techniques to carry out recycling of rubber waste in an effective manner. In our work, devulcanization of rubber was carried out by mechanochemical and thermal methods using various devulcanization aids and conditions. The devulcanized rubber was again vulcanized with other conventional additives. This saves the natural sources as well as processing time. Such reclaiming is expected to occur by scission of sulfur–sulfur and carbon–sulfur crosslink bonds. Final process flow chart is given below (Fig. 10.19).

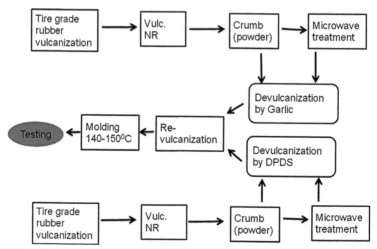

FIGURE 10.19 Process flow chart.

10.5.2 EXPERIMENT

10.5.2.1 MATERIALS

NR (RMA 1X) was supplied by Birla Tyres (India); N-cyclohexyl-2-ben-zothiazyl sulfonamide (Pilcure CBS), 2,2,4-trimethyl-1,2-dihydroquinoline (Pilnox TDQ), and tetramethylthiuram disulfide TMTD (Pilcure TMT) by Nocil Ltd., Mumbai; steric acid (Lubsteric 995) by Godrej Industries, Thane; garlic oil by Allin Exports; DPDS by Alfa Aesar, England; zinc oxide (White Seal), Gujarat; carbon black N330, aromatic oil by Panama Petrochem.

10.5.2.2 PREPARATION OF VULCANIZED NATURAL RUBBER

Initially, the rubber was vulcanized like conventional sulfur vulcanization method. The compounding formulation of rubbers with various ingredients is given in Table 10.3. Accelerator, antidegradants, and other additives were used in this process. The mixing was done in open two roll mixing mill at a friction ratio of 1:2. The rubber compound was cured at 150°C for 5 min by compression molding. It was devulcanized by thermal means using microwave radiation followed by mechanochemical process. This rubber sample was revulcanized and characterized for mechanical properties and sol-gel content.

TABLE 10.3   Compound Formulation for Vulcanizing natural rubber (NR).

Material	PHR	Material	PHR	Material	PHR
NR	100	TDQ	1	Paraffin wax	1
Peptizer	1	Pilflex	0.5	Pilcure CBS	0.7
ZnO	7	N330	50	Pilcure TMT	0.5
Steric acid	2.5	Aromatic oil	6	Sulfur	6

10.5.2.3   THERMO-MECHANOCHEMICAL DEVULCANIZATION PROCESS

Initially, the rubber was vulcanized as per the compounding formulation of rubbers with a mixture of ingredients as given in Table 10.3 and then ground to 20–40 mesh size. The vulcanized rubber compound was ground for 10 min at 40–50°C by using two roll mills.

This crumb was devulcanized mechano-chemically as per formulation given in Table 10.4 mentioned as (W/o Microwave). In this, the devulcanization has been carried out only by shearing action which generates the free radicals and by using devulcanizing agent.

In the next experiment, devulcanization was carried out for this NR crumb by thermal means using microwave radiation followed by mechanochemical process. The microwave which was used was domestic type microwave. The mixing and mastication was done with a friction ratio of 1:2 in an open two roll mill. This crumb was milled again at 40–50°C temperature for 10 min in a two roll mill with instantaneous addition of garlic oil and process oil, and then revulcanizing it. Compound formulation is given in Table 10.4. The batch is cured in compression mold at 145°C for 4 min.

10.5.2.4   CHARACTERIZATION

10.5.2.4.1   Sol-Gel Percentage

The vulcanized and devulcanized samples were immersed in 100 ml of toluene for 72 h to attain equilibrium swelling. Subsequent to taking out samples, solvent was removed from the sample surface using blotting paper. These samples were immediately weighed using electronic balance followed by drying at 80°C for 5 h.

$$\text{Sol Fraction}(\%) = \frac{weight\ before\ swelling - weight\ of\ dried\ sampls}{weight\ before\ swelling} \times 100$$

$$Gel\ fraction(\%) = 100 - Sol\ fraction(\%)$$

TABLE 10.4   Compound Formulation for Revulcanizing NR.

Mat	W/o microwave		With microwave		Mat	W/o microwave		With microwave	
PHR	NG0	NG01	NG1	NG2	PHR	ND0	ND01	ND1	ND2
RNR	100	80	100	80	RNR	100	80	100	80
NR	0	20	0	20	NR	0	20	0	20
Garlic oil	6	6	6	6	DPDS	2	2	2	2
Oil	6	6	6	6	Oil	6	6	6	6
CBS	1	1	1	1	CBS	1	1	1	1
Sulfur	2.2	2.2	2.2	2.2	Sulfur	2.2	2.2	2.2	2.2
PVI	1	1	1	1	PVI	1	1	1	1
Total	116.2	116.2	116.2	116.2	Total	112.2	112.2	112.2	112.2

10.5.2.4.2   Hardness Measurement

Hardness measurement was carried out as per ASTM D2240 using hardness durometer Shore A tester. Table 10.5 shows that there is increase in hardness or it remains the same as vulcanized rubber. It is found that there is minor effect on hardness change in the recycled vulcanizate.

TABLE 10.5   Properties of NR.

Samples	Tensile strength (Kg/cm²)	Elongation (%)	Sol content (%)	Hardness A⁰ shore
NR	138.64	273.03	13.63	55–60
NR	149.39	286.31	–	55–60
NG0	42.73	214.15	27.84	51–55
NG1	33.68	144.61	26.10	51–56
NG01	73.154	237.85	14.92	50–54
Ng2	92.61	318.46	18.60	52–54

TABLE 10.5 *(Continued)*

Samples	Tensile strength (Kg/cm²)	Elongation (%)	Sol content (%)	Hardness A⁰ shore
ND0	48.06	176.76	31.95	50–52
ND1	48.48	187.07	13.76	50–54
Nd01	36.99	211.07	24.85	52-55
ND2	89.38	335.07	31.2	50-53

10.5.2.4.3 Mechanical Properties

The samples were tested for tensile strength and percent elongation as per ASTM D 412-2002 standards using universal tensile testing machine. The initial gauge length of the specimen was 6.5 cm and tensile speed was 300 mm/min. And the tensile strength of the revulcanized rubber is poorer than the original formulation. This maybe attributed to chain scission throughout thermal and mechanical shearing. But improvement in properties was seen for crumbs heated by microwave radiation.

$$\frac{L - Lo}{Lo} \times 100$$

where L is the stretched length.
   Lo is the original length.

10.5.2.4.4 Fourier-transform Infrared Spectroscopy

Original and revulcanized rubber samples were characterized at room temperature using attenuated total reflectance (ATR) technique on an Opus Fourier transform infrared spectroscopy (FTIR) instrument. Characteristic infrared (IR) bands at 886 and 1374 cm^{-1}·(vs) for NR are found much similar in vulcanized and revulcanized rubber samples (refer Figure 10.20(a, b)).

   The FTIR spectrum of all NR vulcanizates showed strong bands in the series of 3015 2744 cm^{-1} for stretching frequency of the C–H groups (refer Fig. 10.21). The bands at 1076, 1029, and 1006 cm^{-1}, are related to C–S bonds for different reclaimed sample.

Advanced Polymeric Materials for Sustainability and Innovations

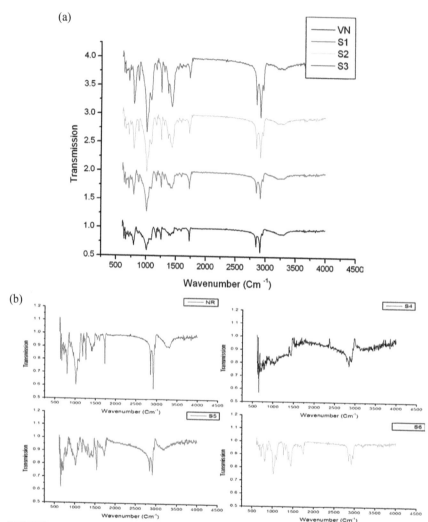

FIGURE 10.20 Fourier-transform infrared spectroscopy (FTIR) spectra of natural rubber (NR) with devulcanization by (a) garlic oil, VN: NR spectra, S1: garlic oil without microwave, S2: garlic oil with microwave, S3: garlic oil with microwave + 20% NR and (b) DPDS, NR: NR spectra, S4: DPDS without microwave, S5: DPDS with microwave, S6: DPDS with microwave + 20% NR.

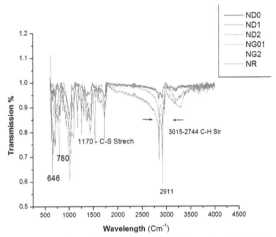

FIGURE 10.21 Attenuated total reflectance (ATR) spectra for NR vulcanizate with different reclaimed rubber.

10.6 RESULT AND DISCUSSION

Increase in sol fraction indicates the increase in devulcanization percentage. Table 10.5 shows the sol percentage of the samples. Sample with 80/20% gives more sol fraction than 100% waste sample. In NR DPDS shows more degree of devulcanization.

It is found that there is minor effect on hardness change in recycled vulcanizate. In some cases there is increase in hardness or it remains the same as vulcanized rubber. Hardness may vary because of curing conditions and time, and it also depends on mastication time. Mastication improves the elasticity of rubber. If there is difference in mastication duration the properties may vary.

The tensile properties of the revulcanized rubber are poorer than the original formulation. This may be attributed to chain scission throughout thermal and mechanical shearing. But improvement in properties was seen for heating of crumb by microwave radiation.

Tensile strength of 80/20 combination is much better. Here we found that there is improvement in tensile strength when rubber undergoes microwave radiation.

In NR, 80/20 combination gives better properties than 100% waste reclaim samples for both DPDS and garlic oil. For 80/20 NR elongation percentage is almost 15–21 % more than virgin rubber.

The FTIR spectrum of all NR vulcanizates showed strong bands in the series of 3015 2744 cm^{-1} for stretching frequency of the C–H groups. The bands at 1076, 1029 and 1006 cm^{-1}, are related to C–S bonds for a different reclaim sample.

10.7 CONCLUSION

Although DPDS gives relatively better result than garlic oil, garlic oil is less expensive. It is obtained from natural source and is not harmful at all. The results of garlic oil studies indicate improvement in results with increase in its percentage in compound. In general, mechanical properties decrease after revulcanization, but when the three devulcanization methods are combined together much better properties in terms of increase in tensile strength, percent elongation, sol content, whereas decrease in cure time can be obtained (Refer to Table 10.5).

Garlic oil can be the best option for devulcanization. Few more experiments are required to be conducted with varying percentage of garlic oil with other conventional additives. Recycling and reuse also results in a decrease in CO_2 gas emissions and increases carbon credits earned.

REFERENCES

1. http://www.statista.com/statistics/275399/world-consumption-of-natural-and-synthetic-caoutchouc/. (Accessed April 2016)
2. http://www.waste-management-world.com/articles/2003/07/scrap-tire-recycling.html. (Accessed Jan 2016)
3. CalRecovery, Inc. Evaluation of Waste Tire Devulcanization Technologies, December 2004.
4. Liang, T. Continuous Devulcanization of Ground Tire Rubber of Different Particle Sizes Using an Ultrasonic Twin Screw Extruder. Thesis, May 2013.
5. Fainleib, A.; Grigoryeva, O. *Recent Developments in Polymer Recycling*; 2011.
6. Rajan, V. V.; Dierkes, W. K.; Joseph, R.; Noordermeer, J. W. M. *J. Appl. Polym. Sci.* **2006,** *102,* 4194.
7. Dijkhuis, K. A. J.; Babu, I.; Lopulissa, J. S.; Noordermeer, J. W. M.; Dierkes, W. K. *Rubber Chem. Technol.* **2008,** *81,* 190.
8. Choi, J.; Isayev, A. I. Ultrasonic Aided Extrusion of Carbon Black-Filled SBR Compounds. *Rubber Chem. Technol.* **2011,** *84,* 55.
9. Sitisaiyidah, S. Post-Consumer Tires Back into New Tires. Thesis, May 13.
10. Maxwell, B. U.S. Patent 4,146,508, 1979.

11. De, D.; Ghosh, A. K.; Maiti, S.; Adhikari, B. Reclaiming of Rubber by a Renewable Resource Material (RRM). II. Comparative Evaluation of Reclaiming Process of NR Vulcanizate by RRM and Diallyl Disulfide. *J. Appl. Polym. Sci.* **1999,** *73,* 2951–2958.

12. Ball, J. M. *Reclaimed Rubber;* Rubber Reclaimers Association Inc.: New York, 1947.

13. Bateman, L. *The Chemistry and Physics of Rubber Like Substance;* Maclaren: London, 1963. (Chapter14).

14. Singleton, R.; Davies, T. L. *Rubber Technology and Manufacture;* Butterworth: London, 1982; 237–242.

15. Hader, R. N.; Le Beau, D. S. Chemistry of Reclaimed Rubber. *Ind. Eng. Chem.* **1951,** *43,* 250–263.

16. Harshaft, A. A. Solid Waste Treatment Technology. *Environ. Sci. Technol.* **1972,** *6,* 412–421.

17. Kohler, R.; O'Neill, J. New Technology for the Devulcanization of Sulfur-Cured Scrap Elastomers. *Rubber World* **1977,** *216,* 32.

18. Sekhar, B. C. European Patent Application EP 0,690,091 AL, 1995.

19. Makrov, V. M.; Drozdovski, V. F. *Reprocessing of Tires and Rubber Wastes;* Ellis Horwood: New York, 1991.

20. Tyler, K. A.; Cerny, G. L. (USA: Goodyear Tire and Rubber). U.S. 4,459,450, 1984.

21. Pelofsky, A. H. Rubber Reclamation Using Ultrasonic Energy. US Patent 3,725,314, 1973.

22. Okuda, M.; Hatano, Y. (Yokohama Rubber Co. Ltd., [JP: 62121741]. Service Oil Co.), 1987.

23. Yun, J.; Isayev, A. I. Superior Mechanical Properties of Ultrasonically Recycled EPDM Rubber. *Rubber Chem. Technol.* **2003,** *76,* 253–270.

24. Meyer, K. H.; Hohenemser, W. Contribution to the Study of the Vulcanization Reaction. *Rubber Chem. Technol.* **1936,** *9,* 201–205.

25. Le beau, D. S. Science and Technology of Reclaimed Rubber. *Rubber Chem. Technol.* **1967,** *40,* 217.

26. Knorr, K. Reclaim from Natural and Synthetic Rubber Scrap for Technical Rubber Goods. *Kautsch. Gummi Kunstst.* **1994,** *47,* 54–57.

27. Anderson, E., Jr. U.S. Patent 4,544,675, 1985.

28. Bilgili, E.; Dybek, A.; Arastoopour, H.; Bernstein, A. New Recycling Technology: Compression Molding of Pulverized Rubber Waste in the Absence of Virgin Rubber. *J. Elastomers Plast.* **2003,** *35,* 235.

29. Yamashita, S.; Kawabata N.; Sagan S.; Hayashi K. Reclamation of Vulcanized Rubbers by Chemical Degradation. V. Degradation of Vulcanized Synthetic Isoprene Rubber by the Phenylhydrazine–Ferrous Chloride System. *J. Appl. Polym. Sci.* **1977,** *21,* 2201.

30. Kasai, K.; Watanabe, T.; Harada, K. *Japan Kokai* **1977,** *77*(105), 951. *Chem. Abst.* **1978,** *88,* P51796n.

31. Martinez, D. F. U.S. Patent 5,304,576, 1994.

32. Bryson, J. G. U.S. Patent 4,148,763, 1979.

33. DEO 4042009 (German Patent Appl. 25.6.1992, Merseburg, TH, Inv. W. Neumann et al.).

34. Martin, J. M.; Smith, W. K. *Handbook of Rubber Technology*; Vol 2; CBS Publishers & Distributors, 2004.
35. Tsuchii, A.; Hayasshi, K.; Hironiwa, T.; Matsunaka, H. The Effect of Compounding Ingredients on Microbial Degradation of Vulcanized Natural Rubber. *J. Appl. Polym. Sci.* **1990,** *41,* 1181–1187.
36. Kojima, M.; Tosaka, M.; Ikeda, Y.; Kohjiya, S. Devulcanization of Carbon Black Filled Natural Rubber Using Supercritical Carbon Dioxide. *J. Appl. Polym. Sci.* **2005,** *95,* 137–143.
37. Tzoganakis, C.; Zhang, Q. Devulcanization of Recycled Tire Rubber Using Supercritical Carbon Dioxide. University of Waterloo, 2004.
38. Ciesielski, A. *An Introduction to Rubber Technology;* Rapra technology Ltd., 1999.
39. http://www.mdidea.com/products/new/new00106.html.
40. Bagudo, B. U.; Acheme, O. D. Chemical Analysis of Locally Cultivated Garlic and it's Oil. *Der Chem. Sin.* **2014,** *5*(1), 128–134.
41. https://www.alfa.com/en/catalog/A12586/.
42. https://en.wikipedia.org/wiki/Diphenyl_disulfide.
43. Jana, G. K.; Das, C. K. Recycling Natural Rubber Vulcanizates through Mechano-chemical Devulcanization. *Macromol. Res.* **2005,** *13*(1), 30–38.
44. Jana, G. K.; Mahaling, R. N.; Rath, T.; Kozloswska, A.; Kozloswski, M.; Das, C.K. Mechano-Chemical Recycling of Sulfur Cured Natural Rubber, *Polymery* **2007,** *52*(2).

CHAPTER 11

PREPARATION AND CHARACTERIZATION OF POLYMER MATRIX COMPOSITES WITH REINFORCED FLY ASH AND SILICON CARBIDE

K. ARUNA[1], B. SUDHEER REDDY[2,*], B. R. P. NAYAK[3], and B. HARITHA BAI[4]

[1]Assistant Professor, Department of Mechanical Engineering, Sri Venkateswara University, Tirupati - 517502, India
E-mail: aruna@svuniversity.ac.in

[2]M.Tech Scholar, Department of Mechanical Engineering, Sri Venkateswara College of Engineering, Karakambadi Road, Tirupati, 517502, India. E-mail: sudheerb121@gmail.com

[3]Teaching Faculty, Department of Mechanical Engineering, Sri Venkateswara University, Tirupati - 517502, India
E-mail: raghu536@gmail.com

[4]Assistant Professor, Department of Mechanical Engineering, SIETEK, Puttur, A. P., India

*Corresponding author. E-mail: sudheerb121@gmail.com

CONTENTS

ABSTRACT

Polymer matrix composites (PMCs) have evoked a keen interest in recent times for potential applications in the industries, such as automotive, aerospace, marine, and so forth, owing to their superior strength to weight ratio. Literature shows that there is a lacuna in achieving a uniform distribution of reinforcement of composite materials in existing molding technology by addition of fly ash and silicon carbide. As such, a solemn attempt is made in the present chapter with an objective to develop linear low-density polyethylene based on the fly ash and silicon carbide particulates as reinforcement materials, so as to obtain homogenous dispersion of composite material with possible low costs. Injection molding technique is proposed and analyzed. Experiments are conducted by varying weight fractions of fly ash and silicon carbide. The output results have indicated that the developed method is quite successful to obtain a uniform dispersion of reinforcement in the PMCs. An increasing trend of hardness with density has been observed with weight percentage of fly ash and silicon carbide.

11.1   INTRODUCTION

Manufacturing of parts and their subassemblies leading to the products with quality is the order of the day and also to meet the global competition, Thus, the usage of conventional materials to manufacture the parts/products have certain limitations in achieving good combination of strength, stiffness, toughness, and density followed by the quality levels. To overcome such limitations and to meet the ever increasing demand of modern day technology, the innovation of the composite materials such as polymer matrix composites (PMCs) has been resulted. PMCs possess high strength, hardness, toughness, and good thermal resistance properties in comparison to other alloys. Additional features such as isotropic properties and possibility of secondary processing lead to easy production of components followed by reinforcement with particulates. PMCs can be easily produced by injection molding process at low cost.

11.1.1 COMPOSITE MATERIALS

A composite material is a composed basic material with reinforcement of fibers, particles, flakes, and/or fillers and embedded in a matrix leading to polymer metals or ceramics. The matrix holds the reinforcement to form the desired shape while the reinforcement improves the overall mechanical properties of the matrix. The classification of composites is shown in Figure 11.1.

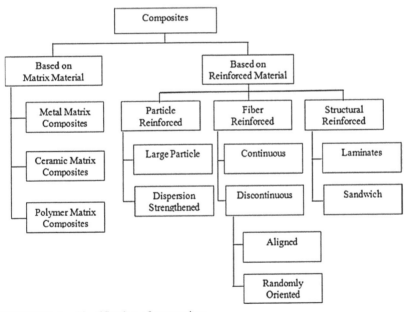

FIGURE 11.1 Classification of composites.

11.2 LITERATURE SURVEY AND OBJECTIVE

A brief literature survey is presented based on the earlier contributions in the following sections.

11.2.1 METAL MATRIX COMPOSITES (MMCs)

A metal matrix composite (MMC) is a composite material with at least two constituent parts: one has to be a metal necessarily, the other maybe a different metal or another material, such as a ceramic or organic compound. When at least three materials are present, it is called a hybrid composite. An MMC is complementary to a cermet.[1]

It consists of at least two chemically and physically different phases. Fibrous or particulate phases in MMC are: aluminum matrix composites; magnesium matrix composite; titanium matrix composite; and copper matrix composites. The automotive industry recognizes that weight reduction and improved engine efficiency will make the greatest contribution to improved fuel economy with the current power trains. This is evidenced by the increased use of aluminum alloys in engine and chassis components. Aluminum and magnesium castings in this sector have grown in leaps and bounds over the past 5 years to help engineers design and manufacture more fuel-efficient cars. The low density and high specific mechanical properties of aluminum metal matrix composites make these alloys one of the most interesting material alternatives for the manufacture of lightweight parts for many types of vehicles.[5] Therefore, the AMCs have been widely used in sporting goods, electronic packaging, aeronautics, and automotive industries. Ceramic matrix composites are designed to improve the toughness of conventional ceramics. The best strengthening effect is provided by chemical vapor deposition of SiC on a substrate made of tungsten or carbon fibers.[8]

11.2.2 POLYMER MATRIX COMPOSITES (PMCs)

The other important category is PMCs. It is a material consisting of a polymer (resin) matrix combined with a fibrous reinforcing dispersed phase; PMCs are very popular due to their low cost and simple production methods. The unreinforced polymers are used as structure materials with low impact resistance. The PMCs are divided into particle reinforced composites (PRCs) which are further divided into large-particle composites (LPCs) and dispersion-strengthened composites (DSCs). As one of the most important fields of current nanoscience, the polymer nanocomposites are promising and efficient way for new generation materials with high performances and multifunctionalities. The incorporation of nanofillers in a polymer matrix may improve mechanical, thermal, electrical, or dielectric properties of the composites.[3]

The concrete and reinforced concrete are the examples of LPCs. In DSCs, the particle size varies from 10–100 nm. The small particles are dispersed throughout the matrix and prevent plastic deformation by blocking the motion of dislocations. Fibers are responsible for high strength and stiffness ratio to weight of the composite. Some examples are the carbon fibers, boron fibers, E-glass fibers, SiC fibers, Al_4C_3 fibers, and so forth. The mechanical properties of the E-glass and carbon fiber specimens were compared from tests including tensile, compressive, flexural, and interlaminar shear strength (ILSS); impact, and head deflection test (HDT). According to American Standard Testing Machine (ASTM),[10] the IPN reinforced carbon fiber specimen showed better results in all the tests than E-glass fiber-reinforced IPN laminate with same thickness of the specimen. Laminate composites having sheets or panels with different fiber orientations are arranged in several layers. These layers are stacked and subsequently cemented together by the orientation such that the high strength can be obtained. A thicker core separates two thin sheets in the sandwich panel. The sheets or faces are bonded adhesively to the core having low density material, such as polymer foam or expanded metal structure, to provide support to the outer faces. The sheets present in outward direction can be made from a strong and stiff material, such as steel, titanium, Al alloys, and so forth, to sustain various stresses due to loading.[4]

11.2.3   CHARACTERISTICS OF COMPOSITES

The characteristics of composites are dependent on the properties of their (i) constituent materials; (ii) geometry of the reinforcement (shape and size); (iii) spherical, cylindrical, or rectangular cross-sectioned prisms or platelets; (iv) concentration distribution and orientation of the reinforcement; (v) size and size distribution (controls texture of the material); and (vi) volume fraction (determines the interfacial area). Then in later stages, the naturally available plants and fibers as alternate reinforcement in the polymer matrix composites because of its excellent characteristics, such as high tensile properties, good damage tolerance, high modulus to its weight, high impact strength, great fatigue behavior, and corrosive properties, produce a significant contribution for traditional materials in several applications, as well as these materials can be converted to any required shape at a minimum cost and acceptable quality.[2,7]

Composites are referred as engineering materials with characteristics similar to artificial materials having two different elements with a well-defined interface and properties, which are influenced by the volume percentage of elements.[6] There has been an increasing interest in composites containing low density and low-cost reinforcements. So far, most of the research work have been carried out by incorporating hard ceramic particles such as Al_2O_3, SiC, fly ash, and graphite particles to a soft matrix, for example, pure aluminum A356, and many more alloys and very few worked on combination of reinforcements (hybrid composites).[9] Therefore, a solemn attempt is made in the present chapter with the objectives of: (i) fabrication of PMC samples by varying the reinforcements; (ii) testing the mechanical properties and; (iii) evaluation of the new PMCs.

11.3   CONSTITUENTS OF PMCs

The main constituents of PMCs are linear low-density polyethylene (LLDPE) as matrix material (base materials) and reinforcement materials in the form of particles. Different matrix materials and reinforcement materials are used to obtain the different composites.

11.3.1   MATRIX MATERIAL

In the present chapter, LLDPE is considered and enriched by the addition of reinforcement material such as silicon carbide and fly ash.

11.3.1.1   SILICON CARBIDE (SiC)

It is also known as carborundum and is a compound of silicon and carbon with chemical formula SiC. Grains of silicon carbide can be bonded together by sintering to form very hard ceramics which are widely used in applications requiring high endurance such as car brakes, car clutches, and ceramic plates in bulletproof vests. Silicon carbide with high surface area can be produced from SiO_2 contained in plant material.

11.3.1.2   *FLY ASH*

Fly ash is one of the residues generated in the combustion of coal. It is an industrial by-product recovered from the flue gas of coal-burning electric power plants. Depending upon the source and makeup of the coal being burnt, the components of the fly ash produced vary considerably, but all fly ash includes substantial amounts of SiO_2, lime, and Al_2O_3. Therefore, larger application of fly ash is in the cement and concrete industries.

11.3.2   *INJECTION MOLDING*

Injection molding is the most commonly used manufacturing process for the fabrication of plastic parts. A wide variety of products are manufactured using injection molding, which vary greatly in their size, complexity, and application. The injection molding process requires the use of an injection molding machine, raw plastic material, and a mold. The plastic is melted in the injection molding machine and then injected into the mold, where it cools and solidifies into the final part.

Injection molding machines are also known as presses, holding the molds in which the components are produced to the required size and shape. The presses are rated by tonnage, which expresses the amount of clamping force that the machine can generate. Such pressure keeps the mold closed during the injection process. The injection molding machine consists of two basic parts: an injection unit and a clamping unit (refer Figure 11.2(a)). The name of the injection molding machine is generally based on the type of injection unit used.

The injection molding machines are classified primarily by the type of driving systems which are: hydraulic, electric, or hybrid. The molding machine used in the present research chapter is shown in Figure 11.2(a) and experimental setup in Figure 11.2(b).

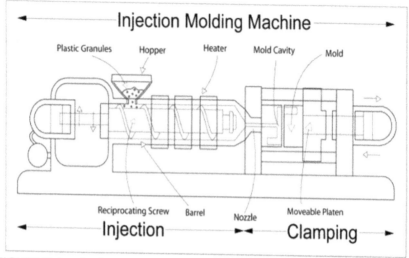

FIGURE 11.2a   Sectional view of machine

FIGURE 11.2b   Experimental setup.

11.4 METHODOLOGY

The methodology consists of two stages. At the first stage, the PMCs are produced with the addition of fly ash and SiC in the injection molding machine. The second stage consists of the testing of the PMCs and discussed the suitability of the application.

11.4.1 PRODUCTION OF PMC

Though the fly ash has so much of important industrial application, the reinforcement is being tried for the first time in the present work to produce PMCs. The preparation starts with four required raw materials of LLDPE, fly ash and SiC as reinforcement with different weight ratios of 0, 2, 4, 6, 8, and 10%.

The process cycle for production of PMCs in injection molding is very short, typically between 2 s and 2 min, and consists of the following four stages:

- **Clamping:** Prior to the injection of the material into the mold, the two halves of the mold must first be securely closed by the clamping unit. Each half of the mold is attached to the injection molding machine and one half is allowed to slide. The hydraulically powered clamping unit pushes the mold halves together and exerts sufficient force to keep the mold securely closed while the material is injected. The time required to close and clamp the mold is dependent upon the machine—larger machines (those with greater clamping forces) will require more time. This time can be estimated from the dry cycle time of the machine.
- **Injection:** The raw plastic material (mixed LLDPE granules), usually in the form of pellets, is fed into the injection molding machine, and advanced toward the mold by the injection unit. During this process, the material is melted by heat and pressure. The molten plastic is then injected into the mold very quickly and buildup of pressure packs and holds the material. The amount of material that is injected is referred to as the shot. The injection time is difficult to calculate accurately due to the complex and changing flow of the molten plastic into the mold. However, the injection time can be estimated by the shot volume, injection pressure, and injection power.
- **Cooling:** The molten plastic that is inside the mold begins to cool as soon as it makes contact with the interior mold surfaces. As the plastic cools, it will solidify into the shape of the desired part. However, during cooling, some shrinkage of the part may occur. The packing

of material in the injection stage allows additional material to flow into the mold and reduce the amount of visible shrinkage. The mold cannot be opened until the required cooling time has elapsed. The cooling time can be estimated from several thermodynamic properties of the plastic and the maximum wall thickness of the part.

- **Ejection:** After sufficient time has passed, the cooled part may be ejected from the mold by the ejection system, which is attached to the rear half of the mold. When the mold is opened, a mechanism is used to push the part out of the mold. Force must be applied to eject the part because during cooling, the part shrinks and adheres to the mold. In order to facilitate the ejection of the part, a mold release agent can be sprayed onto the surfaces of the mold cavity prior to injection of the material. The time that is required to open the mold and eject the part can be estimated from the dry cycle time of the machine and should include time for the part to fall free of the mold. Once the part is ejected, the mold can be clamped shut for the next shot to be injected.

After the injection molding cycle, some post processing is typically required. During cooling, the material in the channels of the mold will solidify attached to the part. This excess material, along with any flash that has occurred, must be trimmed from the part, typically by using cutters. For some types of material, such as thermoplastics, the scrap material that results from this trimming can be recycled by being placed into a plastic grinder, also called regrind machines or granulators, which regrinds the scrap material into pellets. Due to some degradation of the material properties, the regrind must be mixed with raw material in the proper regrind ratio to be reused in the injection molding process. The desired part is shown in Figure 11.3. Thus, the PMCs are produced successfully.

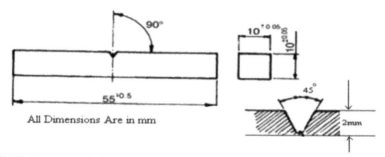

FIGURE 11.3   Sample for testing (as per ASTM D256).

11.4.2   TESTING OF PMC SAMPLES FOR MECHANICAL PROPERTIES

The produced PMCs are tested for their properties so as to determine the suitability of a material for a particular application. The tests conducted are: (i) impact strength on Charpy impact testing machine, (ii) hardness on Rockwell hardness testing machine; and (iii) density.

11.4.2.1   IMPACT STRENGTH

Impact strength signifies the toughness of material which is the ability of material to absorb energy during deformation. The impact test consists of preparation of specimens as per ASTM D256 specification of size $10 \times 10 \times 55$ mm, a 450 V-notch of 2 mm depth and a 0.25 mm root radius, as shown in Figure 11.3. The pendulum of impact testing machine is set at a certain height and released on to the specimen at the opposite end of the notch to produce a fractured sample. The absorbed energy required to produce two fresh fracture surfaces is recorded. The formulae are given below for calculating the Charpy impact test values. For each specimen, the values are recorded as shown in Table 11.1.

TABLE 11.1   Experimental Results.

Exp. no.	Specimen details (%)	Impact strength (N/mm²)	Rockwell hardness number	Density (ρ) (kg/m³) × (10⁻⁴)
1	Fly Ash-0	0.192	24	9.27
2	Fly Ash-2	0.240	72	9.40
3	Fly Ash-4	0.336	73.5	9.52
4	Fly Ash-6	0.36	76.5	9.52
5	Fly Ash-8	0.48	77.5	9.65
6	Fly Ash-10	0.637	84	9.65
7	SiC-0	0.192	24	9.27
8	SiC-2	0.312	61	9.27
9	SiC-4	0.336	70	9.40
10	SiC-6	0.432	74	9.52
11	SiC-8	0.600	75.5	9.65
12	SiC-10	0.673	88	9.65

11.4.2.2   HARDNESS OF PMCs

The Rockwell hardness test is conducted to find the hardness of PMCs according to E18-07 standard test methods. The results are shown in Table 11.1.

11.5   RESULTS AND DISCUSSIONS

The results obtained from the experiments are represented in the form of graphs. The percentage of fly ash and SiC versus the impact strength is 5, shown in Figure 11.4. It can be observed that the impact strength of the PMC is increasing with the increase of percentage of reinforcement. The graph pertaining to percentage of fly ash and SiC versus hardness is shown in Figure 11.5 followed by density in Figure 11.6. From both graphs, it can be seen that with an increase of percentage of fly ash and SiC, both the hardness and density are increasing. To produce the materials with hardness, the PMCs with high percentage of reinforcement material is better suited.

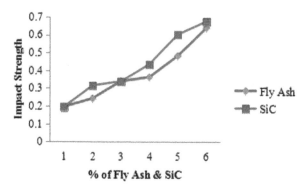

FIGURE 11.4   Percentage fly ash and SiC versus impact strength.

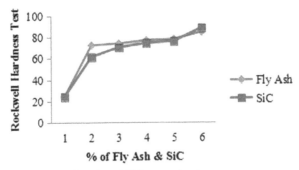

FIGURE 11.5 Percentage fly ash and SiC versus hardness.

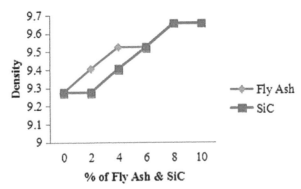

FIGURE 11.6 Percentage fly ash and SiC versus density.

11.6 CONCLUSIONS

An initial attempt is made successfully in the present work and produced new PMCs with the reinforcement materials of fly ash and SiC by injection molding process. Tests are conducted to find their impact strength and hardness and analyze the characteristics of PMCs. The maximum hardness obtained is 84 RHN with the reinforcement of 10% of fly ash followed by 88 RHN with 10% of SiC. The hardness of the PMC is increasing with the increase of the percentage of reinforcement. It is also established that the new materials produced have the suitable mechanical property with strengths for practical applications.

11.6.1   APPLICATIONS

PMCs are used for manufacturing of
- Aerospace structures
- Marine structures
- Automotive bodies
- Sports goods
- Bulletproof vests and other armor parts;
- Chemical storage tanks and pressure vessels
- Biomedical applications
- Bridges
- Electrical: Panels, housing, switch gear, insulators, and connectors

11.6.2   SCOPE FOR FUTURE WORK

The work may be extended by considering high-density polyethylene (HDPE) and LDPE as matrix material and reinforcement materials such as aluminum oxide and boron carbide.

KEYWORDS

- composite materials
- metal matrix composites
- polymer matrix composites
- silicon carbide and fly ash
- molding machine

REFERENCES

1. Callister, W. D. Jr. *Materials Science and Engineering: An Introduction*, 7th ed.; Wiley and Sons Publishing, ISBN-13: 978-0-471-73696-7 (cloth) ISBN-10: 0-471-73696-1 (cloth).
2. Dandekar, C. R., Shin, Y. C. Modeling of Machining of Composite Materials: A Review. *Int. J. Mach. Tools Manuf.* **2012**, *57*, 102–121.

3. Haddadi, M.; Agoudjil, B.; Boudenne, A. Thermal Conductivity of Polymer/Carbon Nanotube Composites. *Mater. Sci. Forum* **2012,** *714,* 99–113. ISSN: 1662–9752.
4. Khalaf, M. N. Mechanical Properties of Filled High Density Polyethylene. *J. Saudi Chem. Soc.* 2015, 19(1), 88–91. ISSN: 1319–6103.
5. Thirumoorthy, A.; Arjunan, T. V.; Senthil Kumar, K. L. Latest Research and Development in Aluminum Matrix with Particulate Reinforcement Composites: A Review. Science Direct, Conference Proceedings, 2016.
6. Mark, J. E. *Physical Properties of Polymers Handbook;* Springer. ISBN: 978-0-387-31235-4.
7. Ramesh, M.; Gopinath, A.; Deepa, C. Machining Characteristics of Fiber Reinforced Polymer Composites: A Review. *Ind. J. Sci. Technol.* **2016,** *9*(42). ISSN: 0974–6846.
8. Rizvi, R.; et al. Development and Characterization of Solid and Porous Polylactide-Multiwall Carbon Nanotube Composites. *Polym. Eng. Sci.* **2011,** *51,* 43–53.
9. Siddique, S. K. A; Nataraj, T. S. C. Deformation Studies on FA/SIC Reinforced Aluminium Composite. *Int. J. Mod. Eng. Res.* **2016,** *6*(11). ISSN: 2249–6645.
10. Suresh, G.; Jayakumari, L. S. Evaluating the Mechanical Properties of E-glass Fiber/Carbon Fiber Reinforced Interpenetrating Polymer Networks. *Polímeros* **2015,** *25*(1). ISSN: 0104–1428.

CHAPTER 12

ION-CONDUCTING POLYMERS

C. S. SUNANDANA*

School of Physics, University of Hyderabad, Hyderabad, Telangana 500046, India

Corresponding author. E-mail: sunandana@gmail.com

CONTENTS

ABSTRACT

A perspective on the problem of ion conduction in polymer electrolytes, focusing on the properties of polymer that favor ion transport is provided. Contemporary issues in polymer electrolytes are briefly discussed next. This is followed by an account of of two fundamental experimental skills: electrical conductivity and positron annihilation. Following up on one of the contemporary issues, recent innovations aimed at enhancing ion conduction are discussed with the aid of most recent work. A fundamental model of ion conduction in polymers based on free volume concept is briefly outlined. Future directions include (1) use of fragile polymers to decouple the ionic conductivity relaxation from segmental relaxation; (2) use of diblock and triblockcopolymers, and (3) polymerization of ionic liquids. Future years would hopefully witness the development and use of new polymer electrolytes andexploration of mechanisms of ion conduction in them.

12.1 WHY POLYMER ELECTROLYTES?

To begin with, let us note that a famous poem "The Dance of Solids" by John Updike that celebrates polymers:

> The *Polymers,* those giant Molecules,
> Like Starch and Polyoxymethylene
> Flesh out, as protein serfs and plastic fools
> The Kingdom with Life's Stuff. Our time has seen
> The synthesis of Polyisoprene
> And many cross-linked Helixes unknown
> To *Robert Hooke;* but each primordial Bean
> Knew Cellulose by heart: *Nature* alone
> Of Collagen and Apatite compounded Bone.

Polymers or more relevantly oxidized polymers which are composed of long alkane chains and segments (for example, polyethylene oxide (PEO)) are electronic insulators. A major finding by Wright and coworkers in 1973[1] was that ionic salts such as LiI may be dissolved in polymers which result in an ion conducting polymer, in this case, a Li^+ ion conducting polymer. Solid-state Li^+ ion batteries have been built using polymer electrolytes[2-4]. Compared to the liquid electrolytes typified by the lead-acid battery, polymer electrolytes possess (1) shape versatility, (2) mechanical strength and rigidity, (3) stable contact at the electrode–electrolyte interface, and (4) much better safety through (a) absence of leakage, (b) nontoxicity, (c) low vapor pressure, and (e) nonflammability.

What is the precise role of the polymer electrolyte membrane in a solid-state Li ion battery such as the LIPO cell in a laptop? It is threefold: (1) enables Li^+ transport between the electrodes, (2) blocks e^- transport between the electrodes, (3) being mechanically rigid prevents direct contact between the electrodes.

A typical sketch of the polymer electrolyte in the context of a Li^+ ion battery is given in Figure 12.1.

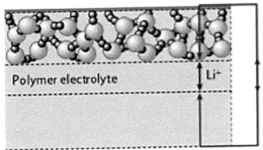

FIGURE 12.1 Polymer electrolyte sits in an all-solid-state Li^+ ion battery made up of porous LiFePO$_4$ cathode and Li anode.

Practically speaking, what polymer characteristics favor ionic conduction? The most important point to note is the fundamental electrochemical compatibility of the inorganic ionic salt with the essentially covalent bonded polymer matrix. Furthermore, we have the following advantages of polymers over other types of electrolytes:

1. Ease of casting into near microlevel thin layers
2. Subambient glass transition temperature T_g (unlike inorganic glasses) and ambient temperature operation of electrochemical devices
3. Manipulation of structure to tailor segmental motion (dynamics)
4. Easy integration into electrochemical devices (battery, fuel cell, sensor)
5. Smooth interface formation between electrode and electrolyte
6. 2D structure facilitates modeling of structure and ionic conduction process(es)/mechanisms.

In an early review,[5] we have considered polymer electrolytes in the general context of theoretical approaches to superionic conductivity. To quote from the abstract "The theory of conductivity in polymer electrolytes—still in its infancy—involves their complex structure and glass transition behavior. Preparative and thermal history, composition, and crystallinity control ionic conductivity. New approaches to the synthesis of optimal polymer electrolytes such as rubbery electrolytes, crystalline polymers and nanocomposites must be considered before achieving a comprehensive theoretical understanding." This understanding is just beginning to be attempted by identifying a few jigsaw pieces of the puzzle.

To focus on the point number (2) above, Angell[6] has given a relaxation time criterion based on the strong coupling between ionic motion and the segmental mobility. The decoupling ratio is the segmental relaxation time/conductivity relaxation time. Conductivity relaxation time derived from experimental dc conductivity (σ_{dc}) data gives a number $\sim 10 \times 10^{-13}/\sigma_{dc}$ s that can be compared to segmental relaxation time of 200 s at the glass transition temperature. The polymer has to relax rapidly to give an increased ionic conductivity. Thus, ionic diffusion is intimately connected to structural relaxation. The experimental situation is that as the polymer electrolyte is cooled such that T approaches T_g, structural relaxation slows down rapidly thereby decreasing σ rapidly. Then how can one increase σ? Addition of more salt is an obvious possibility; thus increasing the mobile ion concentration, one can increase σ. The temperature dependence of σ is controlled by the Vogel–Tamman–Fulcher (VTF) relation for polymer electrolytes:

$$\sigma = \sigma_0 \, e^{-B/(T-T_0)} \qquad\qquad (12.1)$$

Here T_0 is the equilibrium glass transition temperature. T_0 is roughly about 50 K less than the measured T_g. σ_0 is the prefactor and B is the pseudo-activation energy for the redistribution of the free volume in the polymer electrolyte. In Section 12.3, we consider the use of experimental techniques to probe these aspects.

The aim of this brief chapter is to provide an update on the most recent developments in this area from solid-state ionics viewpoint. Organized in five sections, Section 12.2 focuses on contemporary issues while Section 12.3 describes two important experimental probes for polymer electrolytes. Section 12.4 samples a few recent materials innovations. Section 12.5 outlines a fundamental model for ion conduction in polymers while Section 12.6 provides a summary and points at a few future directions.

12.2   CONTEMPORARY ISSUES IN POLYMER ELECTROLYTES

Bonding compatibility between ionic salts and covalent polymers, and possibilities for innovation and processing–structure–property triangle have provided a new impetus and motivation in the field of polymer electrolytes. The coexistence of the major amorphous and minor crystalline phases in these materials and the role of segmental mobility on ionic conductivity have pushed progress far. But questions of amorphous-phase-limited ionic conductivity and alternative strategies to enhance Li^+ (and other) ion conductivity have recently redefined the problem of ion conducting polymers.

The current issues in polymer electrolytes include:

1. High crystallinity and enhanced conductivity,
2. Control of polymerization through mobile ions.

Let us discuss these two issues briefly. The materials innovations that have resulted are discussed in Section 12.4.

The guiding principle of development of polymer electrolytes has been: "reach maximum amorphicity and lowest glass transition temperature." The recent focus is on controlled crystallization to achieve high ionic conductivity. In a recent study, Cheng et al.[7] have made single crystal polymers that exhibit anisotropic ionic conductivity parallel and perpendicular to the plane of the polymer sample, by soaking PEO in $LiClO_4$ pentyl acetate solutions. This has introduced the subject of conductivity anisotropy in polymer electrolytes.

The ethylene oxide segments in PEO/PS polymer blends doped with Li salt are apparently "cross-linked" by the Li^+ ions, thereby affecting their phase behavior. This interesting finding is the result of a mean-field modeling investigation by Ren et al.[8]

The combinatorial entropy that arises in the complexation between the ion-binding polymer and the Li^+ ions seems to provide a driving force for phase separation in the ionic salt–PEO-PS system. The most intriguing prediction of this study is that the asymmetric interaction between the Li^+ ions and PEO-PS results in an unusually asymmetric phase diagram. This study motivates experimental investigations on Li^+ salt polymer blends. Next we describe two important experimental probes for polymer electrolytes and devices.

12.3   BASIC EXPERIMENTAL PROBES FOR POLYMER ELECTROLYTES

We briefly consider two important but connected probes for polymer electrolytes: (1) electrical conductivity and (2) positron annihilation.

12.3.1   ELECTRICAL CONDUCTIVITY

Figure 12.2(a) depicts the apparatus used for measurement of ionic conductivity of polymer samples. Figure 12.2(b) and 12.2(c) illustrates the types of sample cells suitable for AC and DC conductivity measurements, respectively. It is important to note that the AC conductivity measurement by and large avoids the problem of polarization effects and offers signal frequency as an important variable to study conductivity relaxation,[2] (Chapter 4) through complex impedance or electrochemical impedance spectroscopy. Figure 12.3 illustrates the complex impedance plots for poly(propylene oxide) (PPO)-based electrolytes. Figure 12.4 gives the Arrhenius plots for a number of polymer electrolytes.

An important electrical characterization of an electrochemical device such as Li^+ ion battery is the monitoring of the open-circuit voltage as a function of time. This generates the discharge curve. Figure 12.5 illustrates the polymer electrolyte-based Li–air battery and Figure 12.6 gives discharge curves under (a) air and (b) flowing oxygen gas ambient.

What can we learn from ionic conductivity measurements on polymer electrolytes? A simple answer is: ion diffusivity and ion pairing. A recent study by Stolwijk et al.[9] has thoroughly analyzed temperature-dependent conductivity measurements on (a) NaI-PEO and (b) lithium bis (trifluoromethanesulfonyl) imide ($LiN(CF_3SO_2)_2$ or LiTFSI)-PEO. Their approach is to use an extended VTF equation which involves a Boltzmann factor containing pair formation enthalpy. This enthalpy increases upon decreasing salt concentration. While NaI-PEO combines a high pair fraction with high diffusivity of I^- ion, the opposite happens in LiTFSI-PEO (low ion pairing and relatively low mobility of the bulky TFSI ion).

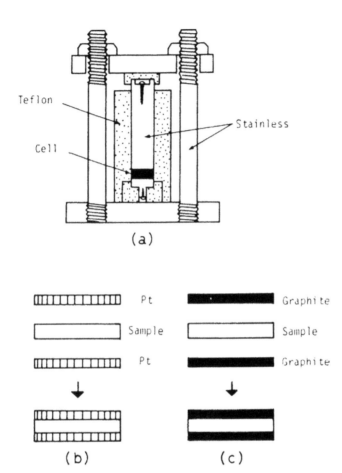

FIGURE 12.2 (a) Apparatus for electrical conductivity measurements on polymer electrolytes; (b) cell for AC conductivity measurement; and (c) cell for DC measurement. *Source:* Watanabe, M.; et al., *Polymer J.* 1982, *14*, 877; reproduced with permission.

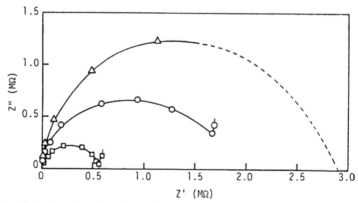

FIGURE 12.3 Complex impedance $Z^* = Z' + iZ''$ plots for poly(propylene oxide) (PPO)-based polymer electrolytes ΔPPO 3000 polycation-LiClO$_4$ salt/polymer = 0.02.

Source: Watanabe, M.; et al., *Polymer J.* **1982,** *14,* 877 with permission.

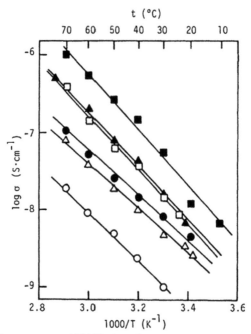

FIGURE 12.4 Log σ versus 1000/T plots for a number of polymer electrolytes. □ Conductivity plot for PPO 3000 polycation-LiClO$_4$ salt/polymer = 0.02. Slope gives activation energy for Li$^+$ ion transport.

Source: Watanabe, M.; et al., *Polymer J.* **1982,** *14,* 877; reproduced with permission.

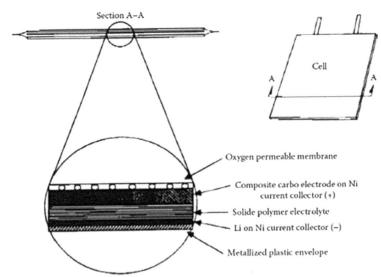

FIGURE 12.5   Schematic representation of a polymer electrolyte-based Li–air battery. The calculated open-circuit voltage of this battery for the idealized electrochemical reaction $4Li + O_2 \rightarrow 2Li_2O$ is 2.01 V. Theoretical specific energies including and excluding O_2 are 5.200 and 22.140 Wh/kg, respectively.

Source: Sunandana, C.S. *Introduction to Solid State Ionics;* CRC Press, 2015.

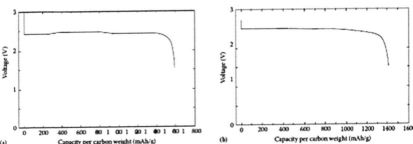

FIGURE 12.6   Probing the performance of a Li/PAN polymer electrolyte/oxygen cell packaged in a metalized plastic envelope. (a) Discharge curve obtained by exposing the carbon electrode to laboratory air, at a current density of 0.1 mA/cm^2 at room temperature. (b) Discharge curve of the same battery (as in (a)) but packaged in a D-shaped cell. O_2 from a tank maintains a flowing O_2 atmosphere.

Source: Sunandana, C.S. *Introduction to Solid State Ionics;* CRC Press, 2015.

12.3.2 POSITRON ANNIHILATION

As will be discussed in Section 12.5 (a fundamental model for ion conduction in polymers), free volume is an important thermodynamic concept involved in the ion conduction process in polymer electrolytes. Measurements of the lifetime of the positron annihilation process by bombarding positrons on the sample give free volume information. Positron annihilation lifetime spectroscopy (PALS) measurements that probe free volume on an atomic scale are carried out using a conventional fast–fast coincident spectrometer. A 15 µCi 22Na source is sandwiched between two identical samples. Typically, the spectrum of positron annihilation in $LiClO_4$/PEO has 1×10^6 gamma photon counts for analysis.[10] A plot of the orthopositronium lifetime or free volume versus temperature gives two straight lines separated by a knee. This brings out the importance of free volume in the ion conduction process to be discussed as a fundamental model in Section 12.5.

The next section focuses on the need for, and most recent examples of, materials innovation in polymer electrolytes.

12.4 RECENT DEVELOPMENTS IN MATERIALS INNOVATION

For over three decades now, the major focus in ion conducting polymers has been the synthesis of mostly amorphous polymers—often plasticized or modified by ceramic fillers. However, only a maximum conductivity of ~ 10 µS/cm is realizable in such polymers at room temperature. Why amorphous polymers? It is because segmental motion (of the Brownian type) occurs in them but not in crystalline polymers and such segmental motion is the main cause of ion transport. Recently, this contention was questioned leading to materials innovations in polymer electrolytes.

To put the problem in perspective, consider the PEO-Li salt polymer composite. Molecular interactions occurring in Li salt-doped polymer are such that most of the Li^+ ions are coordinated with ether oxygen. The negative charges are chemically bonded making them very sluggish compared to Li^+, seriously limiting ionic conductivity. This issue forms the crux of materials innovations.

Polymer electrolytes such as the ones found for some time in laptop batteries based on PEO/Li salt had to work with a conductivity of

~10 μS/cm. Such electrolytes suffer from the problem of chemically bonded negative charges which are sluggish compared to the very mobile Li^+ ions. Thus, there is an urgent need to develop separators for polymer-based Li^+ ion batteries. This together with the need to increase mechanical strength poses a challenge to discover and develop through innovations. Let us illustrate with three examples from recent research.

Bolloli et al.[11] have used nanocrystalline cellulose (NCC) as a reinforcer in a PVdF-based porous nanocomposite membrane. To describe briefly the synthesis, PVdF homopolymer, dry NCC (99%), and ethanol (Sigma Aldrich, 95%) were used as precursors. The reference electrolyte for the electrochemical testing was prepared by dissolving 1 M lithium hexafluorophosphate ($LiPF_6$) in an equivolume mixture of dimethyl carbonate and ethylene carbonate. The dispersion of NCC in dimethylformamide (DMF) was prepared by adding the NCC to the solvent (2 wt.% (standard concentration)) raised up to 3 wt.% to obtain the proper composition for the most loaded PVdF-NCC slurries). Sonicating the whiskers/solvent mixture at room temperature for several cycles of 30 min with a VC 505 Vibra-Cell™ using a 13 mm probe and storing the mixture in an ice bath for 10 min between each cycle solved the problems of excessive solvent evaporation and degradation. The quality of the dispersion was assessed by inspection of the dispersion between two crossed polarizers using birefringence as the criterion for having obtained a good dispersion. In the next step of nanocomposite films processing, PVdF was first dissolved in DMF. The resulting solution was added to the NCC dispersion in order to obtain a concentration of cellulose whiskers in the final membranes of 6, 12, and 20 wt.% and a PVdF concentration of $\approx$11.5 wt.% with respect to the solvent. The obtained suspension was stirred for 8 h at 50°C and degassed under vacuum in order to remove the remaining air.

To obtain dense films, the PVdF-NCC dispersion was cast into Petri dishes. The films were then left to evaporate at 60°C for 2 days. Eventually, they were dried under vacuum for 4 h until constant weight. The thickness varied between 15 and 40 μm, depending on the amount of dispersion cast onto the Petri dish. The samples were then pressed at 180°C in a laboratory hot press for 5 min in order to prepare dense membranes.

To obtain macroporous membranes, the PVdF-NCC dispersion was spread over a glass support with a doctor blade, and then immersed into an ethanol bath for 10 min. The resulting films were then dried in the same way as described for the dense membranes. A thickness of about 20 µm was obtained. These membranes exhibit superior electrochemical performance compared to polyolefin-based microporous separators.

Jo et al.[12] have demonstrated enhancement of cation transport of polymer electrolytes without the deleterious presence of chemically bonded negative charges. How do they circumvent the problem? It is done by ensuring an efficient ion transport along nanoscale PEO domains, by using a block copolymer poly (dithiooxamide) (PDTOA). This hard copolymer (as block or end-linked charged blend) helps stabilize the anions through hydrogen bonding. This trick enhances ion conduction in the block copolymer electrolyte by realizing organized lamellar morphology. A direct experimental evidence is that steady-state current/initial current ratio, 0.67 for the PEO-b-PDTOA electrolytes, surpassing the value of 0.31, is observed for conventional PEO-salt electrolytes. A brief description of their synthesis procedure follows.

A new type of polymer electrolyte based on PEO and PDTOA has high ionic conductivity and improved cation transport, besides improved mechanical stability (PDTOA has a T_g of 130°C). Furthermore, thioamide group of PDTOA helps reduce anion diffusion rate. As a first step, PDTOA homopolymer has to be synthesized.

For this purpose, dithiooxamide(DTOA), 1,1′-thiocarbonyldiimidazo le(TCDI), and triethylamine (TEA) were employed as monomer, linker, and catalyst, respectively. The residual imidazole moiety of PDTOA was converted to an amine group through a further reaction with excess ethylenediamine (EDA). The molecular weight of thus made PDTOA was 6.5 kg/mol. Then, the PEO-b-PDTOA block copolymer was synthesized by reacting carbonylimidazole-capped poly(ethylene glycol) methyl ether (MPEO-CDI, 5 kg/mol) with PDTOA. The Schemes 12.1 and 12.2 below illustrate the synthesis steps.

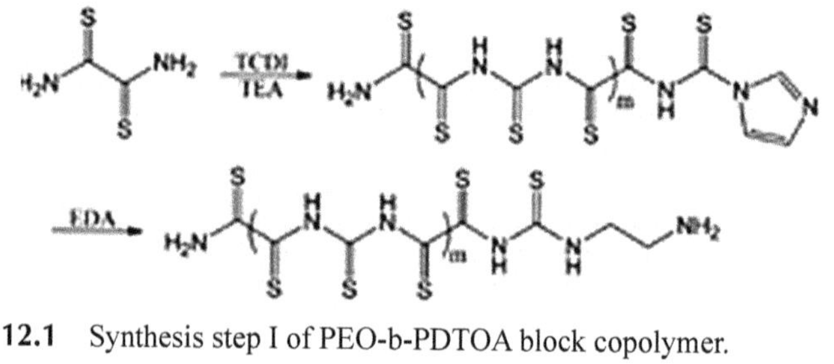

SCHEME 12.1   Synthesis step I of PEO-b-PDTOA block copolymer.

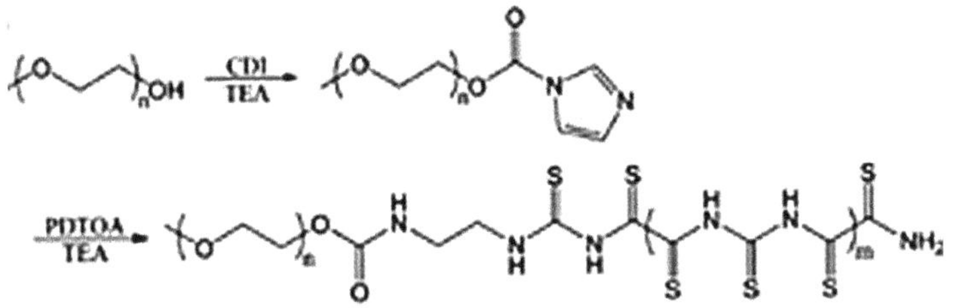

SCHEME 12.2   Synthesis step II of PEO-b-PDTOA block copolymer.

Young et al.[13] have synthesized a triblock copolymer PP-b-PEO-b sPEOP(b: block, s: syndiotactic) by mixing with LiTFSI. PEOP formation ensures strongly segregated in the absence and presence of LiTFSI. Furthermore, LiTFSI inhibits PEO crystallization without affecting PP crystallinity. This copolymer exhibits a nonlinear molecular weight dependence of conductivity with a maximum near 20 kg/mol. Given below is a brief summary of their synthesis procedure.

The first step is purification of relevant reagents. Toluene as a solvent was purified over columns of alumina and copper (Q5). Tetrahydrofuran used for block copolymer synthesis was purified over alumina column and degassed by three freeze–pump–thaw cycles before use. Propylene (Airgas, research purity) was purified over columns of BASF catalyst R3-12, BASF catalyst R3-11, and 4 Å molecular sieves. Polymethylaluminoxane (PMAO-IP, 13 wt.% Al in toluene, Akzo Nobel) was dried in vacuum to remove residual trimethyl aluminum and used as a white solid powder. Sodium azide, PEO polymers (M_n: 3, 8, 16 and 38 kg/mol; polydispersity index PDI $= M_w/M_n = 1.02–1.12$, where M_w is the weight-averaged molecular weight), p-toluenesulfonyl chloride, sodium hydride (60% dispersion in mineral oil), tripropargyl amine (98%), tetrakis (acetonitrile) copper (I) hexafluorophosphate (97%), 2,6-lutidine (>99%), borane-tetrahydrofuran complex (1.0 M solution in THF, stabilized with 0.005 M N-isopropyl-N-methyl-tert-butylamine), propargyl bromide solution (80 wt.% in toluene), and copper bromide were purchased from Sigma-Aldrich and used as received. Acetonitrile (HPLC grade) was obtained from Mallinckrodt Baker and used as received. Benzyl azide (94%) purchased from Alfa-Aesar was used as received. CDCl$_3$ from Cambridge Isotope Laboratories (CIL) was used as received. Dry Tetrahydrofuran (THF) for electrolyte preparation obtained from Sigma-Aldrich was used as received in an argon-filled glove box. Dry LiTFSI obtained from Novolyte under argon was brought into the glove box, and dried under vacuum in the glove box antechamber at 120°C for 3 days prior to use.

An important polymer allyl-terminated sPP (PDI=1.4–1.9) was prepared following a previously reported procedure.[13] The syndiotactic fraction of sPP samples determined from ^{13}C nuclear magnetic resonance (NMR) spectroscopy using the fraction of fully syndiotactic pentads [*rrrr*] was found to be 0.80. Tris-(benzyltriazolylmethyl) amine (TBTA) ligand for alkyne-azide "click" chemistry was synthesized according to the litera-ture procedure (see ref 13 also for supporting information for the synthesis of end-group functionalized polymers). End-group analysis of a[1] H NMR spectrum obtained from the product of the sPP functionalization reaction determined the weight fraction of unfunctionalized sPP chains, $w_{sPP,h}$, which ranged from 0.17 to 0.43 and are given in Table 12.1 of ref 13.

The synthesis of PEOP triblock copolymers is also done in the glove box, into which a 100 ml Schlenk tube was charged with azido-terminated sPP (0.60 g, 0.070 mmol N_3 functional groups), dipropargyl-terminated PEO (0.36 g, 0.085 mmol functional propargyl groups), CuBr (3 mg, 0.02 mmol), and TBTA ligand (11 mg, 0.020 mmol). THF (7.1 ml) was added and the Schlenk tube was heated at 50°C for 24 h. After the reac-tion had completed, the mixture was cooled to room temperature, and the polymer was precipitated in methanol. The resultant light green polymer was thoroughly washed with methanol to remove the copper catalyst and excess PEO. Insoluble polymer was isolated by vacuum filtration, washed with methanol, and dried in vacuum to constant weight (0.85 g, 94% yield). ^1H NMR (400 MHz, $CDCl_3$, 60°C) δ 7.55 (s, 2H), 4.69 (s, 4H), 4.31 (t, J=7.4 Hz, 4H), 3.64 (s, 1010H), and 1.70–0.58 (m, 2664H). ^{13}C NMR (126 MHz, $CDCl_3$, 60°C) δ 70.85, 46.74, 27.74, and 19.97.

The final aspect is the preparation of the electrolyte. For this purpose, PEOP polymers were brought into an MBraun argon glove box for elec-trolyte preparation after drying in the glove box antechamber at 100°C for 1 day. In the glove box, a mixture of PEOP polymer and dry LiTFSI was codissolved in THF. For all samples, the amount of LiTFSI added was predetermined to obtain a molar ratio r of lithium ions (Li^+) to ethylene oxide (EO) moieties equal to 0.063±0.06. This salt concentration is in the vicinity of the optimum salt concentration for PEO-based electrolytes.[13] The solution was stirred for several hours at 90°C until complete dissolu-tion was visually observed, and then the THF was allowed to evaporate to obtain a solid polymer–salt mixture. Subsequently, the electrolyte was dried further in the glove box antechamber under vacuum at 90°C for at least 8 h prior to characterization.

For each electrolyte, the volume fraction of the conducting phase, ϕ_c, is determined assuming that the LiTFSI is located in the PEO domain and that the volume change of mixing was negligible:

$$\phi_c(r) = \frac{V_{EO} + r \cdot V_{LiTFSI}}{V_{EO} + r \cdot V_{LiTFSI} + \frac{2 \cdot M_{n,sPP} \cdot M_{EO}}{M_P \cdot M_{n,PEO}} V_P} \qquad (12.2a,b)$$

where V_{EO}, V_{LiTFSI}, and V_P are the molar volumes of EO monomer units (41.56 cm³/mol) LiTFSI (141.9 cm³/mol) and propylene repeat units (105.99 cm³/mol), respectively, at a reference temperature of 140°C. The values of ϕ_c for each polymer at the desired salt composition are given in Table 12.1 of ref 13 which may also be consulted for additional references.

Polymer electrolytes are usually operated at temperatures above the glass transition temperature at which the polymer makes a transition from rubbery to glassy state. In the glassy state, the polymer is essentially amorphous and from the conductivity viewpoint, the structural relaxation of the polymer introduces a timescale that interferes with the ion diffusion timescale. Agapov and Sokolov[14] have considered the issue of decoupling ionic conductivity from structural relaxation (segmental dynamics) as an innovative strategy for development of solid polymer electrolytes.

Their detailed studies of temperature dependences of ionic conductivity and segmental dynamics in a number of polymers including PEO, P2VP, and P4VP have demonstrated that the ionic conductivity indeed decouples from segmental relaxation in most of the polymers; the degree of decoupling is higher in more fragile polymers. They suggest that use of fragile polymers, that is, polymers with relatively rigid structure, might be a better way to design advanced solid polymer electrolytes. This approach is opposite to the traditional approach in search for polymers with flexible structures, such as PEO and its derivatives.

A novel innovation is to look for ion transport in a polymerized ionic liquid. A polymerized ionic liquid is a solid polymer made up of ionic liquid monomers. The ionic liquid is incorporated into the backbone or side groups of polymers so that the high intrinsic ionic conductivity of the former and the mechanical properties of the latter results in a new class of functional materials. In a recent work,[15] local charge transport and structural changes in films of a PolyIL are studied using an integrated experiment theory-based approach. Experimental data for the kinetics of charging and steady-state current–voltage relations arise from the dissociation of ions under an applied electric field (Wien effect). Onsager's theory of the Wien effect together with the Poisson–Nernst–Planck formalism for the charge transport accounts very well for the experimental results.

We now move on to a free volume-based model for ion conduction in polymers.

12.5 A FUNDAMENTAL MODEL FOR ION CONDUCTION IN POLYMERS

We now outline the Miyamoto–Shibayama[16] free-volume model for ion transport in polymers. The model assumes that ion movement in a polymer is regulated mainly by the extent and distribution of free volume and aims to derive a simplified electrical conductivity equation. Taking account of free volume, and we apply it to T-dependence of electrical conductivity of polymer–ionic salt composites.

Consider a cluster of polymer segments. Say one at the center and six around it (Fig. 12.7a).

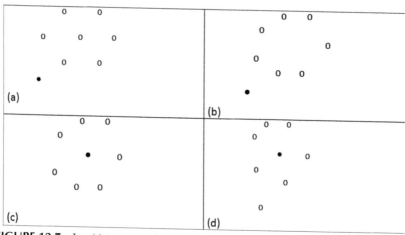

FIGURE 12.7 Ion (•) transport through a polymer made of structured segments (O) via a random walk. (a) Ion just outside a cluster of polymer segments, (b) creation of free

volume inside the segment cluster, (c) ion diffusion into the cluster taking the place of the central segment, and (d) the displaced segment out diffuses (see text).

We need to model ion transport as the moving in of an ion replacing the segment at the center (Fig. 12.7b). How does that happen? It occurs in terms of frequency of transition (f) of an ion from one equilibrium position to another. On what variables does it depend? Note that the free volume should redistribute and create a hole for the ion to jump in.

Thus, two probabilities arise: P_h for hole creation, P_j for ion jump. Therefore,

$$f = va\, P_h\, P_j \tag{12.3}$$

v is vibration frequency of trapped ion, a is a correlation factor $(0 < a < 1)$.

Polymers used in electrochemical devices generally operate at temperatures $T > T_g$ the glass transition temperature of the relevant polymer.

Following Cohen and Turnbull, focus is on molecular transport in liquids. This occurs through the movement of molecule into a hole whose size is greater than a critical value V^*. How is it formed? It is formed by thermal fluctuation of free volume with energy. Now ask for the probability of finding a hole as described above, volume say V_i^*.

$$P_h = exp\left(-gV_i^* / V_f\right) \tag{12.4}$$

Factor g corrects for the overlap of free volume (that is V_f). Ion conduction in a polymer is visualized as a rate process. What is the probability for ion jump into V_i^* by surmounting the potential energy barrier E_j?

$$P_j = exp\left(-E_j / k_B T\right) \tag{12.5}$$

k_B is the Boltzmann constant. With the help of (Eq. 12.4) and (Eq. 12.5), Equation 12.3 becomes

$$f = va\, exp\left(-gV_i^* / V_f\right) exp\left(-E_j / k_B T\right) \tag{12.6}$$

Barrier height E_j decreases along the applied electric field (F) at frequency f^+. Therefore, apparent jump energy is $E_j - (1/2)elF$. l is jump *distance between two equilibrium positions. E_j increases in the opposite direction (frequency f^-).*

Where apparent energy is $E_j + (1/2)elF$. Mean velocity is $v_{mean} = l(f^+ - f^-)$.
Current density,

$$i = nel(f^+ - f^-)$$ (12.7)

where $n = ion$ concentration. In the ionic dissociation theory,

$$n = n_0\, exp\left(-W / 2\varepsilon k_B T\right)$$ (12.8)

n_0 *is a constant, ε is dielectric constant, and W is the ionic dissociation energy.*

Electrical conductivity (σ) is given by

$$\sigma = \sigma_0\, exp[-(gV * i / V_f) + (E_j + W / 2\varepsilon) / k_B T$$ (12.9)

where

$$\sigma_0 = (2v_0\, eval / F)\, sinh(elF / 2k_B T)$$ (12.10)

Note that Equation 12.9 gives both T- and F-dependences of σ.

Unlike Equation 12.7, a simple Arrhenius-type equation $\sigma = (\sigma_0/T)\, e^{(-Ea/kBT)}$ ignores the effect of P_h, whereas Williams–Landel–Ferry equation considers only P_h.[17] Thus, the Miyamoto–Shibayama free-volume model strikes the golden mean between the two extreme approaches.

In the simple Arrhenius picture, a plot of log σ versus $1000/T$ (K) for a typical polymer gives two straight lines with a break at a "critical" temperature. This temperature corresponds to T_g of the polymer. The free volume effects have been directly observed in positron annihilation experiments[18] and more recently, by Jing et al[19] on $LiClO_4/PEO$.

An important finding of the analysis of experimental data on a few polymers is widely linear plots below and above T_g, as log $\sigma + gV*I/2.303V_f$ versus $1/T$ plots.

The "hybrid energy" $E_j + W/2\varepsilon$ equals apparent activation energy at $T < T_g$. This underlines the importance of free volume in conductivity experiments on polymers in the practical and applicable range of temperatures.

Note that a correlation between fractional free-volume $f(T)$ and conductivity σ at temperatures above T_g, log$[\sigma/\sigma(T_g)] = C\ 1[f(T)-f(T_g)]/f[T]$ was first experimentally confirmed using measured positron annihilation results. The comparison of the value of the obtained constant C1 with the universal values for the segmental diffusion of amorphous polymers indicated that the critical free volume required for the ion transport is much smaller than that required for polymer chain segment mobility.

12.6 SUMMARY AND FUTURE DIRECTIONS

This brief review on the ion conducting polymers provides a perspective on the problem of ion conduction in polymer electrolytes, focusing on the properties of polymer that favor ion transport. Contemporary issues in polymer electrolytes are discussed next. This is followed by a discussion of two fundamental experimental probes: electrical conductivity and positron annihilation. Following up on one of the contemporary issues, recent innovations aimed at enhancing ion conduction are discussed with the aid of most recent work. A fundamental model of ion conduction in polymers based on free volume concept is briefly outlined.

Future directions based on important recent work over the last 5 years are as follows: (1) use of fragile polymers to decouple the ionic conductivity relaxation from segmental relaxation; (2) use of diblock and triblock copolymers, and (3) polymerization of ionic liquids. Future years are going to see the development and use of new polymer electrolytes and exploration of mechanisms of ion conduction in them.

12.7 ACKNOWLEDGMENT

I am grateful to the University of Hyderabad for extending full financial support to enable me in making my presentation at ICM 2016. I thank the organizers of ICM 2016 for their kind invitation.

KEYWORDS

- Li-ion conduction
- polymer electrolytes
- polymerization
- ion conduction
- oxidized polymers

REFERENCES

1. Fenton, D. E.; Parker, J. M.; Wright, P. V. *Polymer* **1973,** *14,* 589.
2. Sunandana, C. S. *Introduction to Solid State Ionics—Phenomenology and Applications;* CRC Press: USA, 2015.
3. Hallinan, D. T., Jr.; Balsara, N. P. *Ann. Rev. Mat. Sci.* **2013,** *43,* 503.
4. Golodnitsky, D.; et al. *J. Electrochem. Soc.* **2015,** *162,* A2551.
5. Sunandana, C. S.; Kumar, P. S. *Bull. Mater. Sci.* **2004,** *27*(1), 1–17.
6. Angell, C. A. *Solid State Ionics* **1986,** *18,* 72.
7. Cheng, S.; et al. *Macromolecules* **2014,** *47,* 3978.
8. Ren, C. L.; et al. *Macromolecules* **2016,** *49,* 425.
9. Stolwijk, N. A. *J. Phys. Chem. B* **2012,** *116,* 3065.
10. Jing, G.; et al. *Chin. Phys. B* **2012,** *21,* 107803.
11. M Bolloli; et al. *Electrochim. Acta* **2016,** *214,* 38.
12. Jo, G.; et al. *ACS Macro. Lett.* **2015,** *4,* 255.
13. Young, N. P.; et al. *Solid State Ionics* **2014,** *263,* 87.
14. Agapov, A. L.; Sokolov, A. P. *Macromolecules* **2011,** *44,* 4410.
15. Kumar, R.; et al. *Nanoscale* **2015,** *7,* 947.
16. Miyamoto, T.; Shibayama, K. *J. Appl. Phys.* **1973,** *44,* 5372.
17. Williams, M. L.; Lendel, R. E.; Ferry, J. D. *J. Am. Chem. Soc.* **1955,** *77,* 3701.
18. Peng, Z. L.; et al. *J. Appl. Phys.* **1995,** *77,* 334.
19. Jing, G.; et al. *Chin. Phys. B* **2012,** *21,* 107803.

CHAPTER 13

EFFECT OF MODIFIED CALCIUM CARBONATE FILLER ON PROPERTIES OF POLY(PROPYLENE)/LAYERED DOUBLE HYDROXIDE COMPOSITES

MAHENDRA R. NEVARE and VIKAS V. GITE*

Department of Polymer Chemistry, School of Chemical Sciences, North Maharashtra University, Jalgaon, Maharashtra 425001, India

Corresponding author. E-mail: vikasgite123@gmail.com

CONTENTS

ABSTRACT

The synergetic effect of layered double hydroxide (LDH) and calcium carbonate on poly(propylene) (PP) composites was investigated with respect to mechanical, thermal, morphological, and flame retardant properties. The prepared calcium carbonate was modified with sodium stearate and confirmed by scanning electron microscopy (SEM), Fourier-transform infrared (FTIR) analysis and dispersion stability of filler in solvent. Composites were compounded by melt blending technique using Brabender Plastograph EC and test specimens were prepared using injection molding machine. A comparative study between sodium stearate modified and unmodified calcium carbonate particles on PP composites was also investigated. The modified particles showed improved compatibility with PP matrix which enhanced the mechanical performance of composites. Modified particles well distributed in PP composite had enhanced tensile strength (TS), impact strength, and storage modulus of composites. The synergism between LDH and calcium carbonate resulted in improvement in flame retardancy of composites without losing their mechanical performance.

13.1   INTRODUCTION

The thermoplastic polymers such as poly(propylene) (PP), poly(ethylene), and poly(vinyl chloride) are largely of commercial importance in commodity applications. Such polymers are widely used in household appliances, packaging, automotive, engineering applications, cables, electronic cases, interior decoration, and so forth.[1,2] Due to increasing demand of end users, the performance of polymers is needed to improve with respect to their flammability and mechanical properties. This can be possible by incorporation of reinforcing fillers into polymer matrix to form polymer composites.[3–5] Polymer composites alter properties by incorporating different fillers in varying amount for getting hybrid materials with desirable properties which are not achieved by single polymer alone.[6,7] There are various reports on improved mechanical properties, barrier properties, heat resistance, and dimensional stability of polymers after utilization of fillers in their composites. Significant improvement in performance was observed when modified fillers were used in composites preparation rather than unmodified fillers. This may be due to modified fillers that provide more surface area available for interactions with polymer matrix (hydrophobic), which is the need of the day for getting maximum efficiency of the filler into the polymer composites.[8,9]

Poly(propylene) (PP) is a widely used engineering thermoplastic material as compared to other polymers due to its availability, excellent abrasion, low cost, low density, and good moldability.[10–13] Lately, many engineering polymers are being substituted by PP in applications like automotive, construction, and home appliances due to high performance to price ratio.[14] The performance of PP can be enhanced for engineering applications by enhancing nucleating sites by addition of fillers through higher crystallization in final materials. Reports are available for modification of PP composites to have widespread applications due to improved toughness, enhanced modulus, and barrier properties when filled with fillers.[15,16] Despite many advantages, PP is a highly flammable material because of its hydrocarbon constituent[17] and in many applications PP needs to be modified with better flame retardant properties.[18] Such improvement of flame retardancy of the polymer is a more challengeable research area in today's scenario. Literature has covered miscellaneous and efficient approaches for improving the fire resistance property of polymer.[19] Application of inorganic material is one of the useful methods for flame retardant composites. Inorganic fillers like magnesium hydroxide,[20,21] aluminum hydroxide,[22] montmorillonite (MMT),[23] and layered double hydroxide (LDH)[24] have been tested to improve the flame retardancy of polymers. Sometimes secondary fillers like zinc borate, talc, and zinc phosphate are also employed synergistically for the same purpose.[25,26]

For enhancing flammability of polymers, inorganic clays such as LDH and MMT are gaining attention of researchers due to their low cost, less abrasion, high aspect ratio, and easy availability.[27,28] LDH is an anionic clay-containing metal cation having unique physical and chemical properties.[29–31] LDH is also known as hydrotalcite which has features such as metal hydroxide and layered silicate structure,[32,33] like that of hybrid materials. Incorporation of LDH in polymer matrix alters the mechanical and thermal performance of polymer composites at low filler loading. LDH has been extensively used in polyolefin materials as an environment-friendly halogen-free flame retardant additive because of its endothermic decomposition upon exposure to high temperatures.[34–36] Although it is beneficial for tailoring flammability of polymer composites, its presence in higher percentage deteriorates the mechanical performance of the polymer composites. Hence, in the case of such composites, LDH needs to sacrifice the mechanical properties at the cost of higher flame retardancy or use supplementary filler into composites along with LDH that helps to maintain the mechanical properties of composites to their original or higher level.

Compatibility between the polymer matrix and the filler is one of the key factors that determine the performance of composites. It has been recognized that poor compatibility between filler and matrix results in lower performance of composites than expected performance because of improper dispersion of inorganic fillers into the hydrophobic polymeric matrix. Inorganic fillers have an affinity to agglomerate due to high surface energy and small size.[37] Hence, the surface of such material needs to be modified with a compatible agent (sometimes known as a coupling agent) for the uniform dispersion of inorganic fillers into the polymer matrix in order to have improved performance of composites to a great extent.[38]

Reports are available regarding the use of LDH and calcium carbonate individually as fillers for the preparation of polymer composites.[39] In some cases, specific attention has been given to use LDH synergistically with other fillers.[40] However, limited efforts have been taken for utilization of LDH in association with calcium carbonate. In our previous work, the performance of PP nanocomposites improved with the synergistic effect of LDH and nano zinc phosphate.[41] The present work describes the synergistic effect of LDH and calcium carbonate. The problem regarding weak compatibility between calcium carbonate and PP has been tried to overcome by modifying calcium carbonate by sodium stearate as organically modified filler that can reduce the aggregation tendency. Moreover, polar and nonpolar functional groups present in sodium stearate can help to improve the interfacial adhesion of hydrophilic filler particles with hydrophobic polymer molecules.

13.2 EXPERIMENTAL

13.2.1 MATERIALS

Poly(propylene) (PP) (Repol D120MA) was obtained from Reliance Industries Ltd., India. Calcium chloride and Potassium carbonate (AR grade) were purchased from Loba Chemicals Pvt. Ltd., India. LDH and sodium stearate were purchased from Sigma Aldrich. Poly(ethylene glycol) (PEG) with a molecular weight of 6000 was purchased from S. D. Fine Chemicals, India.

13.2.2 SYNTHESIS OF CaCO₃

The calcium carbonate was synthesized by an in situ precipitation technique as per the method reported previously.[42] The general reaction for synthesis of calcium carbonate is given below.

$$CaCl_2 + K_2CO_3 \rightarrow CaCO_3 + 2KCl$$

The solution of calcium chloride was prepared by dissolving calcium chloride (110 g) in water (100 ml). Simultaneously, separate PEG solution was prepared by dissolving PEG (996 g) in water (450 ml) after gentle heating. Both calcium chloride and PEG solutions were mixed in 1:5 ratios. At the same time, potassium carbonate (124 g) solution was prepared separately in distilled water (200 ml). The complex of calcium chloride and PEG was then added slowly into the above solution and the reaction mixture was kept overnight. The precipitate obtained was filtered and washed with water. Acetone was added to the product and subjected to ultrasonication to break the agglomeration of particles and finally dried at 120°C in vacuum for 2 h.[43]

13.2.3 MODIFICATION OF CaCO₃

The prepared $CaCO_3$ was dried at 120°C for 4 h. Then $CaCO_3$ (300 g) was suspended in a mixture of water and ethanol (1200 ml) with a ratio of 2:1 and the suspension was stirred for 2 h to completely wet the particles. Sodium stearate (3 g) was separately dissolved in the ethanol (100 ml) and gradually added to the above suspension at 80°C for 2 h. The modified calcium carbonate particles were dried in an oven at 120°C at least for 24 h.

13.2.4 MELT COMPOUNDING OF POLY(PROPYLENE)/ LAYERED DOUBLE HYDROXIDE/CALCIUM CARBONATE (PP/LDH/CaCO₃) COMPOSITES

Melt compounding of the PP/LDH/calcium carbonate composites was performed with Brabender Plastograph EC, Germany. The software provided by the manufacturer was used to control processing conditions like temperature (190°C), rotor speed (60 rpm), and blending time (10 min). Only 90% (55 cm³) capacity of a mixing chamber of the equipment was used for compounding of composites. The PP was melted first and after that other ingredients were added to homogenize into it.[44]

Melt compounded samples were coarsely grinded using a laboratory grinder and converted into the specimen for further characterization by molding on a Joy Baby Injection Molding, Ahmedabad, India.

The PP composites were prepared by varying the content of LDH and unmodified as well as modified $CaCO_3$ up to 3 phr. The detailed compositions and coding of samples are given in Table 13.1.

TABLE 13.1   Detailed Compositions and Coding of Poly(propylene) (PP) Composites.

Sample code	Composition	Filler content (phr)
P	PP	100
PL	PP+LDH	100+3
PC1	PP+$CaCO_3$	100+3
PLC2	PP+LDH+$CaCO_3$	100+2+1
PLC3	PP+LDH+$CaCO_3$	100+1.5+1.5
PLC4	PP+LDH+$CaCO_3$	100+1+2
PLmC1	PP+m$CaCO_3$	100+3
PLmC2	PP+LDH+m$CaCO_3$	100+2+1
PLmC3	PP+LDH+m$CaCO_3$	100+1.5+1.5
PLmC4	PP+LDH+m$CaCO_3$	100+1+2

13.3 CHARACTERIZATION

13.3.1 FOURIER-TRANSFORM INFRARED (FTIR) ANALYSIS

Fourier-transform infrared (FTIR) spectra of unmodified and modified $CaCO_3$ particles in powder form were recorded on a Perkin Elmer Spectrum GX in order to study the surface modification of $CaCO_3$ particles. The samples were scanned from 400 to 4000 cm^{-1}.

13.3.2 DISPERSION STABILITY

The effect of particle modification on dispersion stability was determined in an organic solvent (toluene). For this unmodified and modified calcium carbonate particles were ultrasonicated separately into toluene for 30 min. Suspensions were further allowed to stand for 4 h and particle sedimentation behavior was observed.

13.3.3   SCANNING ELECTRON MICROSCOPY (SEM)

A scanning electron microscope (SEM) was used for studying particle size and morphologies of modified and unmodified particles. The morphology of the fractured surface of PP composites obtained from tensile testing was recorded with a field emission scanning electron microscope (FE-SEM, model S-4800, Hitachi High Technologies Corp., Japan). All the samples were sputter coated with a thin layer of gold prior to examination and observed at a voltage range of 0.5–30 kV.

13.3.4   MECHANICAL STRENGTH AND FLAMMABILITY

The determination of tensile strength (TS) of dumbell shaped specimens was carried out on a universal tensile testing machine (UTM) (International Engineering Industries, Mumbai, India) in accordance with ASTM D-638. The analysis was carried out using 1000 kg load cell having speed of 5 cm/min. Izod impact strength was determined on Izod impact tester (International Engineering Industries, Mumbai, India). The hardness of sample was measured on a Shore-D hardness tester (Hiroshima, Apex Enterprises, Mumbai, India). Flammability of all PP composites was studied using UL 94 method.

13.3.5   DYNAMIC MECHANICAL ANALYSIS (DMA)

A dynamic mechanical thermoanalyzer (DMTA) of Anton Paar Instrument, Model MCR 102 was used to perform DMA. The melt behavior of composites was tested by Rheometrics Mechanical Spectrometer (RMS) with an oscillation mode at strain amplitude of 5% and temperature of 200°C.

13.3.6   THERMOGRAVIMETRIC ANALYSIS (TGA)

The thermal stability of composites was studied by a thermogravimetric analyzer (TGA, PerkinElmer, TGA 4000). For each experiment, a sample of about 4–6 mg was kept in an alumina pan and mass loss was recorded during a heating cycle over the temperature range from room temperature to 650°C with a heating rate of 20°C/min under nitrogen atmosphere.

13.4 RESULTS AND DISCUSSION

13.4.1 FOURIER-TRANSFORM INFRARED (FTIR) SPECTRA

Fourier-transform infrared spectra of unmodified and modified $CaCO_3$ fillers are shown in the Figure 13.1. The FTIR spectrum of both types of $CaCO_3$ showed that absorption broad band observed between 3423 and 3429 cm^{-1} was due to –OH stretching. The absorption bands observed in case of unmodified/modified calcium carbonate fillers are due to C–O stretching and C=O stretching at 2888–2864 cm^{-1} and 1540 cm^{-1}, respectively. The characteristic signal of sodium stearate modified calcium carbonate was observed at 2920 cm^{-1} due to the C–H stretching of long stearate chain and 1800 cm^{-1} due to the carbonyl group of sodium stearate. [45] In conclusion, the filler was successfully surface modified by sodium stearate.

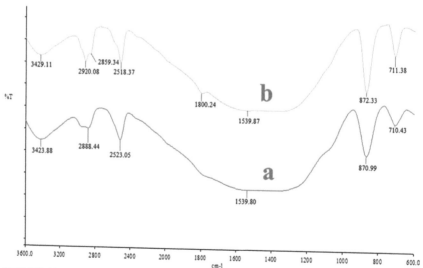

FIGURE 13.1 Fourier-transform infrared (FTIR) spectra of a) unmodified $CaCO_3$ and b) sodium stearate modified $CaCO_3$ (m$CaCO_3$).

13.4.2   DISPERSION STABILITY

The test tube containing modified and unmodified $CaCO_3$ particles dispersed into toluene are shown in Figure 13.2. After 4 h, it was observed that modified solution remained turbid while a solution of unmodified particles was clear with settled particles at the bottom of the test tube. This maybe the result of modified particles that avoid the formation of agglomerates and improves dispersion of inorganic particles in organic media.

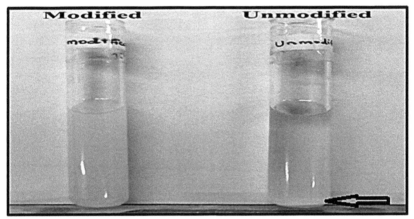

FIGURE 13.2   Dispersion stability test for modified and unmodified calcium carbonate particles.

13.4.3   SCANNING ELECTRON MICROSCOPY (SEM) OF FILLER

The SEM images of unmodified and modified calcium carbonate are shown in the Figures 13.3 and 13.4, respectively.

From these images the rectangular shaped plate-like structure formatting was observed that modified and unmodified $CaCO_3$. These plates have approximate dimensions from 25 to 300 nm. However, it was also observed that unmodified $CaCO_3$ has sharp edges because of its crystalline nature, while in the case of modified $CaCO_3$ particles they do not have sharp ages due to modification with amorphous stearate part. The size of unmodified particles was less than the modified $CaCO_3$. In comparison, unmodified calcium carbonate showed larger size than modified calcium carbonate.

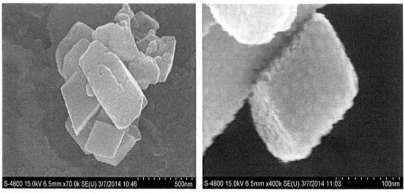

FIGURE 13.3   SEM of unmodified CaCO₃.

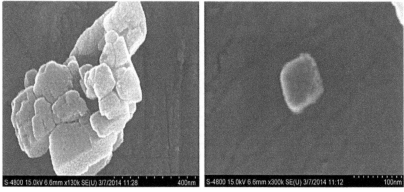

FIGURE 13.4   SEM of modified CaCO₃.

13.4.4   TENSILE STRENGTH (TS)

The results of tensile behavior and other mechanical properties of PP composites are given in the Table 13.2 and tensile behavior of composites is graphically represented in the Figure 13.5.

The results showed that composites containing $CaCO_3$ had higher TS than the pristine PP samples. This maybe due to better compatibility of surface modified particles with PP, which enhances load bearing capacity per cross sectional area of composites. In the category of unmodified $CaCO_3$ highest strength was reported to composites having equal amounts of $CaCO_3$ and LDH. On further increase in LDH it has decreased. In another series, highest strength was observed for composites possessing 2% modified $CaCO_3$ and 1% LDH. Highest values of the TSs in each series were obtained for composites having both types of fillers, which show the synergistic effect of $CaCO_3$ with LDH. On comparing TS of composites with modified and unmodified $CaCO_3$, it was found that composites having modified filler have higher strength than with unmodified filler composites.

TABLE 13.2 Properties of the PP Composites.

Sample code	Tensile strength (MPa)	Impact strength (KJ/m²)	Hardness value (Shore-D)	Flammability (mm/min)
P	30	2.69	57	49
PL	36	2.55	60	43
PC1	39	2.62	58	45
PLC2	38	2.57	59	44
PLC3	42	2.61	59	37
PLC4	39	2.65	59	38
PLmC1	42	2.78	59	38
PLmC2	46	2.63	60	35
PLmC3	44	2.71	59	38
PLmC4	44	2.69	59	37

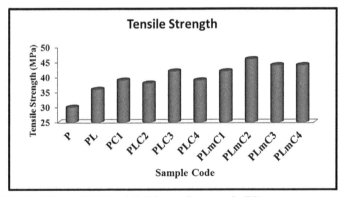

FIGURE 13.5 Effect of LDH and $CaCO_3$ tensile strength (TS).

13.4.5 HARDNESS

The graphical representation on hardness of PP composites is shown in the Figure 13.6. From these values, it was observed that the hardness of all compositions of PP composites was increased with the addition of LDH as well as modified and unmodified $CaCO_3$ as compared to virgin PP. By adding an equal amount of inorganic fillers, there was no major change observed in these values except two samples, that is, PL and PLmC2; this may be attributed to the higher amount of LDH in these samples.

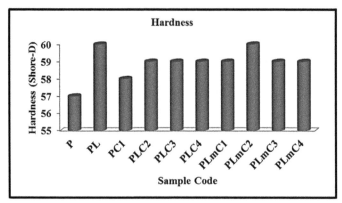

FIGURE 13.6   Effect of layered double hydroxide (LDH) and $CaCO_3$ on hardness.

13.4.6   IMPACT STRENGTH

The data for the impact strength of the PP composites is graphically represented in the Figure 13.7.

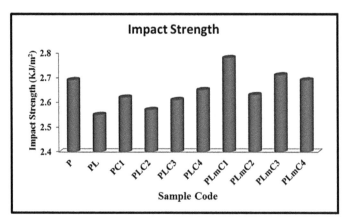

FIGURE 13.7   Effect of LDH and $CaCO_3$ on impact strength.

The results showed that all composites have lower impact strength than pristine polymers and it was least for PP/LDH composite sample. This indicates that addition of LDH decreases the impact strength of PP. However, on the addition of $CaCO_3$, it increased with an increase in the amount of $CaCO_3$. The results of impact strength were found to be better for composites incorporated with modified $CaCO_3$ particles along with LDH. This may be due to the good distribution of modified fillers into the PP matrix than unmodified that may help in the distribution of applied energy throughout the polymer matrix.

13.4.7 FLAMMABILITY

The results of the flammability test conducted on the PP composites are shown in the Figure 13.8.

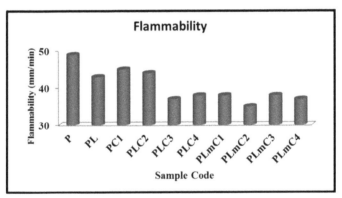

FIGURE 13.8   Effect of LDH and $CaCO_3$ on flammability.

It has been observed that the addition of fillers increased flame retardancy of PP composites than the pristine PP matrix. The resistance to flame increased when LDH and calcium carbonate fillers were used in combination. The sample PLmC2 showed the least flammability than any other sample. The flame retardancy effect was more pronounced when modified calcium carbonate filler was added into the PP composites. The enhancement in flame retardancy may be due to releasing of water molecules from LDH during burning which may have delayed or restricted the burning of composites and it was slower when well dispersed LDH is present in the composites than poorly dispersed LDH.

13.4.8 DYNAMIC MECHANICAL ANALYSIS (DMA)

The storage modulus of pristine PP, PP/LDH composites containing modified and unmodified calcium carbonate are shown in the Figure 13.9. From the results, it was observed that pristine PP has lower storage modulus as compared to the composites. The increase in storage modulus of composites was more pronounced for composites of LDH along with modified calcium carbonate. This suggests that strong interfacial interactions must have occurred between both fillers and PP which has helped to transfer stress from matrix to fillers and vice versa.

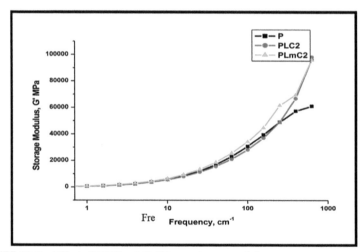

FIGURE 13.9   Effect of LDH and CaCO₃ on storage modulus.

13.4.9 THERMOGRAVIMETRIC ANALYSIS (TGA)

The thermogravimetric curves of PP composites are shown in the Figure 13.10. All samples showed slight weight loss up to 200°C due to the presence of moisture, while the pristine PP has degraded from 250–425°C in a single step. When we compared the thermal stability of composites with unmodified and modified $CaCO_3$, it was observed that the modified PP composites showed better thermal stability than unmodified PP composites. The enhancement in thermal stability of composites with modified filler than composites having unmodified filler maybe due to homogeneous dispersion of particles into the polymer matrix. The residual content of modified material was less than the unmodified; this is because of volatile organic compound present in the modified composites.

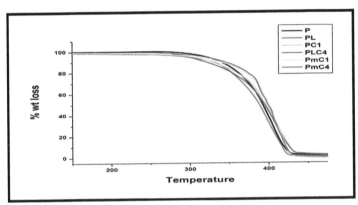

FIGURE 13.10   Thermogravimetric curves of PP composites.

13.4.10   SCANNING ELECTRON MICROSCOPY (SEM) OF PP COMPOSITES

Figure 13.11 shows SEM images of PP/LDH composites. From these images, it was observed that unmodified $CaCO_3$-based PP composites showed more fractured surface than composites containing modified $CaCO_3$ due to weak interactions occurred between the unmodified filler and PP matrix. Conversely modified $CaCO_3$ formed more compatibilized surface with polymer matrix due to the presence of the hydrophobic functional group in the sodium stearate. It was clearly seen in the images that unmodified particles would have poorly wetted with polymer matrix than modified particles. Hence, when composites were fractured, such particles were pulled out easily as compared to that of with the modified particles of $CaCO_3$. The effect of compatibility reflects the mechanical properties of modified and unmodified $CaCO_3$ loaded PP/LDH composites, where mechanical properties of modified $CaCO_3$ composites were more superior to the counterpart.

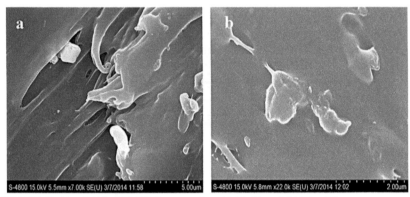

FIGURE 13.11   SEM of PP/LDH composites (a) PLC3 and (b) PLmC3.

13.5   CONCLUSION

The $CaCO_3$ was synthesized by an in situ precipitation technique and modified with sodium stearate. Modification of hydrophilic to the hydrophobic surface of $CaCO_3$ was confirmed by FTIR spectroscopy. The PP/LDH/$CaCO_3$ composites were prepared by melt blending technique using Brabender Plastograph. The effect of unmodified and modified $CaCO_3$ on PP/LDH composites was studied by different mechanical properties. Synergism was noticed for flame retardancy of composites by incorporation of LDH and $CaCO_3$ fillers. The improvement in tensile properties and thermal properties was achieved with the help of modified $CaCO_3$. However, the impact strength of composites was quite lower than pristine PP sample.

KEYWORDS

- polypropylene
- composites
- layered double hydroxide
- calcium carbonate
- mechanical properties

REFERENCES

1. Ahamad, A.; Chaudhari, A. B.; Patil, C. B.; Mahulikar, P. P.; Hundiwale, D. G.; Gite, V. V. Preparation and Characterization of Polypropylene Nanocomposites Filled with Nano Calcium Phosphate. *Polym. Plast. Technol. Eng.* **2012**, *51*, 786–790.
2. Zaman, H. U.; Hun, P. D.; Khan, R. A.; Yoon, K. B. Polypropylene/Clay Nanocomposites: Effect of Compatibilizers on the Morphology, Mechanical Properties and Crystallization Behaviors. *J. Thermoplast. Compos. Mater.* **2014**, *27*, 338–349.
3. Ishida, H.; Campbell, S.; Blackwell, J. General Approach to Nanocomposite Preparation. *Chem. Mater.* **2000**, *12*, 1260–1267.
4. Pinniavaia, T. J. Polymer Clay Nanocomposites; Beal, G. W., Ed.; 2000, ISBN: 978-0-471-63700-4.
5. Vaia, R. A.; Kishnamoorti, R. *Polymer Nanocomposites.* American Chem. Society, 2001, Vol. 804. DOI: 10.1021/bk-2002-0804.
6. Ghugare, S. V.; Govindaiah, P.; Avadhani, C. V. Polypropylene-Organoclay Nanocomposites Containing Nucleating Agents. *Polym. Bull.* **2009**, *63*, 897–909.
7. Schmidt, G.; Malwitz, M. M. Properties of Polymer-Nanoparticle Composites. *Curr. Opin. Colloid Interface Sci.* **2003**, *8*, 103–108.
8. Li, L.; Zhang, M.; Song, S.; Wu, Y. Starch/Sodium Stearate Modified Fly-Ash Based Calcium Silicate: Effect of Different Modification Routes on Paper Properties. *BioResources* **2016**, *11*, 2166–2173.
9. Shimpi, N.; Mali, A.; Hansora, D.; Mishra, S. Synthesis and Surface Modification of Calcium Carbonate Nanoparticles Using Ultrasound Cavitation Technique. *Nanosci. Nanoeng.* **2015**, *3*, 8–12.
10. Wang, L.; He, X.; Lu, H.; Feng, J.; Xie, X.; Su, S.; Wilkie, C. A. Flame Retardancy of Polypropylene (Nano) Composites Containing LDH and Zinc Borate. *Polym. Adv. Technol.* **2011**, *22*, 1131–1138.
11. Wu, L.; Kang, W.; Chen, Y.; Zhang, X.; Lin, X.; Chen, L.; Gai, J. Structures and Properties of Low-Shrinkage Polypropylene Composites. *J. Appl. Polym. Sci.* **2017**, DOI: 10.1002/APP.44275.
12. Ashori, A.; Menbari, S.; Bahrami, R. Mechanical and Thermo-Mechanical Properties of Short Carbon Fiber Reinforced Polypropylene Composites Using Exfoliated Graphene Nanoplatelets Coating, *J. Ind. Eng. Chem.* **2016**, *38*, 37–42.
13. Morlat, S.; Mailhot B.; Gonzalez, D.; Gardette, J. L. Photo-Oxidation of Polypropylene/Montmorillonite Nanocomposites. 1. Influence of Nanoclay and Compatibilizing Agent. *Chem. Mater.* **2004**, *16*, 377–383.
14. Moore, E. P. *Polypropylene Handbook. Polymerization, Characterization, Properties, Processing, Applications;* Hanser-Gardner, 1996. ISBN 1-56990-208-9
15. Onuegbu, G. C.; Igwe, I. O. The Effects of Filler Contents and Particle Sizes on the Mechanical and End-Use Properties of Snail Shell Powder Filled Polypropylene. *Mater. Sci. Appl.* **2011**, *2*, 811–817.
16. He, A.; Wang, L.; Li, J.; Dong, J.; Han, C. C. Preparation of Exfoliated Isotactic Polypropylene/alkyl-Triphenyl Phosphonium Modified Montmorillonite Nanocomposites via in situ Intercalative Polymerization. *Polymer* **2006**, *47*, 1767–1771.

17. Duquesne, S.; Samyn, F.; Bourbigot, S.; Amigouet, P.; Jouffret F.; Shen, K.; Influence of Talc on the Fire Retardant Properties of Highly Filled Intumescent Polypropylene Composites. *Polym. Adv. Technol.* **2008**, *19*, 620–627.

18. Wang, D. Y.; Leuteritz, A.; Kutlu, B.; Landwehr, M. A.; Jehnichen, D.; Wagenknecht, U.; Heinrich, G. Preparation and Investigation of the Combustion Behavior of Polypropylene/Organomodified MgAl-LDH Micro-Nanocomposite. *J. Alloys Compd.* **2011**, *509*, 3497–3501.

19. Laoutid, F.; Bonnaud, L.; Alexandre, M.; Lopez-Cuesta, J. M.; Dubois, P. New Prospects in Flame Retardant Polymer Materials: From Fundamentals to Nanocomposites. *Mater. Sci. Eng.* **2009**, *63*, 100–125.

20. Sain, M.; Park, S. H.; Suhara, F.; Law, S. Flame Retardant and Mechanical Properties of Natural Fibre–PP Composites Containing Magnesium Hydroxide. *Polym. Degrad. Stab.* **2004**, *83*, 363–367.

21. Ahamad, A.; Patil, C. B.; Mahulikar, P. P.; Hundiwale, D. G.; Gite, V. V. Studies on the Flame Retardant, Mechanical and Thermal Properties of Ternary Magnesium Hydroxide/Clay/EVA Nanocomposites. *J. Elastomers Plast.* **2013**, *44*, 251–261.

22. Ramazani, S. A.; Rahimi, A.; Frounchi, M.; Radman, S.; Investigation of flame retardancy and Physical–Mechanical Properties of Zinc Borate and Aluminum Hydroxide Propylene Composites. *Mater. Design.* **2008**, *29*, 1051–1056.

23. Chen, X. S.; Yu, Z. Z.; Liu, W.; Zhang, S. Synergistic Effect of Decabromodiphenyl Ethane and Montmorillonite on Flame Retardancy of Polypropylene. *Polym. Degrad. Stab.* **2009**, *94*, 1520–1525.

24. Wang, D. Y.; Das, A.; Costa, F. R.; Leuteritz, A.; Wang, Y. Z.; Wagenknecht, U.; Heinrich, G. Synthesis of Organo Cobalt-Aluminum Layered Double Hydroxide via a Novel Single-Step Self-Assembling Method and its Use as Flame Retardant Nano-filler in PP. *Langmuir* **2010**, *26*, 14162–14169.

25. Genovese, A.; Shanks, R. A. Structural and Thermal Interpretation of the Synergy and Interactions Between the Fire Retardants Magnesium Hydroxide and Zinc Borate. *Polym. Degrad. Stab.* **2007**, *92*, 2–13.

26. Durin-France, A.; Ferry, L.; Lopez Cuesta, J. M.; Crespy, A. Magnesium Hydroxide/ Zinc Borate/Talc Compositions as Flame-Retardants in EVA Copolymer. *Polym. Int.* **2000**, *49*, 1101–1105.

27. Ding, P.; Qu, B. Synthesis of Exfoliated PP/LDH Nanocomposites via Melt-Intercalation: Structure, Thermal Properties, and Photo-Oxidative Behavior in Comparison with PP/MMT Nanocomposites. *Polym. Eng. Sci.* **2006**; *46*, 1153–1159.

28. Tatiya, P. D.; Nevare, M. R.; Mahulikar, P. P.; Gite, V.V. Melt Processing and Characterization of Nanocomposites of PP/Na+ MMT Blending with Hyperbranched Polyester. *Polym. Plast. Technol. Eng.* **2014**, *53*, 941–951.

29. Li, F.; Duan, X. Applications of Layered Double Hydroxides. *Struct. Bonding* **2006**, *119*, 193–223.

30. Leuteritz, A.; Kutlu, B.; Meinl, J.; Wang, D.; Das, A.; Wagenknecht, U.; Heinrich, G. Layered Double Hydroxides (LDH): A Multifunctional Versatile System for Nanocomposites. *Mol. Cryst. Liq. Cryst.* **2012**, *556*, 107–113.

31. Illaika, A.; Vuillermoz, C.; Commereuc, S.; Taviot-Gueho, C.; Verney, V.; Leroux, F. Reactive and Functionalized LDH Fillers for Polymer. *J. Phys. Chem. Solids* **2008**, *69*, 1362–1366.

32. Penco, M.; Spagnoli, G.; Rahman, M. A.; Passaglia, E.; Coiai, S.; Ciardelli, F. Nonisothermal Crystallization Kinetics of Polypropylene-Layered Double Hydroxide Composites: Correlation with Morphology. *Polym. Comp.* **2011,** *32,* 986–993.

33. Costa, F. R.; Saphiannikova, M.; Wagenknecht, U.; Heinrich, G. Layered Double Hydroxide Based Polymer Nanocomposites. *Adv. Polym. Sci.* **2008,** *210,* 101–168.

34. Jiao, C.; Chen, X. Synergistic Effects of Zinc Oxide with Layered Double Hydroxides in EVA/LDH Composites. *J. Therm. Anal. Calorim.* **2009,** *98,* 813–818.

35. Kang, N. J.; Wang, D. Y.; Kutlu, B.; Zhao, P. C.; Leuteritz, A.; Wagenknecht, U.; Heinrich, G. A New Approach to Reducing the Flammability of Layered Double Hydroxide (LDH)-Based Polymer Composites: Preparation and Characterization of Dye Structure-Intercalated LDH and its Effect on the Flammability of Polypropylene-Grafted Maleic Anhydride/d-LDH Composites. *ACS Appl. Mater. Interfaces* **2013,** *5,* 8991−8997.

36. Shabanian, M.; Basaki, N.; Khonakdar, H. A.; Jafari, S. H.; Hedayati, K.; Wagenknecht, U. Novel Nanocomposites Consisting of a Semi-Crystalline Polyamide and Mg–Al LDH: Morphology, Thermal Properties and Flame Retardancy. *Appl. Clay Sci.* **2014,** *90,* 101–108.

37. Schonhals, A.; Goering, H.; Costa, F. R.; Wagenknecht, U.; Heinrich, G. Dielectric Properties of Nanocomposites Based on Polyethylene and Layered Double Hydroxide. *Macromolecules* **2009,** *42,* 4165–4174.

38. Wan, W.; Yu, D.; Xie, Y.; Guo, X.; Zhou, W.; Cao, J. Effects of Nanoparticle Treatment on the Crystallization Behavior and Mechanical Properties of Polypropylene/Calcium Carbonate Nanocomposites. *J. Appl. Polym. Sci.* **2006,** *102,* 3480–3488.

39. Hanim, H.; Zarina, R.; Fuad, M. Y. A.; Mohd. Ishak, Z. Y.; Hassan, A. The Effect of Calcium Carbonate Nanofiller on the Mechanical Properties and Crystallisation Behaviour of Polypropylene. *Malaysian Polym. J.* **2008,** *12,* 38–49.

40. Jiao, C.; Chen, X. Synergistic Effects of Zinc Oxide with Layered Double Hydroxides in EVA/LDH Composites. *J. Therm. Anal. Calorim.* **2009,** *98,* 813–818.

41. Nevare, M.; Gite, V. V.; Mahulikar, P. P.; Ahamad, A.; Rajput, S. D. Synergism Between LDH and Nano-Zinc Phosphate on the Flammability and Mechanical Properties of Polypropylene. *Polym. Plast. Technol. Eng.* **2014,** *53,* 429–434.

42. Ahamad, A.; Patil, C. B.; Gite, V. V.; Hundiwale, D. G. Evaluation of the Synergistic Effect of Layered Double Hydroxides with Micro- and Nano-CaCO$_3$ on the Thermal Stability of Polyvinyl Chloride Composites. *J. Thermoplast. Compos. Mater.* **2013,** *26,* 1–11.

43. Patil, C. B.; Kapadi, U. R.; Hundiwale, D. G.; Mahulikar, P. P. Preparation and Characterization of Poly (Vinyl Chloride) Calcium Carbonate Nanocomposites via Melt Intercalation. *J. Mater. Sci.* **2009,** *44,* 3118–3124.

44. Tatiya, P. D.; Nevare, M. R.; Mahulikar, P. P.; Gite, V. V. Melt Processing and Characterization of PP/Pristine Montmorillonite (MMT) Nanocomposites: Influence of Compatibilizer and Hyperbranched Polymer. *J. Vinyl. Add. Technol.* **2016,** *22,* 72–79.

45. Tran, H. V.; Tran, L. D.; Vu, H. D.; Thai, H. Facile Surface Modification of Nano Precipitated Calcium Carbonate by Adsorption of Sodium Stearate in Aqueous Solution. *Colloids Surf. A.* **2010,** *366,* 95–103.

CHAPTER 14

DIFFERENT FABRICATION TECHNIQUES OF AEROGELS AND ITS APPLICATIONS

FATHIMA PARVEEN[1] and RAGHVENDRA KUMAR MISHRA[2,3,*]

[1]Department of Engineering Amity University, Dubai

[2]International and Inter University Centre for Nanoscience and Nanotechnology (IIUCNN), Mahatma Gandhi University, Kottayam, Kerala 686560, India

[3]Indian Institute of Space Science and Technology, Trivandrum, ISRO P.O., Kerala, India

*Corresponding author. E-mail: raghvendramishra4489@gmail.com

CONTENTS

ABSTRACT

In the past few decades, aerogels have attracted great interest from the researchers due to their unique structure and associated properties such as insulating capability, extremely low density, high porosity, high specific surface area, low dielectric constant, and low thermal conductivity. This chapter explains and compares different aerogel fabrication techniques such as supercritical drying (SCD), ambient pressure drying (APD), and freeze drying along with their mechanisms, advantages, and disadvantages. The applications of aerogels in various fields are also described in brief.

14.1   INTRODUCTION

Aerogels are a class of materials with various attractive properties. They are open porous networks having very low density. Most of them have more than 90% porosity.

Aerogels are fundamentally derived from gels. However, unlike gels, they are solid, porous, rigid, and dry. Sol-gel synthesis is usually used to prepare gels in which the liquid phase is trapped within the solid network. When the liquid phase is extracted from the gel without collapsing the solid network structure, an aerogel is formed.

However, extracting liquid from the gel is not a simple task because the properties, number of the pores, and pore size are highly dependent on the drying technique used. As a result, a lot of research is being carried out for the successful removal of liquid from solid networks. There are three main drying techniques widely used for the removal of liquid from the gel to form an aerogel: supercritical drying (SCD); ambient pressure drying (APD); and freeze drying. In this chapter, the abovementioned drying techniques for preparing the aerogel along with their mechanisms, pros, and cons are explained in detail.

14.2   DIFFERENT TECHNIQUES FOR AEROGELS

Aerogel structure is formed by sol-gel polymerization followed by drying. In sol-gel polymerization, the sol (solid nanoparticles dispersed in a liquid) agglomerate together to form a gel (continuous three-dimensional network extending throughout the liquid).[1] The continuous phase in a sol is a liquid, and the dispersed phase is a solid. A gel is a wet solid-like material in which nanostructures form a solid network, which is distributed throughout the liquid medium.[1] When the liquid is extracted without damaging the cross-linked network frame using any available drying technique, a solid porous structure called the aerogel is formed. In short, aerogels are the solid framework of a gel isolated from the gel's liquid medium.[1]

The major requisite in fabricating an aerogel is to remove the liquid solvent from the wet gel. Removal of solvent from the wet gel is called *drying*. During drying, factors such as surface tension of the gel, temperature, and pressure play significant roles. But the surface tension at the liquid–vapor interface during conventional drying causes large capillary forces that are difficult for the nanostructures to resist. These capillary forces cause the gels to fracture and shrink as they dry. The capillary forces exerted by the meniscus of the liquid in the pore and the pressure gradient exerted by the large shrinkage of the network are the main reasons for the collapse of the network structure.[2–3]

Drying is governed by capillary pressure, P_c according to the equation,

$$P_c = 2\gamma \cos\theta / r_p;$$
γ-surface tension of the pore liquid
r_p-pore radius

$$r_p = 2V_p / S_p$$
V_p-pore volume
S_p-surface area

Smaller the capillary radius, higher the liquid will rise and higher the hydrostatic pressure will be exerted. Since the pore diameters in the gel are in the order of nanometers, the liquid will exert high hydrostatic pressure. As the liquid in the pores evaporates, the meniscus in the pores and the surface tension forces try to pull the particles together. These combined forces try to collapse the pores and shrink the structure. Hence, during drying, the gels with very fine pores have a tendency to crack and shrink.[4]

So, the solution to this problem is to skip crossing the vapor–liquid boundary and find out other methods to extract the liquid by reducing the capillary forces.[5] In this respect, the main challenge during the drying step is to eliminate the liquid solvent from the gel without collapsing the already existing nanoporous structure, and thereby avoid the subsequent shrinkage and cracking of the dried gel. The method of drying, hence, decides the structure of the final product and is very important in preserving the porous network.

Aerogels are generally prepared using techniques like SCD, APD, or freeze drying. Supercritical drying is the most widely used method to fabricate aerogels. In SCD, the liquid component of the gel is extracted by allowing it to dry off slowly without causing the solid matrix in the gel to collapse from the capillary action like evaporation.[6] The solvent in the hydrogel is exchanged with a water-soluble alcohol (e.g., ethanol). This alcogel is then dried supercritically to obtain the aerogel. In APD, the solvent is removed from the pores by simple evaporation at ambient pressure. But, the shrinkage caused by evaporation is reduced by prolonged aging, aging with an additional catalyst or by using chemical additives before the drying step. In freeze drying/lyophilization, the hydrogel is frozen and dried under high vacuum.[7] These methods are generally able to remove the solvent from the gel without entirely collapsing the network structure.

This chapter discusses the following drying techniques that are used to extract the solvent from the wet gel to form an aerogel in detail:

    i)   Supercritical drying (SCD)
    ii)   Ambient pressure drying (APD)
    iii)   Freeze drying

14.2.1   SUPERCRITICAL DRYING

Supercritical drying is the most widely adopted drying technique for aerogels. Supercritical drying was first introduced by Kistler in 1930s when he successfully replaced the liquid in gel with air without destroying the gel structure by using ethanol as a supercritical fluid in an autoclave. In SCD, the solvent changes its state from liquid to gas at high temperature and pressure by passing through the supercritical region without crossing the liquid–vapor phase boundary. Supercritical method succeeds in forming aerogel by removing the solvent above its supercritical point without crossing the liquid–vapor interface.

Gels are unable to withstand the high drying stress produced during the normal drying process as it passes through the liquid–vapor interface which results in considerable shrinkage. Shrinkage results from the formation of liquid–vapor meniscus and the capillary pressure build-up.[8] The resultant capillary forces in the pores cause gels to shrink and crack due to internal stress.[9] In SCD since the solvent does not pass through liquid–vapor interface, the problem of capillary forces caused by surface tension does not arise. Hence, we can retain the original gel structure in aerogel even after drying.

Advantages

Compared to other drying methods, SCD helps to preserve high porosity and superior textural properties of aerogel to a greater extent.

Disadvantages

Supercritical drying techniques are generally expensive on a large scale. So, the research is being carried out to find alternative ways to reduce the production cost, increase the mechanical integrity, and reduce the shrinkage.

14.2.1.1   GENERAL MECHANISM

In SCD, the solvent in the gel is washed away with a suitable alcohol to form an alcogel. The initial solvent must be completely miscible in the alcohol for this step to be executed. The temperature and pressure of the alcogel are set above the critical point of the solvent in an autoclave. The pressure and temperature of the alcogel are increased till it crosses the critical point of the solvent.

Supercritical fluids are obtained when the liquid in or around the gel is above the critical temperature and pressure. Critical pressure is achieved by heating the liquid within a confined space. When the liquid molecules reach the critical point, their kinetic energy and mean free path increases.[10] Subsequently, the surface tension of the liquid decreases that in turn decreases its capillary stress. At the critical point, the liquid has zero surface tension and, hence, is incapable of exerting any capillary stress on the gel structure.[10] At a critical point, there is no difference between liquid and vapor phases and their densities become equal. Hence, SCD takes place in the absence of both the capillary forces and surface tension. At this supercritical state, for the liquid, the liquid–vapor meniscus responsible for the gels' collapse is eliminated, allowing crack free aerogels to be produced.[8]

Once the solvent is completely turned to its supercritical state, the system is gradually depressurized under isothermal conditions (temperature must be above the critical temperature at all times), allowing the gas to escape from the pores. Once the ambient pressure is reached, the sample is allowed to cool down to room temperature. Often the vessel is pre-pressurized by nitrogen to avoid evaporation of the solvent.[2] Completion of the entire process leaves behind a solid network with gaseous pores in the place of the solvent which is termed as an aerogel. Figure 14.1 shows the P vs T diagram for supercritical drying.

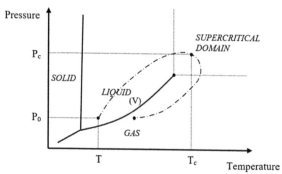

FIGURE 14.1   Supercritical drying procedure.[11]

The gel can be heated and pressurized to reach the supercritical state and then cooled and depressurized to reach the room conditions. During this process, the solvent vaporization curve (V) is not crossed.[11]

14.2.1.2   TYPES OF SCD

The solvent can be removed through SCD process in two different ways:

i)   High-temperature supercritical drying (HTSCD): Solvent in the wet gel is mixed with solvents of high supercritical temperatures such as ethanol, methanol, acetone, and so forth. The wet gel is heated to the supercritical temperature of the alcohol in the pores which is then vented out from the wet gel to form an aerogel.

ii) Low-temperature supercritical drying (LTSCD): The solvent is exchanged with a suitable solvent system (usually liquid CO_2) in which the former solvent must be completely soluble. The solvent system formed is then dried supercritically.

14.2.1.2.1   High-temperature Supercritical Drying

In HTSCD, the solvent in the wet gel is replaced by a suitable organic solvent such as methanol or ethanol. It is then placed in an autoclave and subjected to an increase in temperature, which in turn results in a corresponding increase in pressure. The pressure and temperature are allowed to rise to a value higher than the critical point of the solvent used. This method usually uses high pressures (above 8 MPa) and temperatures (above 260°C).[5] The pressure is kept constant for a period of time until it reaches above the critical pressure of the solvent. The solvent is then slowly vented out at a constant temperature, which results in a gradual pressure drop. The system is then cooled down to room temperature when the ambient pressure is reached. Some of the solvents that are used for HTSCD are mentioned in the Table 14.1 below.

TABLE 14.1   List of Solvent for High-Temperature Supercritical Drying.

Liquid	Critical temperature (°C)	Critical pressure (MPa)
Methanol	240	8.09
Ethanol	243	6.3
Acetone	235	4.66
Water	374	22.04

At the critical point, the alcohol reacts with the hydroxyl groups on the gel surface to form alkoxy groups which make the aerogels partially hydrophobic. High-temperature supercritical drying using organic solvents produces aerogels with minimum shrinkage than LTSCD which helps in drying low-density gels.

One of the risk factors for using this method is that the solvents used here are inflammable and hazardous at high temperature and pressure. In addition to the safety issues, a higher temperature will lead to variations and rearrangement reactions during aging of the gel. Therefore, the resulting aerogels will have small pore sizes, low surface area, and a stiffer network. There is a chance for destruction of organic groups, phase separation, loss of stoichiometry, and crystallization in multicomponent gels while using HTSCD.[2]

14.2.1.2.2 Low-temperature Supercritical Drying

In LTSCD, solvent existing in the wet gel is mixed with another solvent with a low critical point by washing, which is then extracted supercritically. The process that follows the solvent exchange is similar to the HTSCD. Some of the solvents that are used for LTSCD are mentioned in Table 14.2.

TABLE 14.2   List of Solvent for Low-Temperature Supercritical Drying.

Liquid	Critical temperature (°C)	Critical pressure (MPa)
Carbon dioxide	31	7.37
Nitrous oxide	36	7.24

In LTSCD, a solvent which has a critical point near to the ambient temperature of alcohol mixture is used. CO_2 is the most commonly used supercritical fluid for aerogel drying because of its easy availability, lower production temperatures, cost, and toxicity.[8] Nitrous oxide has similar supercritical values and physical behavior of CO_2, but it is a powerful oxidizer in its supercritical state and direct contact will cause health issues such as drowsiness.

CO_2 SCD can be executed at low temperatures (<40°C) and moderate pressure (<80 bar).[12] Therefore, it can be applied even to organic gels which are temperature sensitive because CO_2 has a low critical temperature.

An important criterion that must be met for CO_2 SCD process is the miscibility of solvent in the pores with CO_2. In the first place, solvent existing in the wet gel is mixed with CO_2 through simple washing. The supercritical CO_2 drying is performed in an autoclave. The temperature is raised to the critical temperature of CO_2 which in turn causes the critical pressure to raise consequently. The liquid–vapor meniscus disappears and a supercritical fluid is formed when the carbon dioxide's supercritical point is reached (31°C, 7.37 MPa). Then, the system is depressurized and supercritical CO_2 is vented off slowly and completely from the pores. When ambient pressure is reached, the system is then cooled down to room temperature leaving the aerogel with gaseous pores behind.

The supercritical extraction of CO_2 is much easier and less expensive than HTSCD. But this usually results in the formation of hydrophilic aerogels because they retain the hydroxyl groups. These hydrophilic aerogels can be modified to become hydrophobic through surface functionalization.

For example, in the case of silica aerogel, the dry silica aerogel is placed in tetramethylsilane (TMS) vapors for several hours and hydroxyl groups are replaced by TMS groups.[13]

It is observed that the quality and shape of the aerogel are decided by the diffusion of alcohol to the CO_2 phase.[14] Further, it is reported that rapid drying resulted in cracked samples, whereas slow drying resulted in transparent crack-free aerogels.[9] The drying time strongly depends on the gel thickness, flow velocity of supercritical CO_2, temperature, and pressure. A low flow velocity, a thick gel, a low temperature, and a high pressure are preferred for a crack-free transparent aerogel.[9]

As an example, for nanocellulose suspensions in ethanol solvent, the SCD using CO_2 can be carried as follows. Additional ethanol is added to the nanocellulose suspension placed in the autoclave to prevent air drying. The system is then pressurized to at least 5–6 MPa with CO_2 and cooled to 5–10°C. Ethanol is then extracted by flushing CO_2 through the system. Now, the temperature and pressure of the system are set to reach the critical points of CO_2 by heating and pressurizing. Once the critical points are reached, the system is maintained at constant temperature and pressure for a period, for the supercritical CO_2 to flush through the nanocellulose network to completely remove the ethanol from the pores. After the complete removal of ethanol, the system is depressurized to vent out CO_2 until ambient pressure is reached.[11]

14.2.2   AMBIENT PRESSURE DRYING

Ambient pressure drying is gaining widespread attention due to the fact that it can be easily industrialized for large-scale production of aerogels.

Ambient pressure drying technique is self-defined by its name, that is, the wet gel is dried at ambient pressure. But this technique requires the wet gel to undergo long solvent exchange processes in order to strengthen the structure and is used because it is a cheaper alternative to supercritical and freeze-drying processes. Ambient pressure drying helps to make the gel more flexible to withstand the capillary forces during the drying process.[5] So the wet gel is first chemically modified to manipulate the contact angle and strengthen the network. Pore surfaces inside the gel are passivated in order to prevent the formation of new chemical bonds after condensation reactions when the gel network is compressed under the drying stresses.[15]

In APD, the solvent is removed from the pores by simple evaporation at ambient pressure. The solvent in liquid form is converted to its gaseous form by heating it above its boiling point. Here, drying is carried out along the liquid–vapor interface. When the drying starts three solvent phases coexist in the sample: liquid phase in the solid pores; liquid-gas phase transition regime; and the gas phase.[10] Drying is completed when the liquid phase is completely converted to the gaseous phase. An aerogel monolith can resume its wet size by spring-back effect at the end of the solvent evaporation process as it is no longer submitted to capillary stresses.[15]

Advantages

Ambient pressure techniques for aerogel preparation are still in its infant stages. These techniques are being widely researched now as an alternative to expensive SCD techniques due to its simplicity and cost efficiency. Since APD can reduce the production cost of the aerogels, their relevance has shifted from labs to industries.

Disadvantages

The major drawback of APD is that the solvent evaporation from hydrogels results in large capillary pressures that can easily reach hundreds of bars within nanopores.[10] Hence, the aerogels formed by this method are less porous, form cracks, and show shrinkage. Generally, it is impossible to form aerogels with pore sizes <80 nm using APD.[10] Therefore, it is difficult to preserve the gel structure using APD. To achieve nanoporous high-quality aerogel, the gel must be subjected to pretreatment and the drying process must be optimized.

14.2.3   FREEZE DRYING

Freeze drying is also known as lyophilization or cryodesiccation. Freeze drying technique was invented by Arsène d' Arsonval and F. Bordas in 1906. It is an important technique used for preservation of materials in pharmaceuticals, biotechnology, agriculture, and food industries. In freeze drying, the wet gel is frozen and the frozen solvent in the pores of the gel network is removed through sublimation directly from solid to vapor phase without passing through the liquid phase under vacuum. The remaining unfrozen water is removed by desorption. Under normal conditions, heating a frozen substance is followed by melting to a liquid state and then vaporizing to a gaseous state.

But under reduced pressure, we can transform the solid solvent to gaseous phase directly through sublimation. This helps to obtain materials that are crack free and highly porous by escaping the liquid phase meniscus stresses.[10] This process also provides an alternative to prevent passing through the liquid–vapor phase boundary like SCD.

Freeze drying is a high time and energy consuming process. Freeze drying process requires a low-temperature environment. The structure of the final product after freeze drying is highly dependent on freezing rate, sublimation temperature, and solvent nature.[10] The aging period has to be extended to stabilize the gel network, which makes the process highly time-consuming. The solvent must be exchanged with another solvent with a low thermal expansion coefficient and high pressure of sublimation.[2] It also requires the addition of salts to achieve low freezing rates and freezing temperatures.[16]

Cellulose aerogels are obtained by freezing nanofiber suspensions and sublimating the solvent to obtain a lightweight porous structure. Size of the solvent crystallites is highly dependent on the freezing speed.[17] In the aerogel, the nanocellulose fibrils are connected by hydrogen bonds and entanglements to form an interconnected network which maintains its physical integrity. The water in cellulose nanofiber hydrogel is frozen and sublimated instead of evaporating in order to prevent the collapse of the porous structure due to the capillary forces arising during evaporation.[17] Paakko et al.[18] were the first to report the fabrication of cellulose aerogels from cellulose nanofiber aqueous suspensions by freeze drying. The major step that determines the final aerogel structure is the freezing stage. The cooling speed should be maintained constant throughout the sample for the formation of evenly sized ice grains.[17] When the water (solvent) is sublimated, tiny cavities are formed in the place of the ice grains.

14.2.3.1   GENERAL MECHANISM

14.2.3.1.1   Pretreatment

The concentration of the sample and the solvent to be used in the sample should be determined before freeze drying. The surface texture and density of the aerogels can be tuned by selecting the concentration of the nanocellulose dispersions.[19] The components that will affect the stability, appearance, surface area, and other parameters should also be monitored. Solvents used in freeze drying must be of low expansion coefficient and high sublimation pressure. The solvents that are generally used in freeze drying are mentioned in Table 14.3.

TABLE 14.3   List of Solvent for Freeze Drying.

Solvent	Triple point (°C)
Water	0
Acetone	−95
Methanol	−97
Ethanol	−114

14.2.3.1.2 Freeze Drying

Freeze drying involves three main steps: freezing, primary drying, and secondary drying. The sample is first frozen under atmospheric pressure (freezing), then the frozen water/ice is removed by sublimation (primary drying), and the nonfrozen bounded water molecules are removed by desorption (secondary drying). The conditions under which the freeze drying is carried out will play a significant role in the quality of the dried product.

14.2.3.1.3 Freezing

In the first step, the sample is cooled below its triple point. A triple point indicates the pressure and temperature at which the solid, liquid, and gaseous states of a material can coexist in equilibrium. This has to be done if sublimation is to be carried out instead of melting. Freezing can be done by mechanical refrigeration using dry ice, liquid nitrogen, or alcohols. Larger crystals are easier to freeze dry. Large crystals can be formed by lowering the temperature slowly or by annealing. Figure 14.2 shows the freezing (liquid to solid conversion), condensation and vaporization (vapor to liquid and liquid to vapor state conversion), and sublimation (solid to vapor conversion). In order to, Figure 14.3 shows phase diagram of water with respect to the P vs T curve.

The drying process following the freezing consists of two stages, that is, primary drying and secondary drying. A major part of the water is removed from the product during the primary drying stage by the sublimation of frozen ice crystals.

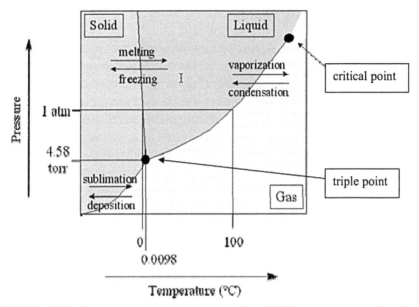

FIGURE 14.2 Pressure versus temperature graph of different states and freezing, condensation, and sublimation.[20]

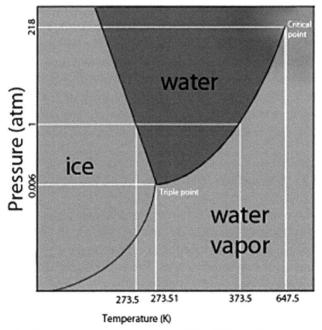

FIGURE 14.3 Pressure versus temperature graph for different phases of water.[21]

14.2.3.1.4   Primary Drying

Primary drying is a slow drying process in which, the latent heat of sublimation required by the frozen ice molecules to convert to vapor, is provided under low pressure and a temperature below the product's critical collapse temperature. Two factors that control the product temperature during the sublimation are system pressure and temperature in the chamber.

During this process, the pressure is lowered by applying partial vacuum. The pressure has to be lowered below the triple point pressure of water in order to avoid the liquid phase during sublimation. Even though sublimation can also take place at atmospheric pressure by passing dehydrated air, vacuum helps to speed up the sublimation process.

Heat energy supplied must be lower than the product's eutectic and has to be supplied slowly and steadily to avoid damage to the structure of the product. All three methods of heat transfer are used for freeze drying a product—conduction, convection, and radiation. If the heat supplied is insufficient, then the water will sublime at a very low temperature making the freeze drying cycle long. A condenser/vapor trap which is at a cooler temperature is used to condense the vapors that are released by sublimation. One of the methods used to determine the completion of primary drying is to use a thermocouple to measure the product temperature.

During the sublimation process, the product uses the heat in the chamber for sublimation which causes the product temperature to be lower than the chamber temperature. When the product temperature is equal to the chamber temperature, it indicates the completion of primary drying. After the completion of primary drying, there will still be moisture content in the range of 5–10% due to the presence of sorbed water molecules in the product.[20]

14.2.3.1.5   Secondary Drying

The secondary drying stage removes the residual water content in the product that remains even after the primary drying. In secondary drying stage, the temperature is increased above the primary drying stage, which in turn increases the desorption rate. The water vapor diffuses and may induce structural changes and agglomeration of cellulose nanofibrils.[22]

Another option to remove the water adsorbed on the internal surface of the product is to increase the pressure. This helps to overcome the capillary forces of water. Secondary drying parameters need to be carefully controlled to avoid product collapse and over drying. After the whole process, the vacuum is broken using an inert gas like nitrogen and the product is sealed. The water content in the final freeze-dried product is restricted around 1–4% as per the product requirements. A schematic diagram for the primary and secondary freeze drying cycle is illustrated in Figure 14.4.

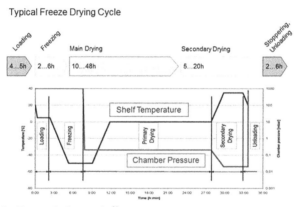

FIGURE 14.4 Freeze drying cycle.[21]

Advantages

Freeze drying is a simple, economic, and environmental-friendly method to obtain good porous aerogels. Lyophilization is an apt way to dehydrate and preserve heat sensitive gels that degrade at high temperatures or are very sensitive to moisture like cellulose, biopolymer gels, etc. without damaging their structures and hampering their activities.[10] Degradation of the lyophilized product by microorganisms is prevented due to low water content. This way, materials can be preserved for many years without refrigeration. So it can be used as a dehydration process used for the preservation of degradable materials. Lyophilization helps to dehydrate materials with minimum damage as it does not cause much shrinkage or toughening of the material. Lyophilized products can also be rehydrated very easily due to the presence of water-free pores.

Disadvantages

On the other hand, lyophilizer installation costs are very high. The process is also highly energy consuming (around 2–3 times compared to other conventional methods) and time consuming (around 24 h). This method also demands prolonged aging for network stabilization compared to other drying processes.[22]

Sublimation of volatile compounds in addition to water may sometimes cause unintended outcomes. Crystallization of the solvent in the pores will damage the network structure of the aerogels and forms cracks. Freeze drying also often produce macroporous aerogels due to the sublimation of ice microcrystals that are formed during the freeze drying.[5] The regular freeze-dried aerogels have highly porous structure consisting of microfibrillar networks, but on close examination, the fibrils were severely coagulated to form film-like masses, especially at lower cellulose concentrations.[16]

14.3 COMPARISON OF VARIOUS DRYING METHODS

The drying method adopted and the starting materials used such as nano-fibrillated cellulose (NFC) or cellulose nanocrystals (CNC) determine the morphological properties of the final dried aerogel.

The agglomeration mechanism occurring among the cellulose nano-fibrils varies depending on each drying method. Both SCD and freeze drying create highly networked structures of cellulose agglomerates having multi-scalar dimensions, including the nanoscale.[16]

Among all the drying techniques, SCD is considered as the best method which can form aerogels with pores at the nanoscale. It is also capable of forming the aerogels with the least shrinkage and collapse of mesopores during drying. On the other hand, use of high temperature and pressure during drying, use of organic solvents at extreme conditions and long-time solvent exchange make it a time-consuming and high-cost method, which severely limits its practical applications.[16]

Freeze drying is simpler and easier than SCD. It is more cost-effective and environmental-friendly method. But at the same time, it is a time-consuming process. The vacuum conditions, freezing temperature, and precursor concentration in the wet gel have to be carefully controlled for the formation of the well-defined aerogel.[16]

Ambient pressure drying is the easiest method to dry wet gels and to form aerogels among all the drying methods available. It has the potential to be industrialized for large-scale aerogel fabrication. Here, the wet gel is dried at ambient pressure through simple evaporation. But in order to prevent the shrinkage and collapse of the network that occur due to conventional drying, the gel is subjected to chemical modifications.

14.4 POTENTIAL APPLICATION OF AEROGEL

Aerogels (e.g., silica aerogel) have already been used in space projects as a contaminant collector.[23–24] They are being used as filters, absorbing media for desiccation, encapsulation media, and for hydrogen fuel storage owing to its high porosity ($>85\%$) and very large surface areas (>400 m^2/g).[25–27]

Several types of aerogel have been fabricated for various electronic and semiconductor applications: metal oxide and organic aerogels as dielectric, superconducting, thermoelectric and piezoelectric materials; the block of aerogel as microwave absorber and high voltage insulators; and conducting carbon aerogel as electrodes for batteries and capacitors.[28–30]

Uncommon acoustic (properties of sound and how sound is transmitted through it) and mechanical behavior are one of the important properties of aerogels. So, the aerogels are used for acoustic impedance matching in order to create more effective ultrasonic devices, sound absorber, and shock absorbing materials.[31–32]

Aerogels are used for making architectural design, piping, heat and cold storage devices, automotive exhaust pipes, transport vehicles and vessels, and thermal insulation due to their very inherent low thermal conductivity.[33–34]

Nowadays, laser glass is fabricated from the lanthanide doped aerogels, radioactive tritium and phosphorus doped silica aerogels for making the radioluminescent light source. It is reported that quantum confinement is possible in nanoparticle-loaded silica aerogels.[23] Recently, solar covers and collectors are made by translucent aerogels; they are also being used in solar windows. Ultra-low density aerogels have been used for preparing lightweight mirror backings.[35–39]

KEYWORDS

- aerogel
- fabrication
- techniques
- drying
- application

REFERENCES

1. Mujumdar, A. S.; Benali, M.; Boumghar, Y. Supercritical Fluid-Assisted Drying. In *Handbook of Industrial Drying*, 4th ed.; Boca Raton, Florids: CRC Press, pp 1261–1270.
2. Mortensen, A. Concise Encyclopedia of Composite Materials, 2nd ed.; Elsevier, 2006.
3. Aravind, P. R.; Shajesh, P.; Soraru, G. D.; Warrier, K. G. K. Ambient Pressure Drying: a Successful Approach for the Preparation of Silica and Silica Based Mixed Oxide Aerogels. *J. Sol-Gel Sci. Tech.* **2010,** *54*(1), 105–117.
4. Gurav, J. L.; Jung, I. K.; Park, H. H.; Kang, E. S.; Nadargi, D. Y. Silica Aerogel: Synthesis and Applications. *J. Nanomater.* **2010,** *2010,* 23.
5. Aegerter, M. A.; Leventis, N.; Koebel, M. M. *Aerogels Handbook Advances in Sol-Gel Derived Materials and Technologies;* Springer Science & Business Media: New York, 2011.
6. Aerogel. https://en.wikipedia.org/wiki/Aerogel. (Accessed May 5, 2017)
7. Xuan Yang B. "Hydrogels and aerogels based on chemically cross-linked cellulose nanocrystals." PhD diss., McMaster University, 2014.
8. Boday, D. J. Silica Aerogel-Polymer Nanocomposites and New Nanoparticle Syntheses. Ph.D.Thesis, The University of Arizona. Materials Science & Engineering, ProQuest. 2009.
9. Mukhopadhyay, M.; Rao, B. S. Modeling of Supercritical Drying of Ethanol-Soaked Silica Aerogels with Carbon Dioxide. *J. Chem. Technol. Biotechnol.* **2008,** *83*(8), 1101–1109.
10. Levy, D.; Zayat, M. *The Sol-Gel Handbook: Synthesis, Characterization and Applications* (3-Volume Set); John Wiley & Sons. 2015, Volume 2. Weinheim, Gennany.
11. Biesmans, G., D. Randall, E. Francais, and M. Perrut. *Polyurethane-based organic aerogels' thermal performance. J.* of Non-Cryst. Solids **1998,** *225,* 36–40
12. Brunner, G. H. *Supercritical Fluids as Solvents and Reaction Media;* Elsevier, 2004.
13. Kowalski, S. J. ed. *Drying of porous materials.* Springer, 2007.
14. Novak, Z.; Željko, K. Diffusion of Methanol–Liquid CO_2 and Methanol–Supercritical CO_2 in Silica Aerogels. *J. Non-cryst. solids* **1997,** *221*(2), 163–169.

15. Maleki, H.; Duraes, L.; Portugal, A. Development of Mechanically Strong Ambient Pressure Dried Silica Aerogels with Optimized Properties. *J. Phys. Chem. C* **2015,** *119*(14), 7689–7703.

16. Zuo, L.; Zhang, Y.; Zhang, L.; Miao, Y. E.; Fan, W.; Liu, T. Polymer/Carbon-Based Hybrid Aerogels: Preparation, Properties and applications. *Materials* **2015,** *8*(10), 6806–6848.

17. Mishra, R. K., H. J . Maria, K. Joseph, and S. Thomas. "Basic structural and properties relationship of recyclable microfibrillar composite materials from immiscible plastics blends: An introduction." In Micro and Nano Fibrillar Composites (MFCs and NFCs) from Polymer Blends, pp. 1–25. 2017.

18. Pääkkö, M.; Vapaavuori, J.; Silvennoinen, R.; Kosonen, H.; Ankerfors, M.; Lindström, T.; Ikkala, O. Long and Entangled Native Cellulose I Nanofibers Allow Flexible Aerogels and Hierarchically Porous Templates for Functionalities. *Soft Matter* **2008,** *4*(12), 2492–2499.

19. Chirayil, Cintil Jose, Jiji Abraham, Raghvendra Kumar Mishra, Soney C. George, and Sabu Thomas. "Instrumental Techniques for the Characterization of Nanoparticles." In Thermal and Rheological Measurement Techniques for Nanomaterials Characterization, pp. 1–36, 2017.

20. *Basic Principles of Freeze Drying;* John Barley, SP Scientific. Retrieved 23 Dec.2016 from http://www.spscientific.com/freeze-drying-lyophilization-basics/.

21. *Freeze Drying (Lyophilization)*, Empower Pharmacy. 23 Dec.2016 from https://empower.pharmacy/freeze-drying-lyophilization.html.

22. Peng, Y.; Gardner, D. J.; Han, Y. Drying Cellulose Nanofibrils: In Search of a Suitable Method. *Cellulose* **2012,** *19*(1), 91–102.

23. Sachithanadam, M.; Joshi, S. C. *Silica Aerogel Composites: Novel Fabrication Methods Engineering Materials.* Springer. 2016.

24. Hotaling, S. P. *Rome Laboratory Report.* RL-TR-93-148, July, 1993.

25. Morris, R. E.; Wheatley, P. S. Gas Storage in Nanoporous Materials. *Angew. Chem. Int. Ed.* **2008,** *47*(27), 4966–4981.

26. Attia, Y. ed. *Sol-Gel Processing and Applications;* Springer Science & Business Media, 2012.

27. Hrubesh, L. W. Aerogel Applications. *J. Non-Cryst. Solids* **1998,** *225,* 335–342.

28. Hrubesh, L. W.; Pekala, R. W. Dielectric Properties and Electronic Applications of Aerogels. In *Sol-Gel Processing and Applications;* Attia, Y., Ed.; Springer: US, 1994; 363–367.

29. Hrubesh, L. W. Silica Aerogel: An Intrinsically Low Dielectric Constant Material. *MRS Proceedings*, Vol. 381. Cambridge University Press, 1995.

30. Hrubesh, L. Technical Applications of Aerogels. *5th International Symposium on Aerogels Montpellier, France September.* 1997.

31. Gronauer, M.; Fricke, J. Acoustic Properties of Microporous SiO_2-Aerogel. *Acta Acust. Acust.* **1986,** *59*(3), 177–181.

32. Gibiat, V. et al. Acoustic properties and potential applications of silica aerogels. *J. Non-Cryst. Solids* **1995,** *186,* 244–255.

33. Fricke, J.; Reichenauer, G. Structural Investigaticn CF SiO_2—Aerogels. *J. Non-Cryst. Solids* **1987,** *95,* 1135-1141.

34. Fricke, J. et al. Optimization of Monolithic Silica Aerogel Insulants. *Int. J. Heat Mass Transfer* **1992,** *35*(9), 2305–2309.
35. Dunn, B.; Zink, J. I. Optical Properties of Sol–Gel Glasses Doped with Organic Molecules. *J. Mater. Chem.* **1991,** *1*(6), 903–913.
36. Tillotson, T. M. et al. Synthesis of Lanthanide and Lanthanide-Silicate Aerogels. *J. Sol-Gel Sci. Technol.* **1994,** *1*(3), 241–249.
37. Renschler, C. L. et al. Solid State Radioluminescent Lighting. *Radiat. Phys. Chem.* **1994,** *44*(6), 629–644.
38. Jensen, K. I. Passive Solar Component Based on Evacuated Monolithic Silica Aerogel. *J. Non-Cryst. solids* **1992,** *145,* 237–239.
39. Rubin, M.; Lampert, C. M. Transparent Silica Aerogels for Window Insulation. *Sol. Energy Mater.* **1983,** *7*(4), 393–400.

CHAPTER 15

ROLE OF NANODISPERSOIDS ON CORROSION INHIBITION BEHAVIOR OF SMART POLYMER NANOCOMPOSITE COATINGS

SARAH B. ULAETO, JERIN K. PANCRECIOUS, RAMYA RAJAN, T. P. D. RAJAN*, and B. C. PAI

CSIR-National Institute for Interdisciplinary Science and Technology, Trivandrum, Kerala 695019, India

Corresponding author. E-mail tpdrajan@rediffmail.com, tpdrajan@niist.res.in

CONTENTS

ABSTRACT

The intelligible driving force of nanotechnology-based coatings for corrosion protection can be attributed significantly to the role of the nanodispersoids. This class of materials has provided value-added functionalities in products meant for corrosion protection of metals, alloys, and metallic surfaces from innovations in the area of coatings and linings technology, as well as novel inhibitors and pigments. The multifunctional role discussed herein exhibited by these nanodispersoids is influenced by their functionalization and mode of dispersion in the polymer matrix. These nanoenhancers have greatly improved efficiency and aesthetics of polymeric coatings along with sustainability in the prevention and control of corrosion on structural and engineering materials of great value. This chapter aims at providing an overview of corrosion-inhibiting polymer nanocomposite coatings with particular attention to the role of nanodispersoids in corrosion resistant coatings and their smart characteristics.

15.1 INTRODUCTION

In the last decade, corrosion-resistant coatings have received a lot of attention due to the interdisciplinary nature of material science and nanotechnology. This has led to significant developments in fabrication of coatings with responsive properties due to the interface control which is a key factor when dealing with materials at the nanoscale. The massive modification experienced on structural and engineering coatings with the unification of nanomaterials as one of the phases in the coating mix is a giant step ahead of the traditional perspective. These nanomaterials also known as nanodispersoids are size-dependent materials with fine grain size in the nanometer range (dimensions within 100 nm) and high surface area to volume ratio. These nanodispersoids are notable for their outstanding mechanical and physical properties which have sparked significant progress in composite coating formulations for corrosion-resistant purposes. The resultant formulations from the union of nanoscale additives in polymeric matrices provide a hybrid material with unprecedented functionality and effects. These hybrid materials referred to as polymer nanocomposite coatings have birthed overtime unique designs of smart and multifunctional coatings with the flexibility of compatibility, generative and dispersion modes.

The resultant smart coatings auto-respond uniquely to changes in their environment depending on tailored objective formulation. Smart coatings consist of nanodispersoids as one phase in the nanorange unified within a continuous phase which is the polymer matrix. These nanodispersoids greatly modify the polymeric matrix and, in turn, the properties at the interphase region even at very low filler loadings. The nanodispersoids may be dispersed as nanocontainers containing loaded inhibitors or may function as inhibitors and pigments. They can also be generated in situ during coating processing. The implications of these functional formulations are experienced in both the active and passive barrier properties of corrosion-inhibiting coatings. Ranging from carbon nanomaterials, such as carbon nanotubes (CNTs) and graphenes to layered double hydroxides (LDHs), porous silicon oxide, halloysites, to metal nanoparticles, clay nanoplatelets, and so forth have all synergistically provided robust properties in various coating systems aimed at extending the service life of structural and engineering materials in challenging and severe environments. The outstanding properties associated with these classes of coatings cannot directly be ascribed to any specific component because the synergy therein ensures new properties and bestows the added functionalities.[19,21,33,75,83,106] Following this brief introduction, there is another brief given on polymer composite coatings. Thereafter, a review on nanodispersoids relevant to corrosion-inhibiting coatings is inscribed. Details on fabrication and characterization of polymer nanocomposite coatings are presented as well. Subsequently, effects of nanodispersoids in smart polymer composite coatings in corrosive environments will be discussed. Applications and commercial developments are not overlooked. Finally, our viewpoint on the role of nanodispersoids will be summarized highlighting their importance in smart anticorrosive coatings from a technology that has come to stay. The purpose of this chapter is not to provide an exhaustive summary of the literature but rather to present significant developments on the role of nanodispersoids in smart polymer composite coatings for corrosion protection of metals and alloys.

15.2 POLYMER COMPOSITE COATINGS

Polymeric materials are extensively used in industry due to their lightweight, ease of production, and often ductile nature. But when applied as coatings on metal substrates, these have found limited practical use in industrial applications for corrosion prevention owing to their overall poor adhesion to underlying substrates. Their lack of molecular compatibility often gives rise to slip occurrences between the metal and polymer coatings in response to mechanical stresses. Furthermore, exposure to wet and corrosive environments mostly results in catastrophic debonding. Effect of low thermal resistance and high coefficient of thermal expansion can further accelerate the loss of interfacial adhesion at the polymer/metal interface resulting in delamination of polymeric coatings. Further, vulnerability to damage by surface abrasion and wear, poor resistance against the initiation, and growth of cracks allow corrosive species to penetrate and attack the underlying metallic substrate resulting in localized corrosion occurrences. These defects compromise their appearance, mechanical properties, performance, durability, and so forth and trigger the need for polymer composite coatings. The polymer composite coatings are a combination of fibers, flakes, laminates, or particulates and an organic matrix material, which binds the incorporated additives together. The fibers, flakes laminates, or particulates are reinforcements in the polymer matrix and enhance the corrosion resistance of the composite coatings.[44,61,73] However, the polymer composite coatings were not permanently impenetrable. As an alternative defense against corrosion, various nanoscale inorganic additives were considered and incorporated into different polymer matrices to generate a series of organic–inorganic hybrid or polymer nanocomposite anticorrosive coatings.[94] The presence of nanodispersoids in polymer coatings has greatly enhanced barrier properties alongside autonomic responses due to certain physical or chemical changes in severe environments.

15.2.1 SMART POLYMER NANOCOMPOSITE COATINGS

The concept of "smart coatings" also referred to as "stimuli responsive," "intelligent," or "environmentally sensitive" coatings is attributed to coatings that will undergo physical or chemical changes in response to external stimuli in its environmental conditions.[12,18,48] The illustration in Figure 15.1 is a response to external stimuli. This includes chemical and mechanical triggering, temperature, redox activity or electric field, complex internal or external triggering based on sensed data, and so forth.[18] When these responses occur from polymeric coatings containing a nano-counterpart, it is designated as a polymer nanocomposite coating. The stimuli response of the coating makes it smart and it is a combined effect although greatly influenced by its functional nanoadditive. Polymer nanocomposites coatings obtained by integrating nanodispersoids of various functionalities within polymeric matrices are of immeasurable relevance due to the opportunities realized with their enhanced properties. The properties of these composite coatings are a combination of the individual contributions of both phases and the role of the inner interfaces which modifies chemical (corrosion, wettability, permeation, temperature insulation, biocompatibility), mechanical (friction, wear), optical (color, transmission, reflection, absorption), and transport (electrical and thermal conductivity) properties of surfaces. The choice of the polymers is determined primarily by their mechanical and thermal behavior.[64,76] Effective dispersion of the nanodispersoids in an organic resin system is a major challenge. Therefore, majority of successful marketable products pre-disperse the nanocounterpart in applicable solvents during formulation.[33] The focus on nanocomposite hybrid coatings is due to intimate dispersion of nanodispersoids within polymer matrices that play a significant role in improving tolerance to differential thermal expansion and moderating electrochemical reactions[73] as well as the immobility of the polymer chains. Thus, when the polymer chains are immobilized onto a solid surface of the nanoadditive, the stimuli-responsive phase-transition from hydrophilic to hydrophobic will collapse the polymer into the solid surface.[48] This influences the glass transition temperature (T_g) of the polymer. Hence in an organic/inorganic composite, T_g of the polymer is known to alter due to the steric and enthalpic effects that modify the segmental mobility of the polymer molecule at a polymer/filler interface.[33] Examples of this class of coatings are hydrophobic coatings, self-healing, fire resistant, antibacterial, anti-graffiti, antifouling, antifriction, corrosion sensing, self-cleaning coatings, and so forth.[27,54]

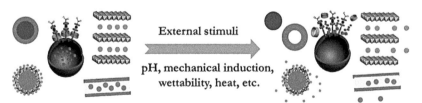

FIGURE 15.1 Scheme of the response of a smart coating to external stimuli.

Source: Reproduced with permission from ref 98. Copyright 2015, Royal Society of Chemistry.

15.3   NANODISPERSOIDS

Development of polymer matrix composite coatings by the incorporation of reinforcements such as fibers, whiskers, platelets, or particles is
an effective approach to improve mechanical properties and the functional behavior of the material. For example, polymers filled with several
synthetic or natural compounds have been used to increase heat and
impact resistance, conductivity, flame retardancy, UV screening, antibacterial activity, corrosion resistance, mechanical strength, or to decrease
electrical conductivity and gas permeability with respect to oxygen and
water vapor due to the nature of reinforcing materials.[34] The reinforcing
materials used in the production of polymer nanocomposite coatings are
classified according to their dimensions and properties. When all the three
dimensions are in the nanometer scale, they are called iso-dimensional
nanoparticles such as spherical silica, metal particles, and semiconductor
nanoclusters. If two dimensions are in the nanometer scale and one larger
forming an elongated structure such as nanofibers, whiskers, or nanotubes
like CNTs and cellulose whiskers, they are put in the second category. The
third category of reinforcement is characterized by only one dimension
in the nanometer scale. Layered materials such as graphite, metal chalco-
genides (MoS_2), clays, and LDHs are in this group. Nanosized particles
show different properties relative to microparticles and bulk materials
thus, providing more effective options for the surface modification, due
to higher surface area to volume ratio. This improves the quality of nanocomposite coatings and adhesion between nanodispersoids and polymer
matrix, such that lower amount of loading achieves equivalent properties.[41,72,102] Therefore, enhanced mechanical and tribological properties,
antibacterial behavior, and corrosion resistance can be achieved by the
addition of different nanodispersoids. Based on their nature, they can be
classified into metal nanoparticles (Au, Ag), ceramic nanoparticles (SiC,
TiO_2 and CeO_2), and other systems such as organics, CNTs, clays, LDHs
and so forth.

15.3.1   METAL NANOPARTICLES

The addition of metal nanoparticles in polymeric resins influences specific
properties of polymer films making them behave differently owing to

the characteristics of particle and complex interface properties between polymer and the nanoparticles. While considering the corrosion resistance of polymer nanocomposite coatings, well dispersed nanoparticles in the polymer matrix could reduce the porosity of the coating, inhibit ions, and water molecules penetrating the coating and subsequent electrochemical reactions between the coating and metal interface. Table 15.1 summarizes the properties of a few metal nanoparticles employed in the formulation of smart polymer nanocomposite coatings.

TABLE 15.1 Summary of Properties and Actions of Metal Nanoparticles Embedded in Polymeric Coating Materials.

Nanomaterial	Particle size	Polymer	Corrosion resistance	Properties other than corrosion resistance	References
Al	300–500 nm	Epoxy	Al–Fe complex oxide barrier layer formation	–	[51]
Si	300–500 nm	Epoxy	Sacrificial effect of Si controls the corrosion	–	[53]
Ag	25–80 nm	–	76.68% of inhibition efficiency on carbon steel in 1 M HCl with 400 ppm silver nanoparticles	–	[4]
Ag	20–30 nm	Poly(ether-amide) (PA)	–	Thermal stability Anti-bacterial properties	[31]

15.3.2   CERAMIC NANOPARTICLES

Ceramic nanoparticles, such as TiO_2, ZnO, SiO_2, SiC, and so forth, have been used to produce polymer nanocomposite coatings in order to enhance the coating performance for a long time with high efficiency. Nano-ZnO and TiO_2 have excellent photo and thermal stability, they do not migrate in the polymer matrix, thus are used as alternative UV-blockers in coatings applications.[109] According to Ramezanzadeh et al., as amount of the nanoparticles increases within the polymeric matrix, the cross-linking density decreases.[69] ZnO's ability to block UV radiation is due to the wide band gap (3.37 eV) and large excitation binding energy of 60 meV, therefore, it can absorb light that matches the UV range of the solar spectrum.[32] Addition of nano-ZnO can provide strength to the polymer composite to prevent it from photodegradation without any disturbance to the optical transparency of the coating. Decrease in curing temperature has been reported for nanocomposite coatings containing nano-ZnO as compared to the neat coating even with small loading level of nano-ZnO in the coating system.[26] Also, nano-ZnO particles can produce reactive oxygen species (ROS), which can induce antimicrobial activity of the composites.[60] Modified TiO_2 in nanocomposite coatings exhibits multi-functionality for various applications.[66] Similarly, in comparison with micro silica-added composites, nanosilica epoxy (EP) composite coating can withstand aggressive environment for a longer time. Furthermore, the mechanical properties such as adhesion, abrasion, and scratch hardness of nano-SiO_2 incorporated coatings exhibited a higher value than that of its micron sized counterpart.[63] Table 15.2 describes the properties and characteristics of some ceramic nanoparticles employed in smart nanocomposite coating formulations.

15.3.3   OTHER SYSTEMS

Incorporation of nanomaterials with tubular structure such as CNTs and platelet structure (nanoclays, graphene, LDHs, etc.) in the polymer matrix has been studied extensively in the recent years. Specific characteristics of each material and their behaviors in different polymer bring desired properties for various applications. Carbon nanotubes have been used to make polymer nanocomposites the most promising material to improve conductivity and other barrier properties. In terms of corrosion resistance, CNTs also reduce the porosity of the coating matrix and inhibit the penetration of corroding ions and water into the coating and subsequent prevention of electrochemical reactions at the coating/metal interface, leading to improved barrier performance of coating. Moreover, they enhance the adherence of polymer coating to the metal surface.[23] Small size effect and high surface area of CNTs are the probable reasons to inhibit the corrosion and reinforcing the barrier effect of the matrix.[24] In a study by TabkhPaz

TABLE 15.2 The Summary of Properties and Actions of Ceramic Nanoparticles Embedded in Polymeric Coating Materials.

Nanomaterial	Particle size	Polymer	Corrosion resistance	Properties other than corrosion resistance	References
ZnO	40 nm	Poly(o-toluidine)/epoxy	Formation of more uniform passive layer that increases the tortuosity of the diffusion pathway of oxygen, water, and chloride ions	–	[42]
	100 nm	Epoxy/PDMS	2 wt.% of ZnO nanoparticles had the most pronounced effect	Hydrophobicity (6 wt.% ZnO)	[3]
	35–40 nm	Waterborne alkyd resin	–	Abrasion resistance Scratch resistance	[26]
	10–18 nm	Chitosan	–	Antimicrobial activity	[60]
TiO_2	10, 50, 100, and 150 nm	Alkyd	Inhibition efficiency increases with decreasing the size of nano-TiO_2 and the temperature	–	[25]
	–	Poly(ester-amide) (PEA)	Elongation of the diffusion path Possible blocking of some pinholes in the PEA coating.	Superhydrophobicity	[28]

TABLE 15.2 *(Continued)*

Nanomaterial	Particle size	Polymer	Corrosion resistance	Properties other than corrosion resistance	References
	18 nm	PVDF (Polyvinylidene fluoride)	—	Reversible switching between superhy-drophobicity and superhydrophilicity through the alternate treatment by UV light irradiation and heating.	[66]
	75 nm	Polyaniline (PANI)	Barrier for both electron and hole transport across the interface. Charge transport from PANI to TiO_2 and vice versa is difficult	—	[67]
SiO_2		Epoxy			[63]
SiO_2 (SiO_2/poly (styrene-co-butyl acrylate))	150–160 nm	Fluoropolymer (JF-3X)	Improvement of resistive properties by polymer network reinforcement		[15]

PDM: polydimethylsiloxane

and coworkers, thermal conductivity of polystyrene was increased by the distribution of CNTs in the matrix. Hybrid composite of hexagonal boron nitride (hBN) nanoparticulates (two dimensional) and CNTs (one dimensional) further improves thermal conductivity due to bridging between the two reinforcements.[86]

Incorporation of functionalized material as reinforcement can enhance interaction with the matrix and the stability of nanocomposites. COO^- functionalized multi-walled CNTs in polypyrrole gave better corrosion protection by formation of COO^- functional group stabilized passive layer.[20] Silane functionalization of fullerene C60 and graphene nanofillers could improve the quality and adherence of the cured EP coating by reducing the porosity of the coating matrix and altering the physiochemical properties of coating-steel interface.[50] Graphene oxide (GO) nanosheets dispersed in polymeric isocyanate cross-linked with hydroxyl-functional acrylic adhesive resulted in impermeable coatings towards diffusion of oxidizing gas and corrosive liquid solution.[84] Epoxy nanocomposite coatings containing self-assembled clay (α-zirconium phosphate nanoplatelets) in smectic order may be used to protect metal surfaces from corrosion.[49] Development of superhydrophobic composite coatings by surface texture modification in the presence of secondary particles is one of the methods to increase the corrosion protection. Nano-silica and CNTs added EP/modified polyvinylidene fluoride (MPVDF)/fluorinated ethylene propylene (FEP) composite coating surface with reticulate papillae structure showed superamphiphobic property. The coatings showed excellent corrosion and wear resistance. The enhanced shielding effect of highly hydrophobic coating by introducing FEP in the presence of CNTs and SiO_2 can decrease the diffusion rate of oxygen and water through the coating. The enhanced wear resistance of the coating was observed due to the designed nano/microstructures and the self-lubrication nature of MPVDF and FEP.[92] Microporous oleophilic-hydrophobic polyvinylidene fluoride (PVDF)/N, N Dimethylformamide (DMF)/candle soot, and PVDF/DMF/camphor soot composite were developed by Sahoo and Balasubramanian (2014).[74] Broccoli-like hierarchical microstructure with micro or nanoscaled roughen surface were obtained for PVDF/DMF/camphor soot and cauliflower like PVDF/DMF/candle soot exhibited superhydrophobicity.[74] In the presence of Mg–Al LDH, EP gave higher mechanical performance and flame-retardant properties.[8] Electromagnetic properties of the graphene nanoparticle (GNP) -based EP composites were studied by Plyushch et al. Annealing of composite above the EP glass transition significantly improved their electromagnetic properties. The electrical conductivity was explained by the result of tunneling between GNP clusters. Since the tunnel barrier decreases after annealing, the composite provided better properties.[65] Table 15.3 illustrates a summary of properties and actions of nanomaterials formulated from other systems including organics.

15.4 FABRICATION OF SMART POLYMER NANOCOMPOSITE COATINGS

The fabrication of polymer-based nanocomposites is achievable for thermoplastics, thermosets, and elastomeric resins irrespective of the conducting or nonconducting nature of the matrix. It is aimed at overcoming the limitations encountered with the use of the macro and microcomposites coatings which is usually early deterioration. The microstructure of the nanocomposites has been categorized based on the interaction between the polymer matrix and integrated nanodispersoids. These microstructures influence the response of the nanocomposite coatings obtained and when triggered by any stimuli reveal its smart qualities. The microstructures could be categorized under the following:

i) Phase-separated composites due to inert fillers inside the organic polymer bulk,

ii) Hybrid-phase composites due to nanoadditives dispersed well in the polymer matrix with occurrence of physical or chemical interactions with the polymer chains,

iii) Intercalated composites due to the presence of nanodispersoids with platelet structures like nanoclays. The polymer chains are sandwiched between the nanodispersoid layers. This permits an increase in spacing between the platelets without a change in the existing regular structure,

iv) Exfoliated composites due to homogeneous dispersion of the individual platelets (particularly clays) in the polymer matrix losing their periodic platelet arrangement arising from total dissolution of electrostatic forces between the platelets.[1,13,48,60]

TABLE 15.3 A Summary of Properties and Actions of Nanomaterials Formulated from Organics and Other Systems.

Nanomaterial	Particle size	Polymer	Corrosion resistance	Properties other than corrosion resistance	References
EP/MPVDF/ FEP CNT and SiO_2	–	Epoxy	Superamphiphobility: Contact angles—($166°$, $155°$) Sliding Angles—($3°$, $5°$) to water and glycerol, respectively. The CNTs and SiO_2 can decrease the diffusion rate of oxygen and water	Designed nano/microstructures and the self-lubrication of modified poly(vinylidene fluoride) (MPVDF) and FEP enhances the wear resistance	[92]
$GO–SiO_2$	<20 nm	Epoxy	$GO–SiO_2$ nanohybrids reduces the coating cathodic delamination	–	[70]
Functionalized MWCNT	Diameter: 10–20 nm Length: 30 μm	Polypyrrole PPy/ MWCNT-COO–	MWCNT creates an obstacle against the path of corrosive ions	"COO–" functional group on stability of the passive layer.	[20]
CNT	diameter: 20–30 nm, length: 1–10 μm, layers: 5–20	polyaniline	CNTs improve the adherence of polyaniline polymer coating to the metal surface	–	[23]

TABLE 15.3 (*Continued*)

Nanomaterial	Particle size	Polymer	Corrosion resistance	Properties other than corrosion resistance	References
CNT	diameter: 20–30 nm, length: 1–10 μm	alkyd	Improvement in adhesion strength of alkyd resin Small size effect and high surface area of CNTs reinforce the barrier effect of alkyd resin film	–	[24]
CNTs and hBN (hexagonal boron nitride)	MWCNT: outer diameter-10–15 nm length—1–10 μm hBN: 70 nm	Polystyrene	–	High thermal conductivity due to bridging between hBN nanoparticulates (two dimensional) by CNTs (one dimensional)	[86]
Silane coupling agent 3-aminopropyltriethoxysine (KH550) modified fullerene C60 graphene		bisphenol-A epoxy	Functionalized nanofillers (FC60 or FG) act as an excellent barrier to corrosive solutions, zigzagging the diffusion path available of corrosive solutions	FC60 showed better tribological and scratch resistance properties compared with that of FG	[50]
Graphene		Epoxy	–	Increase in electromagnetic properties of the GNP-based composites after annealing, above the epoxy glass transition	[65]

TABLE 15.3 *(Continued)*

Nanomaterial	Particle size	Polymer	Corrosion resistance	Properties other than corrosion resistance	References
Carbon soot nanoparticles	average size of 35 nm	PVDF	Micro-porous oleophilic-hydrophobic nature	Designed nano/micro-structures and the self-lubrication of MPVDF and FEP enhances the wear resistance Superhydrophobicity	[74]
Graphene oxide (GO) nanosheets	40–50 nm	Polyisocyanate	Impermeability of ion diffusion of oxidizing gas and corrosive liquid solution	–	[84]

MWCNT: multiwalled carbon nanotubes; *CNT*: carbon nanotubes

The fabrication process results in covalent linkage and physical adsorption based on the interactions between the polymer molecules and the nanodispersoid cores. Immobilization of the polymer molecules on the nanodispersoid atoms occurs in the former process. In the later process, physical adsorption incorporates hydrophobic interior, nonspecific/coordination adsorption, electrostatic interaction, and so forth.[48] Thus, the processing of polymer nanocomposite coatings can be generally classified as in-situ and ex-situ methods. In-situ methods involve generating the nano counterpart inside the polymer matrix. It prevents particle agglomeration, and allows the control of particle size and morphology but its disadvantage is having the unreacted adducts possibly interfere with the properties of the new polymer nanocomposite coating. In-situ polymerization, in-situ intercalative, template synthesis, sol-gel process, etc. are examples of in-situ fabrication techniques. Ex-situ method involves dispersing the already fabricated nanoadditives into the polymer matrix to generate the polymer nanocomposite coatings. Ex-situ methods involve direct mixing, solution blending or dispersion, melt blending or dispersion, etc. and achieving homogeneous dispersions are difficult.[13,37,56,62] In-situ polymerization, most commonly used, suits the preparation of polymer-based nanocomposites with polymers of limited or little solubility. The nanodispersoids in the monomer solution get swollen, this allows the monomer of low molecular weight to diffuse effectively in between the nanodispersoids, and polymerization occurs. The presence of the diffused initiator heat, radiation or a pre-intercalated organic initiator initiates and, at the same time, controls the polymerization reaction. The polymer chains grow between the nanoadditives and may result in exfoliation and uniform distribution of the nanodispersoids in the system. Exfoliation occurring here is influenced by the extent to which the nanodispersoid swells and the diffusion rate of the monomer within. Polymers such as EP, polyurethane, polystyrene, nylon 6, polyethylene oxide, polyethylene terephthalate, and unsaturated polyesters are suitable for this process.[77,102] The sol-gel process is another widely used technique that allows the design of a new generation of advanced materials with unique properties. It is an old technique with an origin that dates back to 17,000 years ago in France when colloids were used for preparation of functional materials seen in cave paintings. With time, it became a fully established technique in the 19th century.[62,95] The use of sol-gel in the preparation of smart polymeric materials grew progressively due to its compatibility with polymers and polymerization processes and allowing the formation of nanoparticles in the presence of organic molecules. The nanoprecursors are mostly ceramic materials.[62] The sol-gel method is a typical hydrolysis–condensation process afterward condensation polymerization occurs upon film application. Although, the evaporation process often times results in formation of voids. This has been overcome with the incorporation of nanodispersoids which lowers porosity and cracking potential as well as increases film thickness. The in-situ sol-gel process has encouraged the preparation of smart coatings such as self-repairing pretreatments coatings.[75]

Blending method with ex-situ nanodispersoids is another widely used fabrication process. The nanodispersoids, mostly inorganic materials, are used as fillers. Among the categories of nanodispersoids, particulate forms are preferable for realizing transparent coatings. These are dispersed either as nanopowders or in the colloidal state. With this technique, the possibility of aggregation is high especially with the powdered nanodispersoids, therefore they are initially pre-dispersed in liquids by any of the following techniques: sonication; shear rate mixing; milling (or grinding); or microfluidic techniques. Among these pre-dispersion techniques, bead milling is the most efficient technique. Ultrasonication and microfluidic techniques are laboratory based and are not suitable for industrial applications.[110] Intercalation method is another fabrication method but particularly suitable for incorporating layered inorganic fillers like clay. The nanophases are generated in situ by the process and dispersion is best achieved with three roll mill and bead mill techniques. These involve solution intercalation and melt intercalation. In solution intercalation, solvents are required for mixing and the polymer chains gradually diffuse into the interlayer of the inorganic filler increasing the interlayer distance. Thereafter, the polymer chains are absorbed onto the nanodispersoids by intercalation displacing solvent molecules between them. Depending on the organic modification of the nanodispersoid used the structure of the nanocomposite produced may appear exfoliated or intercalated. While in melt intercalation, the nanodispersoids are mixed with the polymer matrix directly in their molten state and at temperatures above the polymer's softening point. Diffusion of the polymer chains into the nanodispersoid galleries occurs as electrostatic forces are overcome. An optimum temperature is best for uniformly exfoliated structures. This method is eco-friendly and without the use of solvents.[72,110]

Characterizing smart polymer nanocomposite coatings requires highly sensitive instrumentation for detailed analysis of the underlying behavior of the composite coating material before and after any experimental analysis. The structure of polymer nanocomposite and its process kinetics is determined with wide angle X-ray diffraction technique. In combination with small angle X-ray scattering (SAXS) quantitative structural characterization can be obtained. Transmission electron microscope (TEM) provides direct qualitative information of morphology, spatial distribution of the various components as well as structural defects. Nuclear magnetic resonance (NMR) is important for probing surface chemistry and coordination in exfoliated polymer nanocomposites, such as possibly quantifying the level of clay exfoliation. Fourier-transform infrared spectroscopy (FTIR) and Raman spectroscopy are required to understand the structural formation (due to the presence of functional groups) of the polymer nanocomposite coatings.[102] Studying the contribution of smart properties to corrosion inhibition has been achieved with electrochemical techniques such as electrochemical impedance spectroscopy (EIS), scanning vibrating electrode technique (SVET), scanning electrochemical microscopy (SECM), Potentiodynamic Polarization (PP), localized impedance spectroscopy (LEIS), and so forth.

15.5 EFFECT OF NANODISPERSOIDS ON ANTICORROSIVE SMART POLYMER NANOCOMPOSITE COATINGS

15.5.1 PASSIVE AND ACTIVE BARRIER EFFECTS

The actuality of incorporating nanodispersoids such as carbon nanomaterials, clay nanoplatelets, metal nanoparticles, ceramic-based nanoparticles etc., in polymeric coatings in the area of polymeric nanocomposites for corrosion inhibition has encouraged the emergence of unprecedented opportunities in coatings design and metallic substrate protection. These functional hybrid coatings have found use in the automotive industry, aerospace sector, manufacturing, and other relevant engineering works. The incorporated nanodispersoids provide a toughened homogeneous nanocomposite coating due to cavity filling resulting from particle bridging, crack path deflection, crack bowing, crack pining, polymer disaggregation during curing, and so forth.[61] Aesthetic appearance and corrosion protection is obtained and with the combination of certain inherent and acquired properties on formulation integrity, and durability of coating systems is greatly enhanced. The properties of these classes of materials enable a "smart" response when exposed to severe environments (Fig. 15.2). In general, these coatings exhibit fast drying, high water vapor permeability, increased indentation resistance, high elasticity, no expansion after contact with water, flame retardancy and gas barrier properties, scratch resistance and hardness, superhydrophilic properties if containing polar silanol groups in high concentrations, enhanced barrier effects, wear resistance, self-healing properties, improved mechanical properties, antimicrobial properties, magnetic properties, UV stability, and reduced coating thickness, thereby saving raw materials as well as minimizing toxicity.[21,29,33,38,39,43,71,90] Passive and active barrier effects of these nanodispersoid containing coatings will be considered in two categories based on the functionality as: "nanocontainers" and "nanoinhibitors."

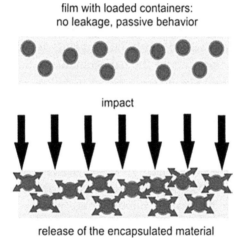

FIGURE 15.2   The schematic of feedback active coatings response mechanism.
Source: Reproduced with permission from ref 82. Copyright 2010, Royal Society of Chemistry.

15.5.1.1   FUNCTIONAL NANOCONTAINERS

These are the nanodispersoids efficient for hosting active species such as corrosion inhibitors, biocides, drugs, and other functional compounds that are released only when triggered by enabling conditions. These functional modifiable materials have yielded smart materials with wide applications in feedback active surfaces, drug targeting, and delivery systems in medicine, energy enriched materials, and so forth. Nanocontainers can be engineered to have controlled uptake and release properties, shell robustness, defined sizes, high loading capacity, and other multifunctional qualities enlarging the capabilities of smart materials. Triggers such as local pH changes, temperature changes, ultrasonic treatment, electromagnetic irradiation, mechanically induced defects, pressure, humidity, ionic strength such as concentrations of NaCl above 0.5 M, electric (electrochemical) potential, and dielectric permeability of the solvent and so forth have been effectively employed in the opening and closing of the nanocontainers. The use of nanocontainers is a successfully proven approach and shows promising trends with different systems.[38,82]

Fabricating nanocontainers on the basis of their shell stability and shell modification versatility is described by three general approaches. First, amphiphilic block copolymers self-assembled into spherically closed nanostructures followed by cross-linking to stabilize the nanocontainer shell.[36] Amphiphilic block copolymers such as poly(ethyleneoxide), poly(N-isopropyl acrylamide), polypyrrolidones, biodegradable polymers like poly(caprolactone), poly(DL-lactide), and so forth have been investigated as components of the nanocontainers. Specifically, poly(N-isopropyl acrylamide) can undergo phase transition at 32 °C and can be used for fabrication of thermosensitive nanocontainers.[83] Second, layer-by-layer assembly of oppositely charged species on the outermost surface of dense template nanoparticles via polyelectrolytes, conductive polymers, biopolymers, CNTs, viruses, lipid vesicles, and nanoparticles as constituents of the nanocontainer shell.[35,78] Third, utilizing ultrasonic waves to assemble inorganic and composite hollow nanospheres.[80]

Nanocontainers for storing inhibitors are acknowledged as a smart method for guaranteeing long-term corrosion protection of metal substrates.[17] Nanocontainers are of different types and influence the mechanical properties of the nanocomposite coating. For encapsulation of active molecules, the nanocontainers can have an organic, inorganic or composite origin.[82,93] Optimum performance of nanocontainers can be achieved when there is good adhesion and compatibility of the nanocontainers and the coating matrix.[10,11,17] The focus is to release and repair damaged zones in the coating matrix more like targeted delivery once triggered by any stimuli as well as minimizing early inhibitor leaching within the matrix. This curbs undesirable interactions between the inhibitors and the polymeric matrix which may adversely affect overall coating performance. Therefore, the nanocontainers should be impermeable yet sensitive to the tailored triggers, chemically and mechanically stable.[61,85,93]

Shchukin and coworkers demonstrated that nanocontainer shells with controlled release properties are useful for fabricating active coatings. The release of corrosion inhibitors encapsulated within the nanocontainers is triggered by the corrosion process.[83] Borosiva et al. designed mesoporous silica nanocontainers of varying concentrations (0.04−1.7 wt.%) loaded with 2-mercaptobenzothiazole, in a hybrid sol-gel (SiOx/ZrOx) layer. 2-mercaptobenzothiazole (at 0.7 wt.%) loaded nanocontainers (MBT@ NCs) containing 0.14 wt.% inhibitor concentration provided the best passive and active corrosion resistance for the coating, as determined with EIS and SVET analysis.[10] Haase et al. introduced nanocontainers based on Pickering emulsions by application of silica-polystyrene-8-hydroxyquinoline (silica-PS-8HQ) composite nanocontainers in a water-based alkyd resin. Electrochemical impedance spectroscopy revealed that inhibiting films with 8HQ doped silica-PS nanocontainers had higher impedances than the films of silica-PS nanocontainers without 8HQ. The nanocontainers possibly acted as barrier pigments decreasing the transport of charges across the film. Whereas, the analysis of scanning vibrating electrode measurements indicated decreased corrosion rates in scratched areas of silica-PS-8HQ doped films.[38] Choi et al. studied the corrosion inhibiting effects of amine loaded polymeric nanocapsules synthesized by multistage emulsion polymerization. The amounts of encapsulated amines increased in the following order: Propylamine, PPA (0.20)<Ethanolamine, ETA (0.90)<Dipropylamine, DPA (1.25) <Diethanolamine, DEA (1.40)<Triethanolamine, TEA (1.79)<5-Amino-1-pentanol, 5AP (2.00) with 5-Amino-1-pentanol of maximum encapsulation behavior amongst the others. With SVET and EIS evaluations DEA, TEA, and 5AP exhibited self-healing anticorrosion performance with recovery of coating resistance, and the others a rapid drop of coating resistance without self-recovery.[17]

15.5.1.2 FUNCTIONAL NANOINHIBITORS

Recent advances in the design of coatings have aimed at combining corrosion inhibiting compounds synergistically within a single coating system for providing enhanced protection of structural components in corrosive environments. Studies of Jin et al. demonstrated that rose-shaped ZnO nanoflowers in a fluoropolymer matrix, resulted in amphiphilic and superamphiphobic textured surfaces enhancing barrier protection.[43] Similarly, exfoliated graphene and CNTs dispersed in N-methylpyrrolidone were evaluated as inhibitors for polyetherimide (PEI) coated low alloy steel substrates. Coating formulation was achieved via dispersion processing of the carbon nanomaterials alongside in-situ imidization of polyacrylic acid (PAA) to form PEI coatings directly on the steel substrate. These ensured exceptional adhesion and provided good dispersion of the carbon nanomaterials within the PEI matrix. The nanoinhibitors were prepared as nanocomposites for the analysis. The first combination was a mixture of graphene and CNTs at 2 wt.% each and the second contained graphene at 20 wt.%. The nanoinhibitors were successfully dispersed into the polymer matrix forming a passive layer on the surface of the base metal protecting the underlying steel substrate up to 3144 h in comparison with the bare polymer. Corrosion rates deduced from the studies were three orders of magnitude lower than bare substrate and one order of magnitude lower than PEI unloaded coating.[22]

Furthermore, carbon nanostructures as reinforcing additives in hybrid anticorrosive coatings were reported by Harb et al.[40] The additives, CNT and GO, improved the scratch resistance, adhesion, wear resistance, and thermal stability of poly(methyl methacrylate) (PMMA)-siloxane-silica (PSS) hybrid coatings. These reinforcement nanoadditives in the hybrid coatings resulted in very efficient anticorrosive barrier, with an impedance modulus about five orders of magnitude higher than the bare carbon steel. In particular, the high corrosion resistance with GO addition was maintained for more than 6 months in saline medium. Figure 15.3 illustrates the Nyquist and Bode plots of unloaded and loaded (CNT and GO) in two different matrices (BPO0.01 and BPO0.05) compared to bare carbon steel, after one day of immersion in the saline medium.

A practical application of carbon nanofibers (CNFs) loaded coatings for corrosion resistance of low carbon steel has been reported by Rout and coworkers. The novel coating was prepared by direct in-situ growth of CNFs onto steel substrates subsequently polyetherimide (PEI) infusion onto the same. The substrate-integrated nanofibers served as a cushioned layer on addition of PEI, locking the polymer layer into the place between nanofibers on the substrate. Strong interfacial bonds were formed that prevented formation of any microcracks and provided a large passivation layer preventing charge-transfer reactions on exposure to severe saline environment. After 30 days exposure and even repeated experiments, there was no evidence of delamination or rust formation whereas PEI coating

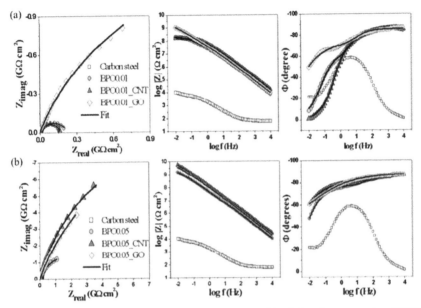

FIGURE 15.3 Nyquist and Bode plots of unloaded and loaded carbon nanotubes (CNT) and graphene oxide (GO) in a) BPO0.01 matrix and b) BPO0.05 matrix compared to bare carbon steel, after one day of immersion in 3.5% NaCl solution.

Source: Reproduced with permission from ref 40. Copyright 2016, American Chemical Society.

alone delaminated in 8 days exposure duration. The nanocomposite coating successfully possessed outstanding hydrophobic and corrosion inhibiting properties.[73] Furthermore, clay nanoplatelets have shown remarkable corrosion resisting properties when coupled with polymers in coating formulations in the past decade. The intense research interest for polymer-clay nanocomposites with the likes of montmorillonite (MMT) clay is because their lamellar elements have displayed high in-plane strength, stiffness, and high aspect ratio. Such that the resultant polymer-clay nanocomposites enhance gas barrier, provide thermal stability, mechanical strength, and impart flame-retardant properties to the polymers alongside anticorrosive properties.[100] The effect of polymer-clay nanocomposites on barrier properties of coatings is partly attributed to the increased path length for molecular diffusion caused by the platy nature of the clay materials. In the preparation of nanocomposites coatings with clays, delamination and exfoliation (random distribution of clay platelets within the polymer) or intercalation (placement of polymer molecules between clay platelets) are involved.[33] Bagherzadeh and Mahdavi, studied the influence of quaternary ammonium-modified MMT clay as a nanoinhibitor in EP-clay nanocomposite coating. Barrier effect was greatly enhanced as incorporation of clay platelets into EP matrix resulted in decreased water permeability for the nanocomposite coating. Increasing the clay concentration simultaneously lead to reduction of degradation and blistering density which improved the anticorrosive performance of the nanocomposite coating.[7] Similar studies confining intercalated polymer chains within the clay galleries and preventing segmental motions of the polymer chains, thereby improving corrosion protection of metals have been reported.[99,101]

Deposition of polyvinyl chloride (PVC) and polystyrene (PS) nanofiber coatings on brass surface has been investigated by El-saheb and coworkers. The study revealed that PVC and PS nanofiber coatings protected the brass surface from corrosion by decreasing its corrosion current and corrosion rate while increasing its polarization resistance. Although brass materials are relatively noble for which good corrosion resistance and machinability, better resistance to biofouling, high thermal and electrical conductivity have been experienced with their use, brass containing more of β-phase accelerates corrosion damage. Coating layers of PVC and PS nanofibers were homogenously distributed on the brass surface by electrospinning technology with an average thickness of about 6.0 μm for PVC and 10.0∼14 μm for PS. These provided outstanding barrier effect in the aggressive chloride solution compared to the uncoated brass alloy.[29] Ag/SiO$_2$ core-shell nanoparticles of 60 nm were investigated for their anticorrosive potentials on steel substrates. For the investigation, Le et al. applied the nanoinhibitors in marine antimicrobial corrosion coatings. Electrochemical noise analysis (ENA) results revealed better antimicrobial corrosion activity with 1 wt.% Ag/SiO$_2$ nanoparticles than that of conventional 40 wt.% Cu$_2$O biocides. With the aid of inductively coupled plasma optical emission spectrometry (ICP), it was observed that very low amounts of Ag ions leached from the acrylic resin, while higher amounts of copper ion leached from the conventional Cu-based coating within the same period of the study.[47]

In Madhankumar et al.'s study, the addition of Si nanoparticles into EP matrix significantly increased the coating film resistance on a coated steel substrate in 0.1 M NaCl solution. The effect of Si nanoparticles on the corrosion protection behavior of EP-coated steel was analyzed with electrochemical impedance spectroscopy (EIS) and scanning electrochemical microscopy (SECM) techniques. Charge transfer resistance (R$_{ct}$)

of the coating sample without nanoparticles was 10 kΩ/sq. and that of Si nanoparticles was as high as 50 kΩ/sq. for the exposure time. While from SECM analysis, the tip current at −0.70 V decreased with the addition of Si nanoparticles in the film. It was inferred that dissolution of the substrate was suppressed due to consumption of dissolved oxygen by the Si nanoparticles from the coating. This process was facilitated by the scratch induced on the coating allowing migration of the nanoparticles to the defected site enriching it and enhancing the corrosion resistance. Furthermore, increase in Si nanoparticles in the coating increased the Vickers hardness of the coating which shows a positive contribution to the mechanical properties of the coating.[52] Furthermore, Ali et al. reported a simple route to synthesize epoxy-based nanocomposite coatings containing silica nanoparticles. The epoxy-modified coatings on AISI 1020 steel substrate demonstrated in Figure 15.4 improved resistance to erosive wear, enhanced hardness and better thermal stability influenced by the nano-silica inclusions.[2]

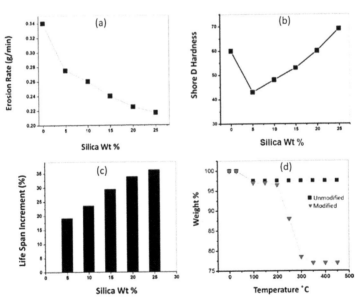

FIGURE 15.4   (a) Erosion rate (g/min), (b) hardness, (c) Life span improvement (%) of the nanocomposite coatings as a function of silica content, (d) Thermogravimetric analysis of nonreactive stearic acid modified silica additives.[2]

The use of silver-decorated polymeric nanoparticles in coating formulations is not left out although challenges such as agglomeration and the cytotoxicity interfere with its meeting up with requirements such as higher concentration. Silver nanoparticles is known to have a strong antimicrobial activity against both Gram-positive and Gram-negative bacteria irrespective of its application and in a non-agglomerated form as agglomeration is known to significantly reduce its antimicrobial effect.[14] The release of Ag$^+$ ion from the coating is possible and the releasing rate describes the antibacterial activity of the coating. Therefore, antibacterial effects are enhanced by the use of silver nanoparticles, owing to their ability to release more Ag$^+$ due to an increased surface area to volume ratio.[55] Atta et al. investigated the effect of silver nanocomposite coatings. High yield dispersed silver nanoparticles (AgNPs) less than 10 nm were synthesized and used in the preparation of hybrid polymer nanocomposites with N-isopropylacrylamide (NIPAm) and 2-acrylamido-2-methylpropanesulfonic acid (AMPS) for the protection of pipeline steel as corrosion inhibitors in acidic media. The addition of the nanocomposite AMPS/NIPAm-Ag NPs to 1 M HCl solution significantly decreased the cathodic and anodic currents and the nanoinhibitor acted as a mixed-type inhibitor. With electrochemical impedance spectroscopy (EIS), the diameter of the capacitive loop increased with increase in inhibitor concentration protecting the steel surface in 1 M HCl solution (Fig. 15.5). At the highest concentration of 250 ppm, inhibition efficiency was calculated to be 81.46%.[5]

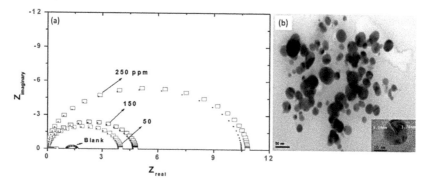

FIGURE 15.5 (a) Nyquist plots for steel with different concentrations of 2-acrylamido-2-methylpropanesulfonic acid (AMPS)/N-isopropylacrylamide (NIPAm)-Ag NPs in 1 M hydrochloric acid solution, (b) silver AMPS/NIPAm hybrid polymer as nanoinhibitor with an inset showing nanoinhibitor size.

Source: From Atta, A. M.; El-Mahdy, G. A.; Allohedan, H. A. Corrosion Inhibition Efficiency of Modified Silver Nanoparticles for Carbon Steel in 1 M HCl. *Int. J. Electrochem. Sci.* **2013**, *8*(4), 4873–4885. Creative Commons Attribution License 3.0.

ZnO embedded epoxy-polydimethylsiloxane (EP/PDMS) coating, developed by Sh. Ammar and coworkers exhibited better corrosion resistance due to the quality of the film by reducing the porosity and zigzagging the diffusion pathways of ionic species through the coating. The surface roughness offered by the uniform distribution of nanoparticles, particularly at 6 wt.%, improved the hydrophobicity of the composite in EP/PDMS matrix by forming air pockets between water and the surface leading to composite solid–liquid–air interface.[3] The corrosion resistance of TiO_2 nanocomposite coating with respect to the particle size (10, 50, 100, and 150 nm) was studied by Deyab and Keera. The inhibition efficiency was found to increase with decreasing the size of nano-TiO_2 particle in alkyd resin. The physical adhesion on the metal surface is higher for smaller particles and hence the O_2 and H_2O permeability of coating decreased with decrease in the nano-TiO_2 size. Depending on the properties of polymer to be used, nanocomposite coatings show smart behavior in the corrosive environment.[25] Likewise, Radhakrishnan and coworkers reported that conducting polyaniline-nano-TiO_2 composites exhibit excellent corrosion resistance much superior to polyaniline (PANI) in aggressive environments. During polymerization in presence of nanoparticles it is expected that the PANI forms around these particles giving a core-shell type structure: the core is TiO_2 and the shell is formed by PANI. Potential barrier for charge transport due to the heterojunction formation improves the corrosion resistance. Since, the charge transport between PANI and TiO_2 is not easy due to the difference in the position of their valence bands and the energy gaps. In addition to this, the p-type PANI offers a large barrier for electron transport while TiO_2 being n-type, gives hindrance to hole transport across the interface.[67] Thus, the barrier properties improved are no doubt due to the nanosize of the PANI–TiO_2. Particularly, the coatings containing PANI–TiO_2 (4.18%) withstood the drastic conditions employed where other coatings completely gave way.

15.5.2 SUPERHYDROPHOBIC EFFECTS

Motivated from nature, hydrophobic surfaces are surfaces having contact angle greater than 90° and superhydrophobic surfaces have contact angle greater than 150°. These surfaces receive continuous attention due to their wide range of applications such as non-wetting, self-cleaning, antifogging, anti-icing, buoyancy, corrosion-resistant, anti-biofouling, oil–water separation, low adhesion and drag-reducing applications and so forth.[95,104] Current research is focused on the super hydrophobization of surfaces for improving corrosion resistance of metallic materials. Superhydrophobic coatings are usually multifunctional. For example, it can prevent contaminants (organic and inorganic) and water molecules from contact with surfaces and act as both corrosion resistant and self-cleaning coatings.[67] The corrosion resistant mechanism of superhydrophobic surfaces is mostly contributed by the existence of air pockets between the substrate and solution, providing an effective blocking to fight against the migration of corrosive ions.[89] Surface roughness required for superhydrophobicity is created by nanoparticles while surface hydrophobicity can be presented by both nanoparticles and organic resins which also serve as binders to connect nanoparticles. This has been observed when ZnO nanoparticles modified by hydrophobic stearic acid were mixed with fluorinated polysiloxane (FPHS) and sprayed on steel substrate. The superhydrophobic coating presented greater corrosion resistant performances.[105] The addition of self-healing functionality to superhydrophobic surfaces would remarkably improve corrosion protection. For example, LDH coating on aluminum alloys showed both corrosion resistance and superhydrophobicity. It provided long-term corrosion protection of the coated aluminum substrate.[103]

Xiao et al. reported a facile and controllable anodization approach to fabricate superhydrophobic copper oxide nano-needle array (CuO NNA) films for the improvement of corrosion resistance of copper substrates. The anodic CuO NNA films were grown on the copper foils by electrochemical anodization in an aqueous potassium hydroxide solution for different anodization times. Superhydrophobic CuO NNA surfaces provided inhibition efficiency higher than 90% in aggressive chloride containing media. The anticorrosion mechanism of superhydrophobic CuO NNA film has been illustrated schematically in Figure 15.6. The nanostructured surface can trap the air inside the valleys between nanoneedles, which leads to the significant increase in surface air fraction in the valleys and the reduction of actual surface area in contact with the aqueous NaCl solution.[96] Nyquist and Bode plots (Fig. 15.7) of pristine Cu and surface-modified Cu coupons after 1 and 7 days of exposure in a 3.5 wt.% aqueous NaCl solution showed better corrosion resistance as reported for the superhydrophobic CuO NNA surfaces.

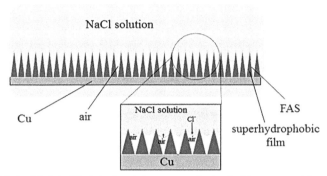

FIGURE 15.6   Model of the interface between superhydrophobic copper oxide nano-needle array (CuO NNA) surface and a 3.5 wt.% NaCl aqueous solution (solid–liquid interface).
Source: Reproduced with permission from ref 96. Copyright 2015, Royal Society of Chemistry.

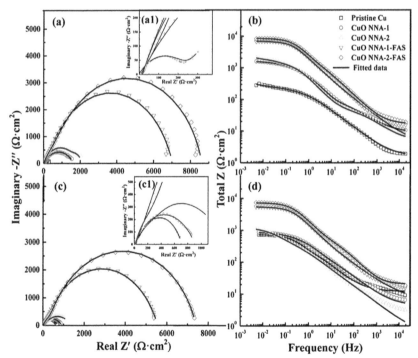

FIGURE 15.7   The graphs (a) and (c) are Nyquist plots of the bare Cu, CuO NNA-1, CuO NNA-2, CuO NNA-1-FAS, and CuO NNA-2- FAS coupons after 1 and 7 days exposure in 3.5 wt.% NaCl solution, (b) and (d) are the corresponding Bode plots. The insets (a1) and (c1) are the magnified Nyquist plots of the pristine Cu coupons in the high frequency range after 1 and 7 days of exposure, respectively.
Source: Reproduced with permission from ref 96. Copyright 2015, Royal Society of Chemistry.

The mechanical stability of superhydrophobic surfaces remains the biggest problem, since any mechanical scratch will dramatically affect the wetting states. For coatings, which require fast recovery of superhydrophobicity (such as repairable surfaces), spraying of hydrophobic materials that spontaneously forms a rough fractal surface with low surface-energy property is a simple, low cost, and effective process.[97] Current researches are focusing on the effective routes to create repairable superhydrophobic surfaces by different encapsulation routes.

15.5.3 SELF-HEALING EFFECTS

Reconstructible polymers and polymer composites are applied in the field of self-healing protective coatings. Self-healing coatings are the class of smart coatings that can repair themselves when exposed to any mechanical damage and recover its functions. Self-healing materials can be classified into intrinsic and extrinsic materials. Intrinsic self-healing materials attain repair due to inherent reversibility of bonding of the matrix polymer. The healing agents are not required for intrinsic self-healing coatings. These types of coatings are based on dynamic covalent or non-covalent chemistries. The covalent approach uses Diels-Alder reaction, transesterification, and so forth, while non-covalent chemistries uses π–π stacking, ligand–metal bonding, hydrogen bonding, and so forth. Polymers can be designed to form strong end group or side-group associations via multiple complementary reversible hydrogen bonds, resulting in a self-healing elastomeric polymer.[9]

The extrinsic self-healing system requires external healing agent. The self-healing process can be initiated by any external stimuli or internal rupture. In extrinsic materials, micro/nanocontainers are used to load the corrosion inhibitors and catalysts. These containers include layered materials, capsules, hollow tubes, and so forth. Our focus hereafter is with respect to the aforementioned nanocontainers. Nanocontainers with the ability to release the inhibitor and catalysts during the time of any damage on the coating surface are required for self-healing coatings. The size of nanocontainers should be less than 400 nm; larger size of the nanocontainers can damage the reliability of the coating by forming large cavities on the coating which reduces the passivity. The corrosion inhibitors and catalysts are preserved in nanocontainers to avoid spontaneous leakage. When the surface is affected by any external impact, the nanocontainers respond to this signal and release the encapsulated active material.[81] Very importantly, the nanocontainers should have good compatibility with the passive matrix, high loading efficiency, and stimuli responsiveness. Depending on the nature of the "smart" components introduced in the container shell, various stimuli can induce reversible and irreversible shell modifications. To obtain stimuli responsive nanocontainers, the outer surface of the inorganic containers can be coated with stimuli responsive polymers such as pH sensitive polyelectrolytes and polyacrylates, and so forth.[79] Tavandashti et al. studied the self-healing effects of cerium nitrate loaded polyaniline (PANI) nanofibers in EP ester coating. They reported that incorporation of polyaniline nanofibers with cerium in EP ester coatings improved corrosion protection.[87] Avila-Gonzalez et al. studied the self-healing effects of a formulated alkyd-based anticorrosive smart coating consisting of silica nanotubes as nanocontainers for loading ferric nitrate as corrosion inhibitor shown in Figure 15.8. The scratched surface containing inhibitor was observed to initially promote corrosion and oxide formation. Nevertheless, the noise resistance obtained for the coating containing silica loaded with inhibitor remained two orders of magnitude higher in comparison with the non-inhibitor condition. Furthermore, at the completion of the experiments after extended immersion times, higher power noise impedance values were obtained for the inhibitor loaded samples, implying lower corrosion rates promoted by the corrosion inhibitor action. The ideal experimental conditions were coatings containing 5% silica particles and 3.1 ppm ferric nitrate inhibitor.[6]

In this chapter, we discuss self-healing effects from three groups of nanocontainers amongst others. These are capsule, vascular, and layered material-based nanocontainers.

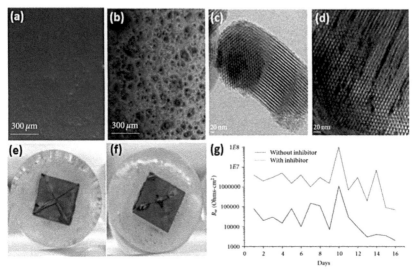

FIGURE 15.8 Optical microscopy micrographs of the coated surfaces (a) without particles, (b) 5.0%; silica particle content (5X), (c) Transmission electron microscope (TEM) micrographs for SBA-15 silica particles without inhibitor, (d) with inhibitor, (e) scratched surface of coated mild steel samples without silica particles and inhibitor after immersion test, (f) with silica particles and 3.1 ppm ferric nitrate inhibitor, (g) electrochemical noise-resistance spectra as a function of time for alkyd coated mild steel without nanoparticles and inhibitor, and with 5% silica particles containing 3.1 ppm ferric nitrate inhibitor after 16 days of immersion in chloride solution.[6]

Source: Ávila-Gonzalez, C., Cruz-Silva, R., Menchaca, C., Sepulveda-Guzman, S., & Uru-churtu, J. (2011). Use of silica tubes as nanocontainers for corrosion inhibitor storage. Journal of Nanotechnology, 2011, doi:10.1155/2011/461313. Creative Commons Attribution License 3.0.

15.5.3.1 CAPSULE-BASED SELF-HEALING

The mechanism of capsule-based self-healing materials sequesters the healing agent in discrete capsules until damage triggers rupture and release of the capsule contents.[9,16] The capsule bond strength with the matrix, capsule volume fraction, and capsule stiffness may affect mechanical properties of the self-healing material such as strength, fracture toughness and elastic moduli. Cerium molybdate hollow nanospheres filled with 2-mercaptobenzothiazole resulted in very good corrosion inhibition on investigation. The corrosion inhibition process may be attributed either to the organic inhibitor or the release of active species, such as cerium ions from the nanocontainers.[59]

15.5.3.2   VASCULAR SELF-HEALING COATINGS

Vascular self-healing materials also sequester the healing agent in a network in the form of capillaries or hollow channels, which may be inter-connected until damage triggers self-healing. Two main techniques used for the assembly of network structures for self-healing have been reported in literature. The first employed hollow glass fibers (HGFs) as channels (~60 μm), filled with the appropriate healing agent. Hollow glass fibers are easily made using existing glass fiber drawing techniques. They are compatible with many standard polymer matrices, and are inert to many popular self-healing agents such as two-part EP resin systems and cyano-acrylates. In addition, these fibers can be integrated into glass and carbon fiber plies for use in composites due to their similar size and shape.[9] For vascular systems, additional connectivity adds numerous performance advantages. Any location in the network has multiple connection points, leading to increased reliability with regard to channel blockages and a larger accessible reservoir for the healing agent(s). Multiple connec-tions between channels also allow for easier refilling of the network after depletion.[81]

15.5.3.3   LAYERED MATERIAL-BASED SELF-HEALING COATINGS

Layered double hydroxide (LDHs) also known as anionic clays or hydro-talcite (HT)-like compounds have been reported to show good corrosion protection with the incorporation of external inhibitors.[59,88] Tedim et al., reported that Zn-Al LDHs intercalated with nitrate anions acted as nano-traps for chloride ions and it reduced the coating permeability of chlo-ride ions.[88] Similarly, in the study by Montemor et al., LDHs loaded with mercaptobenzothiazole (MBT), as a corrosion inhibitor were investigated for self-healing effect. The LDHs loaded MBT significantly delayed corro-sion activity at the early stages because the inhibitor leaching occurred by an exchange mechanism of trapping chlorides ions as inhibitor is released in a sequential manner providing protection on demand.[59] Other layered materials such as nano-clays example montmorillonite nano-clay have been successfully used as nanocontainers in self-healing coatings.[58]

15.5.4 ANTIFOULING AND ANTIBACTERIAL EFFECTS

The antifouling and antibacterial properties are vital for the coating of marine crafts. Tributyltin and tributyltin oxide-like biocides were used for the antifouling coatings. However, these chemicals create serious environmental issues to the marine habitats and these chemicals are no longer in use. Antifouling corrosion resistant coatings are still being explored. Zheng et al. studied the self-healing and antifouling multifunctional system based on pH/sulfide sensitive mesoporous silica nanoparticles. These silica nanoparticles were used as containers for both corrosion inhibitor (benzotriazole) and biocide (benzalconium chloride). This hybrid coating can be used to protect steel parts exposed to sulfur polluted environments.[58,107] Encapsulation of antifouling agents in gel particles is another promising route for introducing functionalities in corrosion protective coatings. The encapsulation of enzymes that produce hydrogen peroxide in polyethylenimine templated silica particles by coprecipitation methods was recently reported.[45] The activity of the enzymes is pH-dependent, with a maximum of near the neutral to slightly acidic range. Thin coatings displaying anticorrosion properties can be prepared based on layer-by-layer assembly of laponites (natural clays) and poly cationic copolymers containing catechol groups and alkyl quaternary ammonium salts.[108] This route can also be used to introduce the antifouling functionality. Also for this purpose, silver-containing micelles were prepared. Effective corrosion inhibition and antifouling ability were demonstrated for galvanized steel and stainless steel substrates, respectively.[58] Encapsulation of natural and environmental-friendly species with antifouling properties in coatings used for corrosion protection is of necessity, but also a challenging task because it demands the proper selection and manipulation of the antifouling agents, its successful encapsulation and controlled release to the coating surface. Encapsulation opens the route to immobilize antifouling agents for prolonged activity and stability, while providing enhanced corrosion protection.[58]

15.5.5 MECHANICAL BEHAVIOR

The interaction between a nanomaterial and the polymer matrix is known to result in the formation of nanocomposite coating influencing cross-linking and adhesion on the coated material, thus, correspondingly an improvement in mechanical properties could be observed.[23,50] Hard ceramic nanoparticles have a great effect on fracture toughness, microhardness, and wear resistance. Nanosilica EP coating has higher mechanical properties such as adhesion, abrasion, and scratch hardness compared to that of its micron sized counterpart and neat polymer.[63] The presence of TiO_2 nanoparticles in polyester has shown greater quasi-static fracture toughness.[30] Nanocomposite coating prepared by incorporation of CNT in polyether ether ketone (PEEK) showed higher hardness and wear resistance, and the uniformly dispersed CNTs in the polymer gave the improvement in the load bearing capacity. In addition to that, effectively dispersed CNTs act as bridges to anchor the polymer chains and preventing the pull out of the polymer during the wear analysis.[57] Also, uniform dispersion of layered materials in polymer coatings, especially as exfoliated or intercalated nanocomposite coating, enhances the mechanical and physical characteristics. Exfoliation of GO in poly (vinyl alcohol) (PVA) matrix occurred owing to the non-covalent interactions of the polymer chains of PVA and the hydrophilic surface of GO layers. The presence of GO in PVA and PVA/starch blends were found to enhance the tensile strength and thermal stability of the formulated nanocomposites.[91] Besides, modified natural clays such as MMT, hectorite, laponite, sepiolite, saponite, halloysite nanotube, synthetic MMT, hectorite, and LDH are some of the most widely accepted reinforcements for high-performance polymer nanocomposites. Even at low loading, they showed higher mechanical, thermal, dynamic mechanical, adhesion and barrier properties, flame retardancy, and so forth.[46] Higher mechanical performance and flame-retardant properties were also observed for EP coating in the presence of Mg–Al layered double hydroxide.[8]

15.6 APPLICATIONS AND COMMERCIAL DEVELOPMENTS

Smart polymer nanocomposite coatings have found wide application in the protection of metallic structures and engineering applications. Among a lot of nanotechnology-based coatings, a few commercialized products are represented in the Table 15.4.

TABLE 15.4 Selected Commercial Smart Anticorrosive Polymer Nanocomposite Coatings.

Products	Characteristics	Applications	Producers
Deletum 5000 Deletum 3000	Anti-graffiti paint- Anti-graffiti paint	Coatings repellant to water and oil	Victor Castaño
2C Marine Sealant PRO	Antifouling coating containing silicon dioxide as nanomaterial	Intended for gel coat marine surfaces with no or only minimal sign of wear	Nanosafeguard
EcoShield® 386 Water Based Coating	Acrylic-based formula with a complex mixture of nanosized nontoxic organic corrosion inhibitors that can provide protection against 1000 h of salt spray performance	For harsh, unsheltered outdoor applications where corrosion and UV protection are required on metals for long periods of time	Cortec Corporation
1K-Nano	Nano paint sealant	Coatings for automobile exterior	Koch-Chemie
NanoChar™	Fire protection coating, VOC-free epoxy intumescent coatings based on nanotechnology	Protects steel amongst other non-metallic substrates	Intumescents Associates Group
Nanotol®	Nano-based sealant that creates a durable water and dirt repellent protective layer and provides deep gloss on all surfaces of the vehicle. Contains nanopolymers that are activated by rubbing with a microfiber cloth during application	Automobile exterior, wheels, and disc	CeNano GmbH and Co. KG

TABLE 15.4 (*Continued*)

Products	Characteristics	Applications	Producers
NanoMate®5810/5820	Functional and decorative metal finishing with anticorrosive properties for metals. Eco-friendly finishing chemical	Pre-painting and pre-bonding applications for metals, stone, ceramics, glass, and so forth	NanoMate Technology, Inc.
NANOMYTE®	Waterborne, solvent-borne, superhydro-phobic, self-healing, anticorrosion, acid resistant, and so forth	Pretreatments, primers, and topcoats	NEI Corporation

Victor Castano: http://www.nanotechproject.org, Nanosafeguard: http://www.nano-techproject.org, Cortec Corporation: http://www.cortecvci.com, Koch-Chemie: http://www.1k-nano.de/en/main.html, Intumescents Associates Group: http://www.advancede-poxycoatings.com, CeNano GmbH and Co. KG: www.nanotol.de, NanoMate Technology, Inc.: http://nanomatetech.com/2014/finishing.html, NEI Corporation:http://www.neicor-poration.com/products/coatings/anticorrosion-paints-coatings/

15.7 CONCLUSIONS AND OUTLOOK

Nanotechnology has brought inspiring potentials which are worthy of note to the methods of mitigating corrosion risk and failures at the metal-electrolyte interface. Incorporation of nanodispersoids in polymeric coatings goes beyond improving barrier performance as it influences other properties as well as imparts new properties to the coating system. Dispersing nanodispersoids, instead of larger particles has allowed increase in the "interfacial material" content significantly. With the dramatic properties at the nanolevel which are better, an impartation is effected in the final product. The investigated novel coating formulations highlighted and their tailored responses have been designed to provide new insight towards understanding the mechanism of corrosion protection of different materials such as aluminum and its alloys, brass, steel, and so forth. This is of great advantage to a wide range of strategic industries ranging from longevity of service life of metal-based products and structures to proactive maintenance provisions minimizing the extent of resource depletion which has tormented all sectors experiencing corrosion for decades. Nano-dispersoids in coating formulations have yielded functional, viable, and outstanding effects for both passive and active corrosion protection in a steady progression from its inception.

KEYWORDS

- nanodispersoids
- polymeric materials
- nanocomposites
- smart coatings
- anticorrosion application

REFERENCES

1. Alexandre, M.; Dubois, P. Polymer-Layered Silicate Nanocomposites: Preparation, Properties and Uses of a New Class of Materials. *Mater. Sci. Eng.: R: Reports* **2000**, *28*(1), 1–63.
2. Ali, S. I.; Ali, S. R.; Naeem, M.; Haq, S. A.; Mashhood, M.; Ali, A.; Hasan, S. A. Synthesis and Characterization of Low-Cost Epoxy-Based Erosion Resistant Nanocomposite Coating. *J. Basic Appl. Sci.* **2016**, *12*, 339–343.
3. Ammar, S.; Ramesh, K.; Vengadaesvaran, B.; Ramesh, S.; Arof, A. K. Amelioration of Anticorrosion and Hydrophobic Properties of Epoxy/PDMS Composite Coatings Containing Nano ZnO Particles. *Prog. Org. Coat.* **2016**, *92*, 54–65.
4. Atta, A. M.; El-Mahdy, G. A.; Allohedan, H. A. Corrosion Inhibition Efficiency of Modified Silver Nanoparticles for Carbon Steel in 1 M HCl. *Int. J. Electrochem. Sci.* **2013**, *8*(4), 4873–4885.
5. Atta, A. M.; El-Mahdy, G. A.; Al-Lohedan, H. A.; Ezzat, A. O. Synthesis and Application of Hybrid Polymer Composites Based on Silver Nanoparticles as Corrosion Protection for Line Pipe Steel. *Molecules* **2014**, *9*(5), 6246–6262. DOI: 10.3390/molecules19056246.
6. Ávila-Gonzalez, C.; Cruz-Silva, R.; Menchaca, C.; Sepulveda-Guzman, S.; Uruchurtu, J. Use of Silica Tubes as Nanocontainers for Corrosion Inhibitor Storage. *J. Nanotechnol.* **2011**, *2011*. DOI: 10.1155/2011/461313.
7. Bagherzadeh, M. R.; Mahdavi, F. Preparation of Epoxy–Clay Nanocomposite and Investigation on Its Anti-Corrosive Behavior in Epoxy Coating. *Prog. Organ. Coat.* **2007**, *60*(2), 117–120.

8. Becker, C. M.; Gabbardo, A. D.; Wypych, F.; Amico, S. C. Mechanical and Flame-Retardant Properties of Epoxy/Mg–Al LDH Composites. *Composites Part A* **2011,** *42*(2), 196–202.

9. Blaiszik, B. J.; Kramer, S. L. B.; Olugebefola, S. C.; Moore, J. S.; Sottos, N. R.; White, S. R. Self-Healing Polymers and Composites. *Annu. Rev. Mater. Res.* **2010,** *40*, 179–211.

10. Borisova, D.; Möhwald, H.; Shchukin, D. G. Influence of Embedded Nanocontainers on the Efficiency of Active Anticorrosive Coatings for Aluminum Alloys Part II: Influence of Nanocontainer Position. *ACS Appl. Mater. Interfaces* **2012a,** *5*(1), 80–87.

11. Borisova, D.; Möhwald, H.; Shchukin, D. G. Influence of Embedded Nanocontainers on the Efficiency of Active Anticorrosive Coatings for Aluminum Alloys Part I: Influence of Nanocontainer Concentration. *ACS Appl. Mater. Interfaces* **2012b,** *4*(6), 2931–2939.

12. Brady, R. F. In *Twenty-First Century Materials: Coatings that Interact with Their Environment.* ACS Symposium Series 957; Oxford University Press, Washington, D.C., USA, 2007; pp 3–11.

13. Camargo, P. H. C.; Satyanarayana, K. G.; Wypych, F. Nanocomposites: Synthesis, Structure, Properties and New Application Opportunities. *Mater. Res.* **2009,** *12*(1), 1–39.

14. Chen, J.; Wang, F.; Liu, Q.; Du, J. Antibacterial Polymeric Nanostructures for Biomedical Applications. *Chem. Commun.* **2014,** *50*(93), 14482–14493.

15. Chen, L.; Song, R. G.; Li, X. W.; Guo, Y. Q.; Wang, C.; Jiang, Y. The Improvement of Corrosion Resistance of Fluoropolymer Coatings by SiO_2/Poly (styrene-co-butyl acrylate) Nanocomposite Particles. *Appl. Surf. Sci.* **2015,** *353,* 254–262.

16. Cho, S. H.; White, S. R.; Braun, P. V. Self-Healing Polymer Coatings. *Adv. Mater.* **2009,** *21*(6), 645–649.

17. Choi, H.; Kim, K. Y.; Park, J. M. Encapsulation of Aliphatic Amines into Nanoparticles for Self-Healing Corrosion Protection of Steel Sheets. *Prog. Organ. Coat.* **2013,** *76*(10), 1316–1324.

18. Cole, I. S. Smart Coatings for Corrosion Protection: An Overview. In *Handbook of Smart Coatings for Materials Protection;* Makhlouf, A. S. H., Ed.; Woodhead Publishing Series: UK, 2014.

19. Cook, R. L.; Myers, A. W. in *Nanoparticle Surface Modification for Advanced Corrosion Inhibiting Coatings.* ACS Symposium Series 1008; Oxford University Press; pp 64–88.

20. Davoodi, A.; Honarbakhsh, S.; Farzi, G. A. Evaluation of Corrosion Resistance of Polypyrrole/Functionalized Multi-walled Carbon Nanotubes Composite Coatings on 60Cu–40Zn Brass Alloy. *Prog. Organ. Coat.* **2015,** *88*, 106–115.

21. Dennis, R. V.; Patil, V.; Andrews, J. L.; Aldinger, J. P.; Yadav, G. D.; Banerjee, S. Hybrid Nanostructured Coatings for Corrosion Protection of Base Metals: A Sustainability Perspective. *Mater. Res. Express* **2015,** *2*(3), 032001.

22. Dennis, R. V.; Viyannalage, L. T.; Gaikwad, A. V.; Rout, T. K.; Banerjee, S. Graphene Nanocomposite Coatings for Protecting Low-Alloy Steels from Corrosion. *Am. Ceram. Soc. Bull.* **2013,** *92*(4), 18–24.

23. Deyab, M. A. Corrosion Protection of Aluminum Bipolar Plates with Polyaniline Coating Containing Carbon Nanotubes in Acidic Medium Inside the Polymer Electrolyte Membrane Fuel Cell. *J. Power Sources* **2014**, *268*, 50–55.

24. Deyab, M. A. Effect of Carbon Nano-tubes on the Corrosion Resistance of Alkyd Coating Immersed in Sodium Chloride Solution. *Prog. Organ. Coat.* **2015**, *85*, 146–150.

25. Deyab, M. A.; Keera, S. T. Effect of Nano-TiO$_2$ Particles Size on the Corrosion Resistance of Alkyd Coating. *Mater. Chem. Phys.* **2014**, *146*(3), 406–411.

26. Dhoke, S. K.; Khanna, A. S.; Sinha, T. J. M. Effect of Nano-ZnO Particles on the Corrosion Behavior of Alkyd-Based Waterborne Coatings. *Prog. Organ. Coat.* **2009**, *64*(4), 371–382.

27. Dolai, S.; Khanna, A. S. Importance of Nanoparticles in the Development of Durable Multi-functional and Smart Organic Coatings. *Nano Res. Appl.* **2016**, *2*(1), 12. http://nanotechnology.imedpub.com/archive.php.

28. Ejenstam, L.; Tuominen, M.; Haapanen, J.; Mäkelä, J. M.; Pan, J.; Swerin, A.; Claesson, P. M. Long-Term Corrosion Protection by a Thin Nano-composite Coating. *Appl. Surf. Sci.* **2015**, *357*, 2333–2342.

29. Es-Saheb, M.; Sherif, E. S. M.; El-Zatahry, A.; El Rayes, M. M.; Khalil, A. K. Corrosion Passivation in Aerated 3.5% NaCl Solutions of Brass by Nanofiber Coatings of Polyvinyl Chloride and Polystyrene. *Int. J. Electrochem. Sci.* **2012**, *7*, 10442–10455.

30. Evora, V. M.; Shukla, A. Fabrication, Characterization, and Dynamic Behavior of Polyester/TiO$_2$ Nanocomposites. *Mater. Sci. Eng. A* **2003**, *361*(1), 358–366.

31. Faghihi, K.; Raeisi, A.; Honardoost, E.; Shabanian, M.; Mirzakhanian, Z. Synthesis, Characterization, and Antibacterial Activity of New Poly(ether-Amide)/Silver Nanocomposites. *Adv. Polym. Technol.* **2016**, *0*(0), 1–10. DOI: 10.1002/adv.21669.

32. Fangli, Y.; Peng, H.; Chunlei, Y.; Shulan, H.; Jinlin, L. Preparation and Properties of Zinc Oxide Nanoparticles Coated with Zinc Aluminate. *J. Mater. Chem.* **2003**, *13*(3), 634–637.

33. Fernando, R. H. Nanocomposite and Nanostructured Coatings: Recent Advancements. In *Nanotechnology Applications in Coatings*; Fernando, R. H., Sung, L. P., Eds.; ACS Symposium Series; American Chemical Society: Washington DC, 2009; Vol. 1008, pp 2–21.

34. Fischer, H. Polymer Nanocomposites: From Fundamental Research to Specific Applications. *Mater. Sci. Eng. C* **2003**, *23*(6), 763–772.

35. Fix, D.; Andreeva, D. V.; Lvov, Y. M.; Shchukin, D. G.; Möhwald, H. Application of Inhibitor-Loaded Halloysite Nanotubes in Active Anti-corrosive Coatings. *Adv. Funct. Mater.* **2009**, *19*(11), 1720–1727.

36. Förster, S.; Plantenberg, T. From Self-Organizing Polymers to Nanohybrid and Biomaterials. *Angew. Chem. Int. Ed.* **2002**, *41*(5), 688–714.

37. Guo, Q.; Ghadiri, R.; Weigel, T.; Aumann, A.; Gurevich, E. L.; Esen, C.; Medenbach O.; Cheng W.; Chichkov B.; Ostendorf, A. Comparison of In Situ and Ex Situ Methods for Synthesis of Two-Photon Polymerization Polymer Nanocomposites. *Polymers* **2014**, *6*(7), 2037–2050.

38. Haase, M. F.; Grigoriev, D. O.; Möhwald, H.; Shchukin, D. G. Development of Nanoparticle Stabilized Polymer Nanocontainers with High Content of the

Encapsulated Active Agent and Their Application in Water-Borne Anticorrosive Coatings. *Adv. Mater.* **2012,** *24*(18), 2429–2435.

39. Hamdy, A. Corrosion Protection Performance via Nano-coatings Technologies. *Recent Pat. Mater. Sci.* **2010,** *3*(3), 258–267.

40. Harb, S. V.; Pulcinelli, S. H.; Santilli, C. V.; Knowles, K.; Hammer, P. A. Comparative Study on Graphene Oxide and Carbon Nanotube Reinforcement of PMMA-Siloxane-Silica Anticorrosive Coatings. *ACS Appl. Mater. Interfaces* **2016,** *8*(25), 16339–16350.

41. Herron, N.; Thorn, D. L. Nanoparticles: Uses and Relationships to Molecular Cluster Compounds. *Adv. Mater.* **1998,** *10*(15), 1173–1184.

42. Hu, C.; Li, Y.; Kong, Y.; Ding, Y. Preparation of Poly (o-toluidine)/Nano ZnO/Epoxy Composite Coating and Evaluation of Its Corrosion Resistance Properties. *Synth. Met.* **2016,** *214*, 62–70.

43. Jin, C.; Li, J.; Han, S.; Wang, J.; Sun, Q. A Durable, Superhydrophobic, Superoleophobic and Corrosion-Resistant Coating with Rose-Like ZnOnanoflowers on a Bamboo Surface. *Appl. Surf. Sci.* **2014,** *320*, 322–327.

44. Jose, J. P.; Malhotra, S. K.; Thomas, S.; Joseph, K.; Goda, K.; Sreekala, M. S. Advances in Polymer Composites: Macro- and Microcomposites—State of the Art. In *Polymer Composites,* 1st ed.; Thomas, S., Joseph, K., Malhotra, S. K., Goda, K., Sreekala, M. S., Eds.; Wiley-VCH Verlag GmbH and Co. KGaA, 2012; Vol. 1.

45. Kharchenko, U.; Beleneva, I. Evaluation of Coatings Corrosion Resistance with Biocomponents as Antifouling Additives. *Corros. Sci.* **2013,** *72*, 47–53.

46. Kotal, M.; Bhowmick, A. K. Polymer Nanocomposites from Modified Clays: Recent Advances and Challenges. *Prog. Polym. Sci.* **2015,** *51*, 127–187.

47. Le, Y.; Hou, P.; Wang, J.; Chen, J. F. Controlled Release Active Antimicrobial Corrosion Coatings with Ag/SiO 2 Core–Shell Nanoparticles. *Mater. Chem. Phys.* **2010,** *120*(2), 351–355.

48. Li, D.; He, Q.; Li, J. Smart Core/Shell Nanocomposites: Intelligent Polymers Modified Gold Nanoparticles. *Adv. Colloid Interface Sci.* **2009,** *149*(1), 28–38.

49. Li, P.; He, X.; Huang, T.; White, K. L.; Zhang, X.; Liang, H.; Nishimurae, R.; Sue, H. Highly Effective Anti-corrosion Epoxy Spray Coatings Containing Self-assembled Clay in Smectic Order. *J. Mater. Chem. A* **2015,** *3*, 2669–2676.

50. Liu, D.; Zhao, W.; Liu, S.; Cen, Q.; Xue, Q. Comparative Tribological and Corrosion Resistance Properties of Epoxy Composite Coatings Reinforced with Functionalized Fullerene C60 and Graphene. *Surf. Coat. Technol.* **2015,** *286,* 354–364. DOI: 10.1016/j.surfcoat.2015.12.056.

51. Madhankumar, A.; Nagarajan, S.; Rajendran, N.; Nishimura, T. EIS Evaluation of Protective Performance and Surface Characterization of Epoxy Coating with Aluminum Nanoparticles After Wet and Dry Corrosion Test. *J. Solid State Electrochem.* **2012a,** *16*(6), 2085–2093.

52. Madhankumar, A.; Rajendran, N.; Nishimura, T. Influence of Si Nanoparticles on the Electrochemical Behavior of Organic Coatings on Carbon Steel in Chloride Environment. *J. Coat. Technol. Res.* **2012,** *9*(5), 609–620.

53. Madhankumar, A.; Rajendran, N.; Nishimura, T. Influence of Si Nanoparticles on the Electrochemical Behavior of Organic Coatings on Carbon Steel in Chloride Environment. *J. Coat. Technol. Res.* **2012b,** *9*(5), 609–620.

54. Makhlouf, A. S. H. Techniques for Synthesizing and Applying Smart Coatings for Material Protection. In *Handbook of Smart Coatings for Materials Protection*; Makhlouf, A. S. H., Ed.; Woodhead Publishing Series (Elsevier): UK, 2014.

55. Mirzaee, M.; Vaezi, M.; Palizdar, Y. Synthesis and Characterization of Silver Doped Hydroxyapatite Nanocomposite Coatings and Evaluation of Their Antibacterial and Corrosion Resistance Properties in Simulated Body Fluid. *Mater. Sci. Eng. C* **2016**, *69*, 675–684.

56. Mittal, V.; Ed. *Synthesis Techniques for Polymer Nanocomposites;* John Wiley and Sons: Weinheim, Germany, 2014.

57. Mohammed, A. S.; Fareed, M. I. Improving the Friction and Wear of Poly-Ether-Etherketone (PEEK) by Using Thin Nano-Composite Coatings. *Wear* **2016**, *364*, 154–162.

58. Montemor, M. F. Functional and Smart Coatings for Corrosion Protection: A Review of Recent Advances. *Surf. Coat. Technol.* **2014**, *258*, 17–37.

59. Montemor, M. F.; Snihirova, D. V.; Taryba, M. G.; Lamaka, S. V.; Kartsonakis, I. A.; Balaskas, A. C.; Kordas G. C.; Tedim J.; Kuznetsova A.; Ferreira, M. G. S. Evaluation of Self-Healing Ability in Protective Coatings Modified with Combinations of Layered Double Hydroxides and Cerium Molibdate Nanocontainers Filled with Corrosion Inhibitors. *Electrochim. Acta* **2012**, *60*, 31–40.

60. Muraleedaran, K.; Mujeeb, V. A. Applications of Chitosan Powder with In Situ Synthesized Nano ZnO Particles as an Antimicrobial Agent. *Int. J. Biol. Macromol.* **2015**, *77*, 266–272.

61. Nazari, M. H.; Shi, X. Polymer-Based Nanocomposite Coatings for Anticorrosion Applications. In *Industrial Applications for Intelligent Polymers and Coatings;* Hosseini, M., Makhlouf, A. S. H., Eds.; Springer International Publishing:Switzerland, 2016; pp 373–398. DOI: 10.1007/978-3-319-26893-4_18.

62. Oliveira, M.; Machado, A. V. Preparation of Polymer-Based Nanocomposites by Different Routes. In *Nanocomposites: Synthesis, Characterization and Applications*; Wang, X., Ed.; Nova Publishers: Hauppauge, New York, USA, 2013; pp 1–22.

63. Palraj, S.; Selvaraj, M.; Maruthan, K.; Rajagopal, G. Corrosion and Wear Resistance Behavior of Nano-silica Epoxy Composite Coatings. *Prog. Organ. Coat.* **2015**, *81*, 132–139.

64. Peng, H. X. Polyurethane Nanocomposite Coatings for Aeronautical Applications. In *Multifunctional Polymer Nanocomposites;* CRC press: Boca Raton, 2011, 337–388.

65. Plyushch, A.; Macutkevic, J.; Kuzhir, P.; Banys, J.; Bychanok, D.; Lambin, P.; Bistarelli, S.; Cataldo, A.; Micciulla, F.; Bellucci, S. Electromagnetic Properties of Graphene Nanoplatelets/Epoxy Composites. *Compos. Sci. Technol.* **2016**, *128*, 75–83.

66. Qing, Y.; Yang, C.; Yu, N.; Shang, Y.; Sun, Y.; Wang, L.; Liu, C. Superhydrophobic TiO_2/Polyvinylidene Fluoride Composite Surface with Reversible Wettability Switching and Corrosion Resistance. *Chem. Eng. J.* **2016**, *290*, 37–44.

67. Radhakrishnan, S.; Siju, C. R.; Mahanta, D.; Patil, S.; Madras, G. Conducting Poly-aniline–Nano-TiO_2 Composites for Smart Corrosion Resistant Coatings. *Electrochim. Acta* **2009**, *54*(4), 1249–1254.

68. Ragesh, P.; Ganesh, V. A.; Nair, S. V.; Nair, A. S. A Review on 'Self-Cleaning and Multifunctional Materials'. *J. Mater. Chem. A* **2014**, *2*(36), 14773–14797.
69. Ramezanzadeh, B.; Attar, M. M.; Farzam, M. A Study on the Anticorrosion Performance of the Epoxy–Polyamide Nanocomposites Containing ZnO Nanoparticles. *Prog. Organ. Coat.* **2011**, *72*(3), 410–422.
70. Ramezanzadeh, B.; Haeri, H.; Ramezanzadeh, M. A Facile Route of Making Silica Nanoparticles-Covered Graphene Oxide Nanohybrids (SiO_2-GO); Fabrication of SiO_2-GO/Epoxy Composite Coating with Superior Barrier and Corrosion Protection Performance. *Chem. Eng. J.* **2016**, *303*, 511–528. DOI: http://dx.doi.org/10.1016/j.cej.2016.06.028.
71. Rathish, R. J.; Dorothy, R.; Joany, R. M.; Pandiarajan, M. Corrosion Resistance of Nanoparticle-Incorporated Nano Coatings. *Eur. Chem. Bull.* **2013**, *2*(12), 965–970.
72. Ray, S. S.; Okamoto, M. Polymer/Layered Silicate Nanocomposites: A Review from Preparation to Processing. *Prog. Polym. Sci.* **2003**, *28*(11), 1539–1641.
73. Rout, T. K.; Gaikwad, A. V.; Lee, V.; Banerjee, S. Hybrid Nanocomposite Coatings for Corrosion Protection of Low Carbon Steel: A Substrate-Integrated and Scalable Active–Passive Approach. *J. Mater. Res.* **2011**, *26*(06), 837–844.
74. Sahoo, B. N.; Balasubramanian, K. Facile Synthesis of Nano Cauliflower and Nano Broccoli Like Hierarchical Superhydrophobic Composite Coating Using PVDF/Carbon Soot Particles via Gelation Technique. *J. Colloid Interface Sci.* **2014**, *436*, 111–121.
75. Saji, V. S.; Thomas, J. Nanomaterials for Corrosion Control. *Curr. Sci.* **2007**, *92*(1), 51–55.
76. Sanchez, C.; Julián, B.; Belleville, P.; Popall, M. Applications of Hybrid Organic–Inorganic Nanocomposites. *J. Mater. Chem.* **2005**, *15*(35–36), 3559–3592.
77. Santos, L. F.; Montiel, A. G.; Alvarado, M. D. B. *U.S. Patent No. 8,344,058;* U.S. Patent and Trademark Office: Washington, DC, 2013.
78. Schneider, G.; Decher, G. From Functional Core/Shell Nanoparticles Prepared via Layer-by-Layer Deposition to Empty Nanospheres. *Nano Lett.* **2004**, *4*(10), 1833–1839.
79. Shchukin, D. G. Container-Based Multifunctional Self-Healing Polymer Coatings. *Polym. Chem.* **2013**, *4*(18), 4871–4877.
80. Shchukin, D. G.; Möhwald, H. Sonochemical Nanosynthesis at the Engineered Interface of a Cavitation Microbubble. *Phys. Chem. Chem. Phys.* **2006**, *8*(30), 3496–3506.
81. Shchukin, D. G.; Möhwald, H. Self-Repairing Coatings Containing Active Nanoreservoirs. *Small* **2007**, *3*(6), 926–943.
82. Shchukin, D. G.; Grigoriev, D. O.; Möhwald, H. Application of Smart Organic Nanocontainers in Feedback Active Coatings. *Soft Matt.* **2010**, *6*(4), 720–725.
83. Shchukin, D. G.; Zheludkevich, M.; Möhwald, H. Feedback Active Coatings Based on Incorporated Nanocontainers. *J. Mater. Chem.* **2006**, *16*(47), 4561–4566.
84. Singh, B. P.; Jena, B. K.; Bhattacharjee, S.; Besra, L. Development of Oxidation and Corrosion Resistance Hydrophobic Graphene Oxide-Polymer Composite Coating on Copper. *Surf. Coat. Technol.* **2013**, *232*, 475–481.

85. Snihirova, D.; Lamaka, S. V.; Montemor, M. F. "SMART" Protective Ability of Water Based Epoxy Coatings Loaded with CaCO 3 Microbeads Impregnated with Corrosion Inhibitors Applied on AA2024 Substrates. *Electrochim. Acta* **2012,** *83*, 439–447.

86. Tabkh Paz, M.; Shajari, S.; Mahmoodi, M.; Park, D. Y.; Suresh, H.; Park, S. S. Thermal Conductivity of Carbon Nanotube and Hexagonal Boron Nitride Polymer Composites. *Composites Part B* **2016,** *100,* 19–30. DOI: http://dx.doi.org/10.1016/j.compositesb.2016.06.036.

87. Tavandashti, N. P.; Ghorbani, M.; Shojaei, A.; Gonzalez-Garcia, Y.; Terryn, H.; Mol, J. M. C. pH Responsive Ce (III) Loaded Polyaniline Nanofibers for Self-Healing Corrosion Protection of AA2024-T3. *Prog. Organ. Coat.* **2016,** *99*, 197–209.

88. Tedim, J.; Kuznetsova, A.; Salak, A. N.; Montemor, F.; Snihirova, D.; Pilz, M.; Zheludkevich M. L.; Ferreira, M. G. S. Zn–Al Layered Double Hydroxides as Chloride Nanotraps in Active Protective Coatings. *Corros. Sci.* **2012,** *55*, 1–4.

89. Tian, Y.; Su, B.; Jiang, L. Interfaces: Interfacial Material System Exhibiting Superwettability. *Adv. Mater.* **2014,** *26*(40), 6871–6897.

90. Tran, T. H.; Vimalanandan, A.; Genchev, G.; Fickert, J.; Landfester, K.; Crespy, D.; Rohwerder, M. Regenerative Nano-Hybrid Coating Tailored for Autonomous Corrosion Protection. *Adv. Mater.* **2015,** *27*(25), 3825–3830.

91. Usman, A.; Hussain, Z.; Riaz, A.; Khan, A. N. Enhanced Mechanical, Thermal and Antimicrobial Properties of Poly(vinyl Alcohol)/Graphene Oxide/Starch/Silver Nanocomposites Films. *Carbohydr. Polym.* **2016,** *153*, 592–599.

92. Wang, H.; Liu, Z.; Wang, E.; Zhang, X.; Yuan, R.; Wu, S.; Zhu, Y. Facile Preparation of Superamphiphobic Epoxy Resin/Modified Poly(vinylidene Fluoride)/Fluorinated Ethylene Propylene Composite Coating with Corrosion/Wear Resistance. *Appl. Surf. Sci.* **2015,** *357*, 229–235.

93. Wei, H.; Wang, Y.; Guo, J.; Shen, N. Z.; Jiang, D.; Zhang, X.; Yan, X.; Zhu, J.; Wang, Q.; Shao, L.; Lin, H. Advanced Micro/Nanocapsules for Self-Healing Smart Anticorrosion Coatings. *J. Mater. Chem. A* **2015,** *3*(2), 469–480.

94. Weng C. J.; Chang C. H.; Yeh J. M. Polymer Nanocomposites in Corrosion Control. In *Corrosion Protection and Control using Nanomaterials;* Saji, V. S., Cook, R., Eds.; Woodhead Publishing: Cambridge, 2012; pp 330–356.

95. Wright, J. D.; Sommerdijk, N. A. J. M. *Sol–Gel Materials: Chemistry and Applications;* Gordon and Breach Science Publishers: Amsterdam, The Netherlands, 2001.

96. Xiao, F.; Yuan, S.; Liang, B.; Li, G.; Pehkonen, S. O.; Zhang, T. Superhydrophobic CuO Nanoneedle-Covered Copper Surfaces for Anticorrosion. *J. Mater. Chem. A* **2015,** *3*(8), 4374–4388.

97. Xue, C. H.; Ma, J. Z. Long-Lived Superhydrophobic Surfaces. *J. Mater. Chem. A* **2013,** *1*(13), 4146–4161.

98. Yang, C.; Wei, H.; Guan, L.; Guo, J.; Wang, Y.; Yan, X.; Zhang, X.; Wei, S.; Guo, Z. Polymer Nanocomposites for Energy Storage, Energy Saving, and Anticorrosion. *J. Mater. Chem. A* **2015,** *3*(29), 14929–14941.

99. Yeh, J. M.; Chin, C. P.; Chang, S. Enhanced Corrosion Protection Coatings Prepared from Soluble Electronically Conductive Polypyrrole-Clay Nanocomposite Materials. *J. Appl. Polym. Sci.* **2003,** *88*(14), 3264–3272.

100. Yeh, J. M.; Liou, S. J.; Lai, C. Y.; Wu, P. C.; Tsai, T. Y. Enhancement of Corrosion Protection Effect in Polyaniline via the Formation of Polyaniline-Clay Nanocomposite Materials. *Chem. Mater.* **2001,** *13*(3), 1131–1136.

101. Yeh, J. M.; Liou, S. J.; Lin, C. Y.; Cheng, C. Y.; Chang, Y. W.; Lee, K. R. Anticorrosively Enhanced PMMA-Clay Nanocomposite Materials with Quaternary Alkylphosphonium Salt as an Intercalating Agent. *Chem. Mater.* **2002,** *14*(1), 154–161.

102. Zeng, Q. H.; Yu, A. B.; Lu, G. Q.; Paul, D. R. Clay-Based Polymer Nanocomposites: Research and Commercial Development. *J. Nanosci. Nanotechnol.* **2005,** *5*(10), 1574–1592.

103. Zhang, F.; Zhao, L.; Chen, H.; Xu, S.; Evans, D. G.; Duan, X. Corrosion Resistance of Superhydrophobic Layered Double Hydroxide Films on Aluminum. *Angew. Chem. Int. Ed.* **2008,** *47*(13), 2466–2469.

104. Zhang, Y. L.; Xia, H.; Kim, E.; Sun, H. B. Recent Developments in Superhydrophobic Surfaces with Unique Structural and Functional Properties. *Soft Matter* **2012,** *8*, 11217–11231.

105. Zhang, D.; Wang, L.; Qian, H.; Li, X. Superhydrophobic Surfaces for Corrosion Protection: A Review of Recent Progresses and Future Directions. *J. Coat. Technol. Res.* **2016,** *13*(1), 11–29.

106. Zheludkevich, M. L.; Tedim, J.; Ferreira, M. G. S. "Smart" Coatings for Active Corrosion Protection Based on Multi-functional Micro and Nanocontainers. *Electrochim. Acta* **2012,** *82*, 314–323.

107. Zheng, Z.; Huang, X.; Schenderlein, M.; Borisova, D.; Cao, R.; Möhwald, H.; Shchukin, D. Self-Healing and Antifouling Multifunctional Coatings Based on pH and Sulfide Ion Sensitive Nanocontainers. *Adv. Funct. Mater.* **2013,** *23*(26), 3307–3314.

108. Zhou, C. H.; Zhang, D.; Tong, D. S.; Wu, L. M.; Yu, W. H.; Ismadji, S. Paper-Like Composites of Cellulose Acetate–Organo-Montmorillonite for Removal of Hazardous Anionic Dye in Water. *Chem. Eng. J.* **2012,** *209*, 223–234.

109. Zhou, S.; Wu, L. Development of Nanotechnology-Based Organic Coatings. *Compos. Interfaces* **2009,** *16*(4–6), 281–292.

110. Zhou, S.; Wu, L. Transparent Organic–Inorganic Nanocomposite Coatings. In *Functional Polymer Coatings: Principles, Methods, and Applications;* Wu, L., Bagdachi, J., Eds.; John Wiley and Sons, Inc.: Hoboken, New Jersey, 2015; pp 1–70.

CHAPTER 16

THERMAL AND DYNAMIC MECHANICAL ANALYSIS OF JUTE POLYPROPYLENE COMPOSITES

HARICHANDRA U. CHANDEKAR[1,*], VIKAS V.CHAUDHARI[2], SACHIN D. WAIGAONKAR[2], and NARENDRA N. GHOSH[3]

[1]Department of Mechanical Engineering, Goa College of Engineering, Farmagudi, Ponda, Goa 403401, India

[2]Department of Mechanical Engineering, BITS Pilani, K. K.Birla Goa Campus, Zuarinagar 403726, Goa, India

[3]Department of Chemistry, BITS Pilani, K. K.Birla Goa Campus, Zuarinagar 403726, Goa, India

*Corresponding author. E-mail: harichandekar@gmail.com

CONTENTS

ABSTRACT

The thermomechanical analysis of jute fabric-reinforced polypropylene (jute–PP) composite has been investigated. Thermogravimetric analysis (TGA) and differential scanning calorimetry (DSC) techniques have been used to quantify the thermal characteristics of PP, jute, and jute–PP composite. TGA results show improvement in the thermal stability of jute–PP composite over PP. Dynamic mechanical analysis (DMA) technique has been used to obtain thermomechanical properties of PP and jute–PP composite. The addition of jute in PP enhanced the operating temperature range of jute–PP composite as compared with PP. The storage modulus of jute–PP composite is found to be 27% higher than PP at room temperature.

16.1   INTRODUCTION

Natural fiber-reinforced polymer composites are being explored mainly for nonstructural applications as they are environment-friendly when compared with glass fiber-reinforced composites. However, more recently they have found applications in the automotive and construction indus-tries.[1,2] The natural fiber composites possess the advantages of being light-weight, low-cost, and also undergo less wear during the processing stage. The jute fiber along with recyclable thermoplastic matrix, polypropylene (PP) has been considered in the present investigation.

The major constituents of jute fibers are cellulose (61–71%), hemicel-lulose (13.6–20.4%), and lignin (12–13%).[3] Chemical treatment of natural fibers is necessary to improve the interfacial adhesion between the natural fibers and PP matrix. Among the several chemical treatments available, the most common and economical method is mercerization or alkali treat-ment.[4] Thermal analysis of the composite constituents is essential to study the thermal stability from a processing viewpoint. The use of high temper-ature beyond the limiting temperature of natural fibers results in their degradation. The thermal analysis of natural fibers and their constituents is carried out by various authors.[5,6] Dynamic mechanical analysis (DMA) is a powerful technique to determine the viscoelastic properties of the composite as a function of temperature and frequency.[7] The use of DMA to study the efficacy of the fiber–matrix bonding of natural fiber-based composites has been carried out.[8] More recently, the DMA technique applied to various natural fiber reinforced composites has been reviewed.[9]

In the present study, to determine the processing parameters for jute–PP composite, thermal analysis has been carried out of its constituents, namely jute and PP using the analysis (TGA) and differential scanning calorimetry (DSC), whereas the DMA has been used to investigate the storage modulus (E'), loss modulus (E''), and the ratio of loss modulus to storage modulus (tan δ) of the jute–PP composite.

16.2 EXPERIMENTAL

16.2.1 MATERIALS AND SAMPLE PREPARATION

Polypropylene (PP) homopolymer in granular form (REPOL H110MA, Reliance Ltd, India) of density 0.910 g/cc (MFR of 11 g/10 min; 230°C/2.16 kg) and commercial plain weave jute fabric (370 g/m²) have been used for the study.

After washing with distilled water, the jute fabric is soaked in 5% NaOH solution for 8 h at room temperature. Then it is neutralized with dilute acetic acid. The fabrics are allowed to dry for 12 h at room temperature followed by oven drying at 100°C for 4 h to ensure removal of moisture.

PP sheets of 1.5-mm thickness are prepared from PP granules by compression molding using a hydraulic press. The PP sheets thus prepared and alkali-treated jute fabrics are stacked alternately and hot pressed in a compression molding machine at 165°C and 10 MPa for 20 min to get the jute–PP composite laminate. A set of three PP sheets of size 250 × 150 mm along with two alkali-treated jute fabrics result in 25 wt.% jute–PP composite. The schematic representation of the materials and the experiments conducted is shown in Figure 16.1.

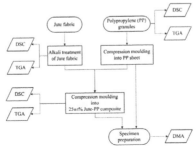

FIGURE 16.1 Schematic representation of jute–polypropylene (PP) composite preparation and its analysis.

16.2.2   DIFFERENTIAL SCANNING CALORIMETRY (DSC)

DSC study has been performed by using DSC-60, Shimadzu, Japan, for the thermal characterization of PP, alkali-treated jute and jute–PP composite samples weighing 1.725, 2.18, and 1.54 mg, respectively. These samples were placed in an aluminum pan in the air atmosphere. The temperature range set was 30–230°C with a temperature ramp rate of 10°C/min.

16.2.3   THERMOGRAVIMETRIC ANALYSIS (TGA)

The TGA has been performed on Shimadzu DTG-60. The temperature range chosen was from 30 to 500°C at a temperature ramp rate of 5°C/min. The sample mass for PP, alkali-treated jute, and jute–PP composite samples was 3.242, 2.245, and 1.501 mg, respectively. All the samples were placed in an aluminum crucible in air atmosphere at a 50 ml/min flow rate.

16.2.4   DYNAMIC MECHANICAL ANALYSIS (DMA)

The viscoelastic properties, namely storage modulus (E'), loss modulus (E''), and mechanical damping parameter (tan δ) as a function of temperature are measured using TA instruments Q800 dynamic mechanical analyzer. Rectangular specimens of dimensions $35 \times 12.5(\pm 0.1) \times 3(\pm 0.1)$ mm have been cut from the PP and jute–PP composite plates and mounted in a single cantilever configuration. The test has been performed at a fixed frequency of 1 Hz. Amplitude of 30 μm which ensured operation in linear viscoelastic region for the polymer was chosen. The temperature range used for the test was from –50 to 160°C, at a heating rate of 2°C/min. All measurements were performed under nitrogen atmosphere.

16.3   RESULTS AND DISCUSSIONS

16.3.1   DIFFERENTIAL SCANNING CALORIMETRY (DSC)

The DSC curves of PP, jute, and jute–PP composite are shown in Figure 16.2. A sharp endothermic peak is observed at 163.83°C which represents the melting temperature (T_m) of PP. For the treated jute, an endothermic peak is observed below 100°C, which is attributed to the evaporation of absorbed moisture. The jute–PP composite also exhibits an endothermic peak indicating the T_m, which is same as that of PP. The DSC test can also reveal the glass transition temperature; however, in the present study, the glass transition temperature for PP and jute–PP composite has been observed to be 11.6 and 6.5°C, respectively, from the DMA test as discussed in Section 16.3.3.

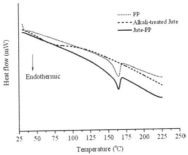

FIGURE 16.2 Differential scanning calorimetry (DSC) heating curves of PP, jute, and jute–PP composite.

16.3.2 THERMOGRAVIMETRIC ANALYSIS

TGA is carried out by observing the process of weight loss occurring in the sample as the temperature increases. The TGA and derivative thermogravimetric (DTG) plots for PP, jute, and jute–PP composite are shown in Figures 16.3 and 16.4, respectively. The DTG curve shows the peaks where there is a sudden drop in the TGA plot.

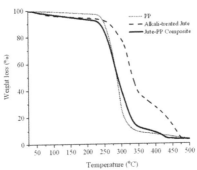

FIGURE 16.3 Thermogravimetric analysis (TGA) plots of PP, jute, and jute–PP composite.

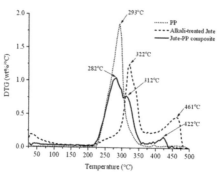

FIGURE 16.4 DTG plots of PP, jute, and jute–PP composite.

The degradation of PP occurs in a single stage as seen in the TGA curve (Fig. 16.3), which begins at 228°C and ends at 345°C, the weight loss for which is 87.24%. The maximum temperature in this transition for PP occurs at 293°C as seen from the single peak in the DTG plot in Figure 16.4. Eventually, the residual weight for PP is 3.67% at the end of the test.

For the jute fiber, 5% weight loss is observed till 100°C, which is attributed to the evaporation of absorbed moisture. The treated jute fiber exhibits two-stage degradation: first stage from 250 to 350°C and the second stage from 350 to 475°C as seen in the TGA curves (Fig. 16.3). The less stable constituent of the jute fiber, namely hemicellulose and α-cellulose is degraded initially[10] in the first stage (temperature range 250 to 350°C) which shows 53% weight loss and the peak temperature for jute fiber in this range as depicted in Figure 16.4 is 322°C. In the second stage, lignin degradation takes place. The peak temperature for this range is 461°C. The residual weight of jute fiber at the end of 500°C is 3.34%.

The degradation of jute–PP composite takes place in three stages as shown in Figure 16.3. The transition temperature ranges for jute–PP composite are 230–295, 295–345, and 345–440°C; the peak transition temperature in these regions as depicted in Figure 16.4 are 282, 312, and 422°C, respectively. The corresponding weight loss in these stages is 46, 36, and 10%. Based on the above wide temperature range (from 230 to 440°C), the stability of the jute–PP composite is more enhanced as compared with PP (i.e., 228 upto 345°C). The residual weight percent for the composite is 3.39%.

16.3.3 DYNAMIC MECHANICAL ANALYSIS

The viscoelastic properties as a function of temperature for PP specimen are shown in Figure 16.5. The temperature range selected for the test is from −50 to 160°C to facilitate the determination of primary transition, that is, the glass transition temperature for PP.

As seen from Figure 16.5, the peak of tan δ at 11.66°C is interpreted as the primary transition of PP. This is referred to as β transition temperature in the literature, whereas the peak at 90.75°C is the α transition temperature. The β transition is interpreted as the relaxation of amorphous chains in PP, whereas the α transition mainly indicates the relaxation of the restricted crystalline chains of PP. The onset of drop in E′ is also interpreted as the glass transition temperature by some researchers.

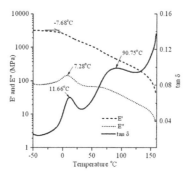

FIGURE 16.5   Viscoelastic properties of PP.

The storage modulus represents the elastic component of the stored energy, it is also referred as in-phase or the real modulus. The storage modulus (E″) for PP and 25 wt.% jute–PP composite as a function of temperature is shown in Figure 16.6. In comparison with PP, the jute–PP composite storage modulus is higher over the entire temperature span. With increasing temperature, E′ values of PP and composite system decrease; this is attributed to the matrix softening. At room temperature, the storage modulus values for PP and 25 wt.% jute–PP composite are 1583 and 2010 MPa, respectively. The storage modulus is higher in the jute–PP composite as compared with the PP due to the contribution of jute fiber stiffness.

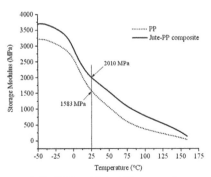

FIGURE 16.6   Storage moduli of PP and Jute–PP composite.

The tan δ is a dimensionless number, which is the ratio of loss modulus to storage modulus. The loss modulus or the viscous modulus represents the energy lost to friction and internal motions. The tan δ plot, therefore, portrays important information about the composite system such as the damping due to matrix and fiber, the bonding between them, and so forth. It is beneficial to have lower tan δ values. Figure 16.7 shows the trend of mechanical damping parameter (tan δ) with temperature. It is observed that the jute–PP composites tan δ values drop in comparison with the PP tan δ plot as the temperature increases. This indicates a good interface in the jute–PP composite, which results in lower energy dissipation and therefore low tan δ values. The polymer transition temperatures are observed for PP at 11.66 and 90.75°C, respectively. It can also be seen that the difference between the transition peaks for PP is 79°C; this is termed as the operating temperature range. Owing to the reinforcement of the PP matrix by jute fibers the operating temperature range is enhanced to 117°C.

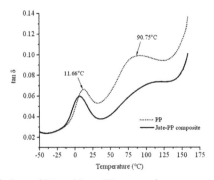

FIGURE 16.7   Tan δ plots of PP and Jute–PP composite.

16.4 CONCLUSIONS

The Tm of PP and the jute–PP composite is the same which is 164°C as seen on the DSC curve, whereas it is not the case for glass transition temperature where the Tg for PP is 11.66°C and that of jute–PP composite is 6.5°C. The TGA study reveals that the thermal stability of the jute–PP composite is better as compared with PP. From the DMA study of PP and jute–PP composites investigated in the temperature range between −50 and 150°C, it can be concluded that:

I) The storage modulus of the jute–PP composite drops at the glass transition temperature of PP, that is, at 11.6°C. The storage modulus at room temperature for 25 wt.% jute–PP composite is higher by 27% in comparison with that of unreinforced PP. At high temperatures up to 120°C the reinforcing effect of jute fibers is more pronounced.

II) The peaks noticed in the tan δ curve for jute–PP composite have reduced peak height in comparison with the peak heights of PP, this indicates lower energy dissipation in the jute–PP composite, which suggests a better interfacial bonding.

KEYWORDS

- jute–PP composite
- alkali treatment
- thermogravimetric analysis
- differential scanning calorimetry
- dynamic mechanical analysis

REFERENCES

1. Pickering, K. L.; Efendy, M. G.; Le, T. M. A Review of Recent Developments in Natural Fibre Composites and Their Mechanical Performance. *Composites Part A* **2016**, *83*, 98–112.

2. Kiruthika, A. V. A Review on Physico-Mechanical Properties of Bast Fibre Reinforced Polymer Composites. *J. Build. Eng.* **2017,** *9,* 91–99.

3. Bogoeva-Gaceva, G.; Avella, M.; et al. Natural Fibre Eco-Composites. *Polym. Compos.* **2007,** *28,* 98–107.

4. John, M. J.; Anandjiwala, R. D. Recent Developments in Chemical Modification and Characterization of Natural Fibre-Reinforced Composites. *Polym. Compos.* **2008,** *29,* 187–207.

5. Martin, A. R.; Martins, M. A.; da Silva, O. R. F.; et al. Studies on the Thermal Properties of Sisal Fibre and its Constituents. *Thermochim. Acta.* **2010,** *506,* 14–19.

6. Bhaduri, S. K.; Mathew, A.; Day, M. D.; Pandey, S. N. Thermal Behavior of Jute Fibre and its Components I. DSC studies. *Cellul. Chem. Technol.* **1994,** *28,* 391–399.

7. Menard, K. P. *Dynamic Mechanical Analysis: A Practical Introduction;* 2nd ed.; CRC press: Boca Raton, 2008.

8. Sreenivasan, V. S.; Rajini, N.; Alavudeen, A.; Arumugaprabu, V. Dynamic Mechanical and Thermo-Gravimetric Analysis of Sansevieria Cylindrica/Polyester Composite: Effect of Fibre Length, Fibre Loading and Chemical Treatment. *Composite Part B* **2015,** *69,* 76–86.

9. Saba, N.; Jawaid, M.; Alothman, O. Y.; Paridah, M. T. A Review on Dynamic Mechanical Properties of Natural Fibre Reinforced Polymer Composites. *Constr. Build. Mater.* **2016,** *106,* 149–159.

10. Doan, T. T. L.; Brodowsky, H.; Mader, E. Jute Fibre/Polypropylene Composites II. Thermal, Hydrothermal and Dynamic Mechanical Behaviour. *Compos. Sci. Technol.* **2007,** *67,* 2707–2714.

INDEX

Milton Keynes UK
Ingram Content Group UK Ltd.
UKHW022045141024
449569UK00022B/808